Federated Learning for Digital Healthcare Systems

Intelligent Data-Centric Systems

Federated Learning for Digital Healthcare Systems

Edited by

Agbotiname Lucky Imoize
Department of Electrical and Electronics Engineering, Faculty of Engineering, University of Lagos, Lagos, Nigeria

Mohammad S. Obaidat
The King Abdullah II School of Information Technology, The University of Jordan, Amman, Jordan

Houbing Herbert Song
Department of Information Systems, University of Maryland Baltimore County (UMBC), Baltimore, MD, United States

Series Editor: **Fatos Xhafa**
Universitat Politècnica de Catalunya (UPC), Barcelona, Spain

Associate Series Editor: **Houbing Herbert Song**
Department of Information Systems, University of Maryland Baltimore County (UMBC), Baltimore, MD, United States

ACADEMIC PRESS

An imprint of Elsevier

ELSEVIER

Academic Press is an imprint of Elsevier
125 London Wall, London EC2Y 5AS, United Kingdom
525 B Street, Suite 1650, San Diego, CA 92101, United States
50 Hampshire Street, 5th Floor, Cambridge, MA 02139, United States

Notices
Knowledge and best practice in this field are constantly changing. As new research and experience broaden our understanding, changes in research methods, professional practices, or medical treatment may become necessary.

Practitioners and researchers must always rely on their own experience and knowledge in evaluating and using any information, methods, compounds, or experiments described herein. In using such information or methods they should be mindful of their own safety and the safety of others, including parties for whom they have a professional responsibility.

To the fullest extent of the law, neither the Publisher nor the authors, contributors, or editors, assume any liability for any injury and/or damage to persons or property as a matter of products liability, negligence or otherwise, or from any use or operation of any methods, products, instructions, or ideas contained in the material herein.

ISBN: 978-0-443-13897-3

For Information on all Academic Press publications
visit our website at https://www.elsevier.com/books-and-journals

Publisher: Mara Conner
Editorial Project Manager: Emily Thomson
Production Project Manager: Swapna Srinivasan
Cover Designer: Vicky Pearson Esser

Typeset by MPS Limited, Chennai, India

Working together
to grow libraries in
developing countries

www.elsevier.com • www.bookaid.org

Contents

CHAPTER 2   Architecture and design choices for federated learning in modern digital healthcare systems.........37

Konstantinos A. Koutsopoulos, Christoph Thümmler,
Angelica Avila Castillo, Alice Abend, Stefan Covaci,
Benjamin Ertl, Giannis Ledakis, Stéphane Lorin,
Vincent Thouvenot, Sahar Haddad, Gouenou Coatrieux,
Reda Bellafqira and Alessandro Bassi

**CHAPTER 3 Curation of federated patient data: a proposed
landscape for the African Health Data Space** 59

*Mirjam van Reisen, Samson Yohannes Amare,
Ruduan Plug, Getu Tadele, Tesfit Gebremeskel,
Abdullahi Abubakar Kawu, Kai Smits, Liya Mamo Woldu,
Joëlle Stocker, Femke Heddema, Sakinat Oluwabukonla
Folorunso, Rens Kievit and Araya Abrha Medhanyie*

*Agbotiname Lucky Imoize, Mohammad S. Obaidat and
Houbing Herbert Song*

List of contributors

Abdulhakeem Oladele Abdulfatai
Department of Electrical and Computer Engineering, Kwara State University, Malete, Nigeria

Alice Abend
6G Health Institute GmbH, Markkleeberg, Germany

María Libertad Aguilar Carlos
Information System Department, Autonomous University of Aguascaleintes, Aguascalientes, Mexico

B. Akoramurthy
Department of CSE, National Institute of Technology, Puducherry, India

Adam A. Alli
Department of Computer Science, Islamic University in Uganda (IUIU), Mbale, Uganda

Emmanuel Alozie
Department of Information Technology, Sule Lamido University, Kafin Hausa, Jigawa, Nigeria

Samson Yohannes Amare
Leiden University Medical Center (LUMC), Leiden University, Leiden, the Netherlands; School of Humanities and Digital Sciences, Tilburg University, Tilburg, the Netherlands

Francis Ifeanyi Anyasi
Department of Electrical and Electronics Engineering, Faculty of Engineering and Technology, Ambrose Alli University, Ekpoma, Nigeria

Baguma Asuman
Department of Computer Science, Islamic University in Uganda (IUIU), Mbale, Uganda

Joseph Bamidele Awotunde
Department of Computer Science, Faculty of Information and Communication Sciences, University of Ilorin, Ilorin, Nigeria

Oluwatobi Banjo
Artificial Intelligent Systems Research Group (ArISRG), Department of Mathematical Sciences, Olabisi Onabanjo University, Ago-Iwoye, Nigeria

Alessandro Bassi
Eurescom GmbH, Heidelberg, Germany

Reda Bellafqira
Inserm, Cyber Health, LaTIM UMR1101, IMT Atlantique, Brest, France

Angelica Avila Castillo
6G Health Institute GmbH, Markkleeberg, Germany

Subrata Chowdhury
Department Of CSE, Sreenivasa Institute of Technology and Management Studies, Chittoor, Andhra Pradesh, India

Gouenou Coatrieux
Inserm, Cyber Health, LaTIM UMR1101, IMT Atlantique, Brest, France

Stefan Covaci
Agentscape AG, Berlin, Germany

K. Dhivya
Department of CSE, Pondicherry University, Puducherry, India

Benjamin Ertl
Agentscape AG, Berlin, Germany

Nasir Faruk
Department of Information Technology, Sule Lamido University, Kafin Hausa, Jigawa, Nigeria; Directorate of Information and Communication Technology, Sule Lamido University, Kafin Hausa, Jigawa, Nigeria

Sakinat Oluwabukonla Folorunso
Artificial Intelligent Systems Research Group (AISRG), Department of Mathematical Sciences, Olabisi Onabanjo University, Ago-Iwoye, Nigeria

Salisu Garba
Department of Software Engineering, Sule Lamido University, Kafin Hausa, Jigawa, Nigeria

Anastasius Gavras
Eurescom GmbH, Heidelberg, Germany

Tesfit Gebremeskel
Leiden University Medical Center (LUMC), Leiden University, Leiden, the Netherlands

Ramya Govindaraj
Department of IT, Vellore Institute of Technology, Vellore, Tamil Nadu, India

Pushpanjali Gupta
Institute of Biomedical Informatics, National Yang Ming Chiao Tung University, Taipei, Taiwan; Institute of Public Health, National Yang Ming Chiao Tung University, Taipei, Taiwan; Health Innovation Center, National Yang Ming Chiao Tung University, Taipei, Taiwan

Sahar Haddad
Inserm, Cyber Health, LaTIM UMR1101, IMT Atlantique, Brest, France

Pascal Haigron
Univ Rennes, Inserm, LTSI - UMR 1099, Rennes, France

Femke Heddema
Leiden University Medical Center (LUMC), Leiden University, Leiden, the Netherlands

Agbotiname Lucky Imoize
Department of Electrical and Electronics Engineering, Faculty of Engineering, University of Lagos, Lagos, Nigeria; Department of Electrical Engineering and Information Technology, Institute of Digital Communication, Ruhr University, Bochum, Germany

Segun Ezekiel Jacob
Department of Electrical and Computer Engineering, Kwara State University, Malete, Nigeria

Kassim Kalinaki
Department of Computer Science, Islamic University in Uganda (IUIU), Mbale, Uganda; Borderline Research Laboratory, Kampala, Uganda

K.V.N. Kavitha
Department of Communication Engineering, School of Electronics Engineering, Vellore Institute of Technology, Vellore, Tamil Nadu, India

Abdullahi Abubakar Kawu
Technological University Dublin, Dublin, Ireland; Department of Computer Science, Ibrahim Badamasi Babangida University, Lapai, Nigeria

Rens Kievit
Leiden Observatory, Leiden University, Leiden, the Netherlands

Konstantinos A. Koutsopoulos
Qualtek Sprl., Brussels, Belgium

Vinoth Babu Kumaravelu
Department of Communication Engineering, School of Electronics Engineering, Vellore Institute of Technology, Vellore, Tamil Nadu, India

Daphne Teck Ching Lai
School of Digital Science, Universiti Brunei Darussalam, Gadong, Brunei

Giannis Ledakis
UBITECH, Chalandri Athens, Greece

Stéphane Lorin
Thales SIX GTS France SAS, Campus Polytechnique, Palaiseau cedex, France

Owais Ahmed Malik
School of Digital Science, Universiti Brunei Darussalam, Gadong, Brunei

Araya Abrha Medhanyie
School of Humanities and Digital Sciences, Tilburg University, Tilburg, the Netherlands

Abolfazl Mehbodniya
Department of Electronics and Communication Engineering, Kuwait College of Science and Technology (KCST), Doha, Kuwait

Sulagna Mohapatra
Health Innovation Center, National Yang Ming Chiao Tung University, Taipei, Taiwan; Division of Gastroenterology, Taichung Veterans General Hospital, Taichung, Taiwan

Pooja Mohnani
Eurescom GmbH, Heidelberg, Germany

Jaime Muñoz-Arteaga
Information System Department, Autonomous University of Aguascaleintes, Aguascalientes, Mexico

Abdulwaheed Musa
Department of Electrical and Computer Engineering, Kwara State University, Malete, Nigeria; Centre for Artificial Intelligence and Machine Learning Systems, Kwara State University, Malete, Nigeria; Institute for Intelligent Systems, University of Johannesburg, Johannesburg, South Africa

Mohammad S. Obaidat
The King Abdullah II School of Information Technology, The University of Jordan, Amman, Jordan; School of Computer and Communication Engineering, University of Science and Technology Beijing, Beijing, P.R. China; Department of Computational Intelligence, School of Computing, SRM University, SRM Nagar, Kattankulathur, Tamil Nadu, India; School of Engineering, The Amity University, Noida, Uttar Pradesh, India

Hope Ikoghene Obakhena
Department of Electrical and Electronics Engineering, Faculty of Engineering and Technology, Ambrose Alli University, Ekpoma, Nigeria; Department of Electrical and Electronics Engineering, Faculty of Engineering, University of Benin, Benin, Nigeria

Hawau I. Olagunju
Department of Information Technology, Sule Lamido University, Kafin Hausa, Jigawa, Nigeria

Daniel Favour Oluyemi
Department of Electrical and Computer Engineering, Kwara State University, Malete, Nigeria

Ruduan Plug
Leiden University Medical Center (LUMC), Leiden University, Leiden, the Netherlands

José Rafael Rojano-Cáceres
Faculty of Statistics and Informatics, University of Veracruz, Xalapa, Mexico

Wasswa Shafik
School of Digital Science, Universiti Brunei Darussalam, Gadong, Brunei Darussalam; Dig Connectivity Research Laboratory (DCRLab), Kampala, Uganda

Antoine Simon
Univ Rennes, Inserm, LTSI - UMR 1099, Rennes, France

Kai Smits
School of Humanities and Digital Sciences, Tilburg University, Tilburg, the Netherlands

Houbing Herbert Song
Department of Information Systems, University of Maryland Baltimore County (UMBC), Baltimore, MD, United States

Joëlle Stocker
Research Advisers and Experts Europe (RAEE), Brussels, Belgium

Samarendra Nath Sur
Department of Electronics and Communication Engineering, Sikkim Manipal Institute of Technology, Rangpo, Sikkim, India

B. Surendiran
Department of CSE, National Institute of Technology, Puducherry, India

Getu Tadele
School of Humanities and Digital Sciences, Tilburg University, Tilburg, the Netherlands

Vincent Thouvenot
Thales SIX GTS France SAS, Campus Polytechnique, Palaiseau cedex, France

Christoph Thümmler
6G Health Institute GmbH, Markkleeberg, Germany

Rasha Tolba
6G Health Institute GmbH, Markkleeberg, Germany

Orazio Toscano
Ericsson, Genova, Italy

Friday Udeji
Department of Mechanical Engineering, Faculty of Engineering, Ambrose Alli University, Ekpoma, Nigeria

Mirjam van Reisen
Leiden University Medical Center (LUMC), Leiden University, Leiden, the Netherlands; School of Humanities and Digital Sciences, Tilburg University, Tilburg, the Netherlands

Julian L. Webber
Department of Electronics and Communication Engineering, Kuwait College of Science and Technology (KCST), Doha, Kuwait

Liya Mamo Woldu
School of Humanities and Digital Sciences, Tilburg University, Tilburg, the Netherlands

Chun-Ying Wu
Institute of Biomedical Informatics, National Yang Ming Chiao Tung University, Taipei, Taiwan; Institute of Public Health, National Yang Ming Chiao Tung University, Taipei, Taiwan; Health Innovation Center, National Yang Ming Chiao Tung University, Taipei, Taiwan; Division of Translational Research, Taipei Veterans General Hospital, Taipei, Taiwan; Department of Public Health, China Medical University, Taichung, Taiwan

Umar Yahya
Motion Analysis Research Laboratory (MARL), Islamic University in Uganda (IUIU), Kampala Campus, Kampala, Uganda

Rufai Yusuf Zakari
Department of Computer Science, School of Science and Information Technology (SSIT), Skyline University Nigeria, Kano, Nigeria

Preface

Modern healthcare systems support the collection of critical medical data for statistical evaluation and inference using machine learning (ML) algorithms. However, the application of ML in healthcare data analytics has not been gainfully exploited due to the proliferating security and privacy issues, among other critical concerns. Specifically, the potential of ML is limited in the case of insufficient data, posing a significant impediment to the transition from research to clinical practice. In recent times, federated learning (FL) has been introduced to enhance the performance of ML algorithms. In FL, artificial intelligence models are trained with data from multiple sources, ensuring data anonymity, security, privacy, and integrity, thus removing potential barriers to data availability and sharing. Additionally, FL-trained models have shown favorable performance in agreement with the ones obtained from centrally hosted datasets.

A successfully implemented FL model can produce unbiased decisions to facilitate good judgment in precision medicine. The primary focus of this book is to critically examine the key factors that contribute to the limitations of applying traditional ML in healthcare systems and investigate how FL is employed to address these limitations. The book explores the potential of FL in digital healthcare systems. It highlights technical considerations, examines the potentials and prospects of FL, analyzes governmental and economic policies, evaluates legal frameworks and regulations, and addresses the critical challenges of FL integration in healthcare systems.

This book discusses the broad applications of FL in digital healthcare systems. First, the book critically looks at the existing ML models and suggests FL-based solutions to revamp the existing ML models. Specifically, the book examines the application of FL solutions toward addressing critical security and privacy concerns in digital healthcare systems. More so, these solutions are critically analyzed and compared in terms of the required resources, computational complexity, and system performance. Additionally, the book addresses the need to design efficient FL-based algorithms to tackle the proliferating security and patient privacy issues in digital healthcare systems. The key highlights of the book are outlined as follows:

- The book provides researchers and academicians with new insights into the real-world scenarios of the design, development, deployment, application, management, and benefits of FL integration in digital healthcare systems.
- The book discusses the critical FL applications to resolve security and privacy issues that affect all stakeholders in the healthcare ecosystem and provides practical FL-based solutions to these issues.

The book features a well-organized structure comprising 15 chapters, summarized as follows:

Chapter 1 gives an overview of digital healthcare systems in an FL setting. The chapter noted that FL uses emerging technologies to improve health care by combining data while it is being computed. As the amount of health care (HC) data increases, more computation resources are needed for data manipulation. FL further involves training the shared data on a generic model using a main server while protecting sensitive information in a local facility. This procedure has a terrific potential to integrate patchy HC data databases, taking privacy into account. The chapter discusses the general idea, terms, possibilities, and challenges of FL in digital healthcare systems. In addition, a detailed explanation of how FL stands within the HC ecosystem is depicted. Finally, the chapter presents identified research questions that need to be answered for FL to be effectively employed in the medical field.

Chapter 2 focuses on the architecture and design choices for FL in modern digital healthcare systems. The study reveals that the adoption of data space practices by digital healthcare domains can create great potential for the evolution of ML solutions due to the high availability of data sources. Also, the authors have remarked that privacy protection is a key aspect to be addressed before any stakeholder engages with the concept of FL in modern healthcare systems. Thus, the chapter proposes an approach for the establishment of trust among participants by addressing all the phases of the data life cycle based on the continuous attestation of the trustworthiness of all involved functional components. The details of the attestation methodology are presented in the context of continuous provision and the availability of immutable proofs. The mechanisms for timely detecting and averting misuse of data are elaborated, and finally, a proof-of-concept implementation is presented.

In Chapter 3, the curation of federated patient data using the landscape of the African health data space is examined. The chapter analyzes new aspects of data curation in a federated, machine-actionable, and semantic format. The work presents the results of use cases developed by the Virus Outbreak Data Network research group in relation to a distributed data infrastructure for FL and privacy-preserving analytics. Specifically, the infrastructure has been deployed across 9 African countries, connecting clinical and research data from 88 health facilities, as well as data from population groups that lack access to health clinics. The chapter sets out the curational aspects, choices, tools, techniques, and prospects of the federated data platform. The findings are relevant to decisions about what parameters are needed for an African health data space and beyond.

Chapter 4 dwells on the recent advances in FL for digital healthcare systems and demonstrates how federated systems are perceived in recent times and what the important considerations are in FL development and deployment. The study notes that the FL approach helps in understanding the need for privacy preservation in a scalable and reliable way. In addition, FL facilitates the integration of cutting-edge technologies and recent innovations in patient information management, enabling the implementation of novel patient-centric care models and services. The study concludes that these advancements have the potential to expedite the digital transformation of both the economy and society.

Chapter 5 conducted a performance evaluation of FL algorithms using a breast cancer dataset. The authors noted that cancer is a raging storm and one of the leading causes of death among persons worldwide, the most dangerous being breast cancer, claiming lives at an alarming rate. The chapter focuses on the performance evaluation of two FL algorithms based on the classification of breast cancer disease as benign or malignant on a geographically dispersed data, leveraging the FL scheme. The authors aim to develop an evolving and comprehensive BC classification model, employing two edge node devices, also known as clients, to train the convolutional neural network (CNN) model used in the study on their own localized dataset, deploying the CNN model from the server. These two clients subsequently send their newly mastered parameter value and weightiness to the central server, where it totals and modifies the comprehensive CNN model. The federated averaging and federated match averaging algorithms were deployed using the BreakHis breast cancer dataset, which contains 7,783 BC binary labels for benign and malignant lumps. Each client possesses a different number of images, for example, client 1 has 6000 images, while client 2 has 2000 images. There is no communication or association between datasets belonging to different clients in the network, and the CNN algorithm visits the data on the client premises, further bolstering data privacy.

Chapter 6 presents a taxonomy for FL in digital healthcare systems. The fundamentals of FL, the underlying technologies, system architectures and challenges, and privacy-preservation methods are described. The chapter provides a comprehensive taxonomy covering critical aspects of FL. These include comparisons between centralized ML and FL, qualitative assessments of existing FL surveys, distinctions between cloud computing and edge computing, assumptions and characterizations specific to FL, discussions on security, and privacy, FL architectures, challenges encountered in FL systems, privacy considerations in FL frameworks, and an extensive exploration of applications and trending use cases. Using the faceted taxonomy method, the chapter presents a robust classification and clustering of the state-of-the-art in FL in numerous application scenarios in the healthcare domain. Finally, future open directions and challenges in FL within the healthcare system are delineated. Overall, the chapter serves as a reference point for researchers and practitioners to explore FL applications in healthcare settings.

Chapter 7 deals with the modeling of the Internet of Health Things (IoHT) using FL to support remote therapies for children with psychomotor deficits. The work emphasizes the need for a systematic approach that applies model-based solutions to develop an explainable technological innovation for the rehabilitation system that delivers remote therapy services online. Children with psychomotor deficits need therapist-assisted rehabilitation services, even if the therapy has to be online. The study advocates an IoHT architectural model that uses FL to support therapies for children with psychomotor disabilities to specify the components, services, and resources. In particular, the interactions of the main actors, such as therapists, physicians, parents, and children, are considered. The model integrates monitoring online therapies for children in conjunction with parents and specialists. The proposed model contributes to the development of a digital system for remote therapies, designed to achieve sustainability while allowing for scalability through necessary adaptations. The system can provide digital resources and services through an architecture based on intelligent software. The framework facilitates the production, organization, consumption, and evaluation of resources based on user profiles and their activities of daily living. A real scenario presents different levels of support given by the proposed model. Finally, the chapter discusses the appeal of implementing remote psychomotor therapies and assesses their impact on therapists and participants.

Chapter 8 considers blockchain-based FL in IoHT applications. The work proposes a framework employing blockchain-based FL that manages authentication, training, distribution, reputation, and trust for digital healthcare systems. The framework provides dataset encryption and inferencing processes. The results revealed that encryption techniques are utilized on every health node and hyperparameters of the model. The framework supports lightweight synthetic data to endorse the complete deidentification and confidentiality of the IoHT data. The authors employ several deep-learning applications created for medical studies involving COVID-19 patients to test the designed framework, and the results show great promise for the more extensive espousal of safe IoHT-based health management.

Chapter 9 is on the integration of FL paradigms into electronic health record systems. The authors noted that the current centralized data collection and storage methods, as well as the high scalability of modern healthcare systems, present several operational challenges that can compromise patient data security by allowing unauthorized entities to alter their medical records. Thus, the chapter provides context and details on the integration of the FL paradigm into electronic health record (EHR) systems to improve the planning, management, and integrity of valuable hospital biomedical data. First, a foundational background on this collaborative technique, its motivations, technical requirements, and potential for confidential medical informatics is presented. Next, innovative solutions to unified records, statistical concerns, communication efficiency, data security, and privacy issues are presented. Finally, the implications and possibilities for digital healthcare systems, as well as future research directions, are highlighted.

Chapter 10 examines the technical considerations of FL in digital healthcare systems. The authors observed that implementing federated networks in hospitals poses new challenges that must be addressed, citing the case of medical imaging data possessing different characteristics from FL algorithms. The authors added that the decentralized nature of the data adds to the complexity and challenges of verifying the completeness and quality of the FL-based healthcare data analysis. In addition, the chapter stated that data heterogeneity can prevent model convergence. To fully grasp these challenges and enhance the potential of FL-assisted tools, the chapter takes a cursory view of the technical considerations and challenges of FL models in digital healthcare systems and discusses technical issues in implementing FL models in practical scenarios. The critical challenges examined include bias between healthcare centers, heterogeneity of clinical data, client management and selection issues, as well as proliferating privacy and security concerns.

Chapter 11 focuses on FL challenges and risks in modern digital healthcare systems. The study emphasizes how FL enables the training of ML models using decentralized data dispersed across a spectrum of IoHT devices, encompassing smartphones, wearables (e.g., fitness trackers, smartwatches), and implantable healthcare devices (e.g., pacemakers). The authors added that FL assures the privacy and security of these devices and raw data throughout the learning process. However, integrating FL into contemporary digital healthcare systems raises challenges and risks that warrant meticulous consideration to ensure the ethical and secure utilization of sensitive patient information. Specifically, the chapter explores the multifaceted challenges, problems, and risks involved in integrating FL into digital healthcare systems. The study underscores potential solutions and effectively outlines future directions for mitigating these challenges and risks. The insights presented in the chapter serve as invaluable guidance for researchers, students, and diverse stakeholders navigating the intricate landscape of FL in digital healthcare systems, with a steadfast commitment to upholding ethical principles and security standards.

Chapter 12 presents case studies and recommendations for designing FL models for digital healthcare systems. The authors stressed the need for healthcare systems previously focused on centralized working to collaborate with the use of FL technology. The authors added that multiple collaborations can be achieved in a distributed manner where collaborators can communicate when desired, in a centralized fashion, without sharing the raw data. This is because FL deals with challenges and concerns related to data confidentiality and privacy, thus helping to minimize the risk of confidential data leakage. The chapter presents a survey and discusses various literature related to the use of FL in digital healthcare systems. Finally, the work highlights the various challenges that arise in the digital healthcare system and provides case studies and solutions to the identified problems using FL.

Chapter 13 sheds light on government and economic regulations on FL in emerging digital healthcare systems. The study reiterates that FL is undoubtedly one of the most exciting and trending technologies in intelligent healthcare systems, and regulatory policies on FL in this field are yet to be fully developed. Further, the authors noted there have been no sufficient studies of the market potential and investment opportunities for FL in smart healthcare systems. Against this premise, the chapter presents government and economic regulations on FL in emerging digital healthcare systems. The chapter investigates the market potential and investment opportunities for FL in emerging digital healthcare systems. Finally, the chapter proposes a commercialization and cost-benefit analysis.

Chapter 14 examines the legal implications of FL integration in digital healthcare systems. The authors state that governments at all levels must be explicit about the laws regulating the usage of FL in healthcare data management. The chapter highlights the legal guidelines for guaranteeing healthcare data security and privacy. The study captures the data protection programs provided by some selected jurisdictions and analyzes critical concepts such as data harvesting, ownership, custodianship, stewardship, privacy, security, legislation, and sovereignty. Additionally, the chapter discusses the legal implications of noncompliance with the regulatory requirements by stakeholders in the healthcare ecosystem. Finally, the chapter highlights critical open research issues and suggests possible recommendations for addressing the legal issues of FL integration in digital healthcare systems.

Finally, Chapter 15 presents secure FL in the IoHT for improved patient privacy and data security. The authors have observed that FL, an emerging distributed and collaborative technique, appears as a potential solution to address security and privacy concerns associated with conventional artificial intelligence methodologies. The study focuses on elucidating the prevailing security and privacy challenges encountered when deploying FL in the context of IoHT while exploring various techniques for implementing and enhancing the security and privacy of FL within the healthcare realm. The authors have explored privacy-preserving techniques such as differential privacy, homomorphic encryption, secure multiparty computation, and other methodologies that bolster the security and privacy of FL in the healthcare sector. Finally, the chapter sheds light on promising avenues for future research in secure FL within the IoHT landscape.

The book will be an ideal reference for practitioners and researchers in digital healthcare systems, as well as a good textbook for graduate and senior undergraduate courses in digital healthcare systems and bioinformatics.

We would like to thank the reviewers of the original book proposal for their constructive suggestions. Special thanks go to all the authors of the chapters for their contributions and cooperation. Many thanks go to the editors and editorial assistants for their cooperation and fine work.

The Book Editors
Agbotiname Lucky Imoize
Mohammad S. Obaidat
Houbing Herbert Song

Digital healthcare systems in a federated learning perspective

1

Wasswa Shafik[1,2]

[1]*School of Digital Science, Universiti Brunei Darussalam, Gadong, Brunei Darussalam*
[2]*Dig Connectivity Research Laboratory (DCRLab), Kampala, Uganda*

1.1 Introduction

The more technology advances, the more it is used to save the public in all aspects of healthcare (HC). Due to the increasing availability of HC data from clinical facilities, medical-related institutions, individual patients, insurance firms, and pharmaceutical companies, HC data analytics has grown in popularity in academia and industry. Drones are being used to support medicine deliveries, particularly in areas hard to reach (Shafik et al., 2020). This is a unique opportunity to avail simple and user-friendly techniques for data-driven HC delivery improvement (Frost et al., 2023).

The complexity of HC systems and operations is more data-distributive. Health facilities may only see clinical records for their patients. In most cases, these documents are classified as extremely sensitive and are commonly and medically referred to as protected health information (Shafik, 2024). The Health Insurance Portability and Accountability Act (HIPAA) safeguards patients' privacy at different sensitivity levels by regulating data access and analysis (Miao et al., 2022). Contemporary machines, such as AI and deep learning, require complex computation due to the exponential data growth in health facilities during the training and testing of these techniques (Maskeliūnas et al., 2023; Shafik, 2023a). These complexities pose significant challenges, necessitating the exploration of alternative learning methods such as federated learning (FL) (Kumar et al., 2023).

FL has become much more popular recently because it can be used to learn a lot from both aggregated and sensitive data. It lets a central (main) server train a shared generic model while protecting the local HC data from which the data comes (Imteaj et al., 2023; Jun et al., 2021; Yang et al., 2019). This is healthier than the traditional discovery and replication architecture, which involves collecting data from many places simultaneously. In contrast, it does not rely on the design of data consolidation at all.

The concept of FL has not only reached academia but has also made its way into the industry. Professor Patrick Hill taught philosophy and is responsible for

Federated Learning for Digital Healthcare Systems. DOI: https://doi.org/10.1016/B978-0-443-13897-3.00001-1

creating an FL Community in 1976 to help researchers in prominent academies deal with the ambiguity and isolation of sparsely distributed data from different data sources (Amon, 2023). Several federations of learning and content repositories were created afterward (Shafik Matinkhah & Etemadinejad & Sanda Shafik Matinkhah Etemadinejad et al. 2020, 2023). According to Peng et al. (2023), a reference model for assembling an interoperable repository structure by federating repositories and collecting information into a central archive (referring to the registry in data science) with a discovery-specific data point has been accessed since 2005. This model enables content repository learning. FL community or search service techniques have helped model FL algorithms.

According to Min et al. (2019), FL enhances HC analytics in modeling a standard for predicting the risk of facility readmission based on patient electronic health records (EHRs) and in developing patient-centered applications using smart technology (Dobischok et al., 2023). At this point, sensitive patient data can be kept at a local facility or with authorized individuals, protecting patients' privacy. This study analyzes FL's setup and issues and visualizes HC applications using FL. This study provides a more in-depth introduction to FL and proceeds to discuss the most significant concerns and recent advances in this research paradigm. Following the demonstration of the success of recent studies, we illustrate the potential application of FL approaches in the HC industry.

Over the past few years, there have been a limited number of reviews specifically addressing FL. Pontillo (2023) presented one of the initial FL studies, a survey that summarizes the general strategies that can be applied to FL to preserve patients' privacy. Some researchers conducted a review of the subproblems that are associated with FL. These concerns include customization approaches (Kumari et al., 2023), semisupervised learning (Lukauskas & Ruzgas, 2023), portable edge computation (Zhu et al., 2023), and data risk protection (Wang, Li, et al., 2023a). Other studies further emphasized contemporary advancements and comprehensively compiled open research concerns that need attention to ensure proper application in digital HC (DHC) (Díaz & García, 2023).

The viewpoint of Li et al. (2020), who researched FL from a system perspective, is in contrast to prior reviews. The current studies were explored, showing the possibility of FL being utilized in the HC industry. Moreover, problems in the FL scenario in DHC and a survey of many representative FL methods for the HC industry are presented. In the previous section of their analysis, we reviewed some aspects of collaborative learning for HC, including guidance or open-ended questions that, once solved, can be implemented without privacy or security concerns. Their draft of these findings is demonstrated in Shafik et al. (2024).

1.1.1 Chapter organization

The remainder of this chapter is structured as follows. Section 1.2 broadly describes FL in general, involving statistical challenges of FL and possible resolutions such as consensus pluralistic solutions. The section further explains security and privacy concerns such as differential privacy (DP), secure multiparty (MP) computation, FL communication efficiency, model compression, update reduction, and P2P (peer-to-peer) learning. In Section 1.3 different applications of FL in HC are demonstrated. Section 1.4 illustrates the future trends of FL as it is currently utilized in the medical sector. These include increased use of wearable technology, personalized medicine, improved patient outcomes, increased data privacy, remote patient monitoring, collaborative HC, real-time decision-making, improved drug discovery, predictive analytics, improved clinical trials, enhanced medical imaging, continuous learning, improved resource allocation, increased patient engagement, and improved public health. In Section 1.5 the most prominent challenges of FL are demonstrated in data privacy and security, heterogeneity of data, data quality, communication overhead, regulatory compliance, and resource constraints. Section 1.6 presents the open research questions that include model precision, personalization, incentive mechanisms, expert knowledge incorporation, and data quality. Section 1.7 involves the conclusion and future research direction.

1.2 Federated learning

The primary objective of FL is to effectively train a high-quality generic model by means of a main (central) server using distributed data, which involves uploading a substantial amount of diverse user data, as shown in Fig. 1.1. To simplify the explanation of FL, we present it mathematically. Let us assume there are x

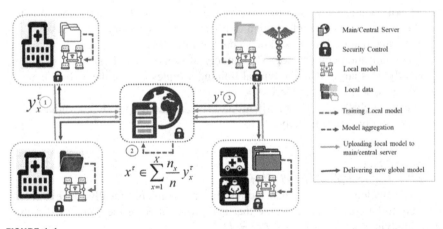

FIGURE 1.1

Federated learning schematic framework overview. Healthcare facilities upload of local raw data using their local model to the central server (main server) to train the global model. The main server merges updates and returns global model parameters to all local models for better computation.

active users, considering that the data is found in different forms and on several devices (for example, it could be on a smart device or system or in a clinical institution, among other applications). D_x represents the data distribution associated with client$x$; n_x represents the available number of samples from that particular HC facility; $n = \sum_{x=1}^{X} n_x$ represents the total number of usable samples. The FL problem can be summed up as the solution to an experiential risk reduction previously discussed by Gao et al. (2023) and Šajina et al. (2023), which is expressed as follows:

$$\min_{y \in \mathbb{N}^k} F(y) := \sum_{x=1}^{X} n_x F_x(y) \text{ given that } F_x := \frac{1}{n_x} \sum_{x=1 \in D_x} f_i(y) \tag{1.1}$$

where y represents the parameter of the model that needs to be learned. The function f_μ is defined by a loss f_x, which is determined based on a set of input−output (I/O) data pairs $\{P_i, Q_i\}$. Given that $P_i \in \mathbb{N}^k$ and $Q_i \in \mathbb{N}$ or $y_i \in \{-1, 1\}$.

Sample applications for linear and logistic regressions are expressed in Eqs. (1.2) and (1.3), respectively, and for support vector machines in Eq. (1.4).

$$f_i(y) = \frac{1}{2} \left(P_i^T y - yi \right)^2, \quad y_i \in \mathbb{N}; \tag{1.2}$$

$$f_i(y) = -\log\left(1 + \exp\left(-yi P_i^T y\right)\right), \quad y_i \in (-1, 1); \tag{1.3}$$

$$f_i(y) = \max\left(0, 1 - y_i P_i^T y\right), \quad y_i \in (-1, 1) \tag{1.4}$$

According to Zhuang (2020) and Li et al. (2023a), FL algorithms encounter a range of concerns, which include the following:

- Security and privacy concerns: Implementing additional privacy safeguards for clients who cannot be relied upon to participate is necessary. It is hard to guarantee that each user is as dependable as the others.
- Communication competence: The number of users x is high and may vary greatly; it is larger than the typical training samples kept in the active clients' storage spaces, for example,$x \gg (n/x)$.
- Statistical challenge: The distribution of the data across all clients varies substantially. For simplicity,

$\forall_x = \tilde{x}$,
Therefore,
$ZP_i \sim D_x[f_i(y; P_i)] \neq ZP_i \sim D_x[f_i(y; P_i)]$.
The situation is that the data points that are available locally are extremely unlikely to be considered a sample that is representative of the complex distribution, implying that $ZP_i \sim D_x[f_i(y; P_i)] \neq f_i(y)$.

Furthermore, we conduct a comprehensive analysis of the existing studies on FL strictly related and relevant to managing the three identified concerns from the DHC perspective.

1.2.1 **Statistical challenges of federated learning**

Federated averaging (FedAvg) addresses the FL challenge in a simple manner (Lee, 2022). Requiring all users to use the same model makes it function with independent identical distribution data. FedAvg does not address the statistical challenges arising from distorted data distributions. Weight divergence can lower the performance of FedAvg-trained convolutional neural networks. Studies on consensus resolution and multiethnic solution directly address the statistical complexity of FL.

1.2.1.1 *Consensus solution*

The overwhelming majority of mainstream models are trained using aggregated training samples obtained from experiments involving local patients (Zhang, 2023). The primary model is trained with the inherent objective of reducing the relative distribution loss (Sattler et al., 2021), as expressed in Eq. (1.5).

$$\tilde{D} \in \sum_{x=1}^{X} \frac{n_x}{n} D_x; \qquad (1.5)$$

Given that, $\tilde{D}$ represents the target distribution data that the learning technique aims to achieve. Nevertheless, this particular sort of uniform distribution is not an appropriate answer for the vast majority of situations.

The most recent studies have been examining how to solve this problem by obliging the data to conform to a uniform distribution (Sattler et al., 2021). This allows the problem to be resolved. In particular, a minimax optimization strategy, also known as agnostic FL (AFL), has been introduced. The approach involves augmenting the main standard for potential target distributions resulting from the mixing of user distributions. The approach has never been implemented on a larger scale. In contrast to AFL, q-Fair FL (q-FFL) was proposed by Shafik et al. (2023b), which involves providing a higher weight to gadgets with poor performance. This is done to reduce the variance in the accuracy distribution across the network. They provide quantitative evidence demonstrating how q-FFL is superior to AFL in terms of flexibility and adaptability.

Generic data exchange is also considered prevalent (Razavi-Far, 2023). The main server and clients must share a common subset. Sharing small fractions of dependability and noise configurations can assist local medical users in choosing a miniature training subset. As users become aware of adjusting selected data samples, it enhances the performance of the generic model during test operation (Miao et al., 2022). Sharing this information helps solve the nonindependent complexity.

1.2.1.2 *Pluralistic solution*

In the case of multitask learning (MUL), it is technologically possible to accommodate various distributions. It involves a generic model based on nonindependent and imbalanced data. Target sparsity, graph-based relatedness, and low-ranking structure. An analytical method that proves this assertion using real-time federated dataset(s) and recommends a unique methodology (MOCHA) to tackle a common convex MUL issue while addressing system restrictions is demonstrated by Zhao (2023).

Furthermore, an established virtual nonconvex MUL-FL approach involves training the main server-client federation as a Bayesian network using estimated variational inference. This work relates to federated and transfer/continuous learning (Jiang et al., 2023). The MUL approach is based on the assumption of relatedness. Pluralism handles heterogeneous data without using MUL. Also, a basic pluralistic strategy can minimize data heterogeneity in block-cyclic training. Pluralism outperforms "perfect" independent data when component distributions diverge (Wojtowytsch, 2023).

1.2.2 Security and privacy

FL involves thousands or millions of customers (phones, automobiles, clinical institutions, etc.). No user is trustworthy regarding privacy and security (Lu, 2023; Shafik et al., 2021). FL prevents the risk of direct leakage of data by internally training the model without sharing input or output data with users. $f(y)$depicts the shared prediction model; y will possibly reveal particular data about an additional user's private data stores. For this purpose, numerous efforts have been made across various HC perspectives to focus on confidentiality at both the MP and individual level, particularly in the context of social media, where MP privacy concerns have significantly aggravated (Amon, 2023; Shetty, 2023), as presented in Fig. 1.2.

1.2.2.1 Differential privacy

DP technically refers to an alternative data security model employed for privacy protection (Pei et al., 2023; Tominaga & Masataka, 2023), such as support vector machine (Sirisha & Chandana, 2023), principal component analysis (Kamila et al., 2023), and other DL studies (Cheng et al., 2021; Hongbin & Zhi, 2023). It maintains against ancillary leakages; lossy DP prevents patients (users and medics) data from leakage only partially and may reduce prediction accuracy (Valente et al., 2023). Accordingly, some studies employ DP and secure multiparty computation (SMPC) to limit noise input and guard against extraction assaults and collusion (Rahman et al., 2023).

1.2.2.2 Secure multiparty computation

FL clients jointly compute a function using SMPC and oblivious transmission (Nazir & Kaleem, 2023). In homomorphic encryption (Alicherif, 2023), any event can encrypt its information using defined public keys and calculate with data coded by others using similar keys. FL studies enhance the success of cloud computing (Shanmugarasa et al., 2023). SMPC cannot prevent an adversary from

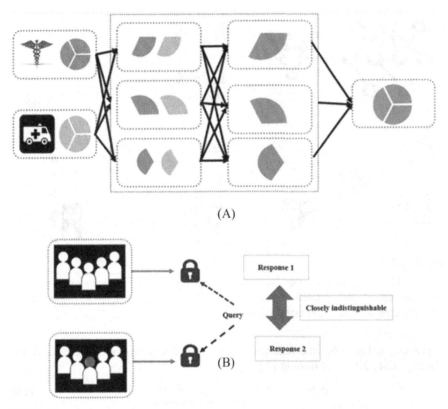

(A)

(B)

FIGURE 1.2

Privacy-preserving systems. (A) The developed models divide security shares (yellow and blue pie). No model can learn the original value or output (green pie). Any node can reclaim its shares. (B) Ensures that differentially private assessments provide the same result (response 1 and response 2 are nearly homogeneous).

gaining access to specific information, such as determining whether the absence of a client may affect a classifier's decision threshold. For fundamental challenges, SMPC strategies require repeated encryption or decryption and participant communication regarding encrypted findings (Elmi et al., 2023).

1.2.3 Federated learning communication efficiency

In an FL setup, training data often continues to be dispersed across a large number of users, each of which will have an unpredictable and somehow sluggish nature in network connectivity. The absolute amount of bytes required for communication uplinking and downlinking by every $x$user during their training (as

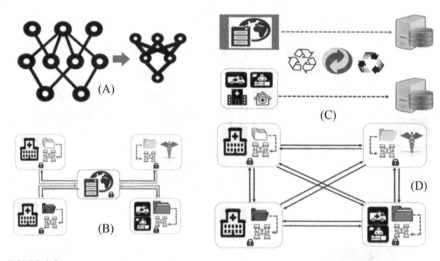

FIGURE 1.3

Effective FL communication methods; studies on communication efficiency include (A) approach compression. (B) User selection. (C) Update reduction. (D) P2P learning. *FL*, Federated learning; *P2P*, peer-to-peer.

expressed in Eq. 1.6) is calculated when using a synchronous protocol for FL (Huang et al., 2023; Jahromi et al., 2023).

$$\lambda^{up \; or \; down} \in \theta^* \psi^* (\xi^* \underbrace{|y|^* H(\Delta_y^{up \; or \; down})}_{updating\_size} + \beta))$$

(1.6)

where ξ is the aggregate summation of updates accomplished by every member; $|y|$ represents and consists of the dimensions of the algorithm; $H(\Delta_y^{up \; or \; down})$ denotes the updated weights of entropy that are traded while the transmission is taking place; and finally, β shows the discrepancy between the actual size of the update and the smallest possible size of the update as demonstrated by Batool et al. (2023).

Three approaches can be used to lower communication complexities, such as reducing$x$, Ψ, and updating size. From these three aspects, it is practically possible to allocate the communication-efficient FL approaches into four clusters: approach compression, user selection, update reducing, and P2P learning; as demonstrated in Fig. 1.3, each approach is detailed.

1.2.4 User selection

Restricting users in the HC domain or updating a portion of parameters in each round reduces communication costs. A study by Obermeyer et al. (2019) demonstrated a discriminatory stochastic protocol that can pick all parameters, or simply those with more significant gradients, thus reducing execution. Anh et al. (2019) presented a dubbed FedCS protocol. In this protocol, the main server controls the sources of various users and evaluates each user's stored information (data) to select which users ought to contribute to the electric current task training.

Relevant resource data (information) examples include wireless channel statuses, processing capabilities, and the number of relevant data resources. At this point, the main server is supposed to evaluate precisely how considerably energy central processing unit (CPU) data resources are available for the mobile devices to control energy utilization, training life span, and bandwidth expenses while still meeting the needs of training activities at the local level and subsequent levels. Subsequently, researchers presented the Deep Q-Learning method (Akter et al., 2022) suggesting techniques that facilitate the main servers to verify the relevant energy and data needed by smart devices participating in mobile phone Crowd-ML (machine learning) via FL without prior network dynamics experience. This would permit the server to locate the ideal solution to the issues.

1.2.5 Model compression

Some approach compression reduces uplink or downlink communications complexities by compressing server-to-the-client interactions. Structured updates are promptly discovered from a constrained space parameterized by a reduced variable, for example, sparse and low ranking or lopping the minor significant networks quantization of weights and distillation of the existing models (Hu et al., 2023). Another compression approach involves FL, a complete approach that updates and compacts using quantization or randomized rotation and subsampling before transmitting responses to the main server. Decoding main server updates can be done by aggregating all incoming data uploads.

1.2.6 Updates reduction

Jin et al. (2023) suggested that each user can train a submodel instead of the generic model. Because these submodels are subsets of the generic model, the processed local updates are natural, and technically, that reduces unnecessary updates at the main server. FL dropouts decrease the number of downlink communications and uplink updates. Due to the smaller dimensions of the parameters utilized by the local training technique, the local computation costs are minimized.

Dynamically based communication utility averaging models reduce transmissions to a level comparable to high-tech systems that frequently communicate. This is suitable for large distributed systems with poor communication infrastructure. Partitioned variational inference improved the FL of the Bayesian neural network. After many data runs in one local epoch, the users upload parameters to the main server. One-shot FL uses network data to develop a generic model. Furthermore, achieving learning efficiency in the context of CPUs and discovering an appropriate training momentum become important. To solve this, round-robin and reinforcement learning are often employed to manage computation and communication resources (Wang, Yi, et al., 2023b).

1.2.7 Peer-to-peer learning

In FL, a main server synchronizes generic model training. As a result of the number of clients involved, communication with the main server may be very expensive. Numerous practical P2P networks are dynamic and rarely connect to a main server. Due to the existence of a main server, all users must correspond to a single, trustworthy main body, the failure of which would terminate the training process for all clients. Consequently, researchers (Chen et al., 2023a; Tian et al., 2023) studied fully decentralized frameworks without a main server. Local medical users only connect to a single-hop network with the neighbors. Based on its data, each user updates its local assumption and then aggregates data from its one-hop neighbors.

1.3 Federated learning data processing: as still, stream, and multistream

This subsection demonstrates insights on data processing in FL, encompassing still, stream, and multistream.

1.3.1 Data processing at rest

The collection, storage, and subsequent batch processing of data from numerous sources constitute data processing at rest (DPR), usually referred to as 'Still' processing in the context of FL (de Miras et al., 2023). In this approach, data is gathered over time and processed simultaneously rather than instantly. A brief explanation of DPR is presented in the following sections.

1.3.1.1 Data collection

This is the first stage of the DPR and involves data accumulation and storage. Under the data accumulation phase, the data is continuously generated from various sources in the HC ecosystem. This data can include medical imaging, patient surveys, EHRs, and genomic data, among other HC data types and categories (de Miras et al., 2023). Depending on the architecture of the HC system, the gathered data may be stored in a centralized repository or among multiple distributed data sources. These storage options include data lakes, databases, and safe cloud environments.

1.3.1.2 Batch or interval-based processing

At this stage, the data is collected in batches and then processed in intervals. The interval-based processing, as opposed to processing data as it arrives in real-time, functions on a periodic schedule or at predetermined intervals, for example, regular, weekly, or monthly. The method enables batch processing, in which data is processed in bulk during certain predetermined times through four practical processes, including data preprocessing, medical AI-assisted model training, model aggregation, and improvement (Nair et al., 2023).

1.3.1.2.1 Data preprocessing
Data may go through preprocessing operations to clean, standardize, and transform information into a suitable format for ML before model training starts. Scaling, feature engineering, and handling missing values are examples of these actions.

1.3.1.2.2 Model training
The complete batch of data is used to train ML models such as neural networks, decision trees, and regression models (Zhou et al., 2023). Typically, a central server is used to initialize these models, and training involves tuning model parameters to fit the dataset.

1.3.1.2.3 Model aggregation
ML algorithms are trained utilizing the whole data set, including neural networks, decision trees, and regression models. The initialization of these models typically occurs on a central server, and training involves tuning model parameters to fit the dataset.

1.3.1.2.4 Model improvement
The aggregated model updates contribute to improving the model's universal performance. After each batch processing cycle, model updates (gradients) from various local sources are compiled at the central server (Nagy et al., 2023). Without disclosing patient data, this aggregation technique synthesizes information from all participating institutions.

1.3.1.3 Merits of data processing at rest
In terms of efficient resource utilization, processing data in batches allows for efficient use of computational resources, as optimizations such as distributed and parallelization computing can be applied. Furthermore, FL-assisted models trained on complete datasets exhibit greater model convergence and consistency, contributing to model stability.

1.3.1.4 Considerations on data processing at rest
Amidst the demonstrated benefits, latency, data volume, and scheduling are some of the critical aspects. DPR may cause a delay in getting information or making choices, making it unsuitable for applications that need instantaneous responses (Zhou et al., 2023). Correspondingly, applications needing real-time responses might not be suitable for DPR since it could cause a delay in the acquisition of insights or judgments. Batch processing cycles must be properly scheduled and coordinated to ensure data is processed consistently and dependably.

1.3.2 **Data processing as a stream**

Data processing as a stream is a second method in the FL framework that involves real-time or near-real-time data processing, with data arriving from various sources. The method is particularly suitable for HC applications that require adaptability to changing data patterns and instantaneous responses (Calbimonte et al., 2023). As illustrated below, data processing as a stream involves two main operational phases: data ingestion and stream processing.

1.3.2.1 Data ingestion

This stage has two approaches, that is, continuous and continuous data streams. For a continuous data stream, data is uninterruptedly generated in real-time or near-real-time from various sources in the HC ecosystem. The data could be obtained from various IoT (Internet of Things) device sources, including patient monitoring systems, medical wearable devices, and IoT sensors (Jiang et al., 2023). Data streams are ingested as they are produced, frequently through data pipelines or stream processing frameworks such as Apache Flink or Apache Kafka, and many more. These systems enable gathering, transforming, and routing data to the proper processing parts.

1.3.2.2 Stream processing

Under stream processing, simultaneous processing is considered. Rather than waiting for data to tally over time, data processing as a stream processes data instantly as it enters the FL-assisted system. Complex event processing, data augmentation, and analysis may be a part of this real-time processing (Nair et al., 2023). This is possible through three main procedures: data preprocessing, continuous model updating, and model aggregation.

The data preprocessing stage is the initial preparation step used to clean, validate, and enrich incoming data. This can involve eliminating background noise, dealing with missing data, and combining data from several sources (Qin et al., 2023). This is followed by continuous model updates. In FL, ML models are continuously updated as new data streams in. The local models at participating HC or medical institutions receive updates in real time as data is processed, allowing models to adapt and learn from the most recent data (Chen et al., 2023b). Finally, model aggregation, which combines data from several sources, may still occur periodically, even while model updates are made on local models. However, because this aggregation occurs more often than batch processing, the global model is always current with the most recent information.

1.3.2.3 Advantages of data processing as a stream

Since data processing as a Stream is excellent for applications where prompt replies and real-time judgments are crucial, this technique has shown improved simultaneous decision-making. This includes tools for early disease identification and remote patient monitoring, increasing model adaptivity. The continuous updates enable AI models to rapidly adapt to changing data patterns, emerging trends, or critical events such as remote patient monitoring and telemedicine alerts (Zhao, Mao, et al., 2023).

1.3.2.4 Considerations

Data processing as a stream is a valuable approach in FL, particularly for HC applications where timely responses and adaptability to dynamic data are crucial. As long as data privacy and security are upheld, HC practitioners can make quick judgments, monitor patients in real time, and react quickly to new HC scenarios. Nevertheless, due to its low latency, stream processing is appropriate for real-time applications (Rajagopal et al., 2023). This calls for the optimization of systems for low-latency processing. Real-time data processing can be resource-intensive and necessitates a solid computational architecture to manage data streams effectively (Xu et al., 2021). Managing massive amounts of incoming data streams can also be difficult, and systems need to be scalable to handle growing data flow.

1.3.3 Multistream data processing

Multistream data processing (MSDP) is another strategy used inside the FL framework that involves processing data concurrently from numerous streams, each representing a different data type or source. The method is beneficial in HC applications where combining several data sources is required to provide a thorough knowledge of a patient's health.

1.3.3.1 Data streams

This involves diverse data sources and continuous data ingestion. For diverse data sources, MSDP involves the simultaneous processing of data streams from multiple sources throughout the HC ecosystem. These sources may include wearable device data, medical imaging, genomics data, patient questionnaires, EHRs, and more (Nagy et al., 2023). Furthermore, there is continuous data ingestion. Each source's data is continually and in real-time or almost real-time ingested. These data streams could be produced at various frequencies and speeds.

1.3.3.2 Stream processing

At this level of FL computation, parallel data processing, simultaneous model updates, and integrative analysis are carried out. First, data obtained from several streams is processed in parallel (parallel from each source), with each stream undergoing unique preprocessing steps (Calbimonte et al., 2023). These preprocessing procedures may involve feature engineering, data normalization, and cleansing specific to each data type. Second, depending on the peculiarities of each data type, these preparation stages may also involve feature engineering, normalization, and data cleansing. Lastly, in integrative analysis, the processed data from several data streams is merged and analyzed simultaneously to obtain a comprehensive knowledge of a patient's health or a HC scenario (Jiang et al., 2023). Data combining at different levels, for example, gathering patient records from EHRs, superimposing medical pictures, and correlating genomic data, can be done as part of this integration.

1.3.3.3 Advantages of multistream data processing

MSDP has demonstrated a comprehensive analysis that enables instantaneous incorporation and analysis of various HC data sources. This strategy affords a more all-inclusive view of a patient's health, medical, or HC state of affairs (Qin et al., 2023). Furthermore, in the realm of personalized HC, HC professionals can develop individualized HC interventions that take into consideration an individual's comprehensive health profile, combining clinical, genetic, lifestyle, and environmental aspects by taking data from many streams into account. Some use cases include inclusive patient monitoring that integrates data from wearable devices and drug discovery (Chen et al., 2023b).

1.3.3.4 Considerations of multistream data processing

MSDP is a potent strategy in FL, especially for HC applications that require a thorough and customized understanding of patient health. HC practitioners can make better judgments, adapt treatments, and gain deeper insights into complicated HC scenarios by processing data from multiple sources while adhering to data privacy and security regulations (Zhao, Mao, et al., 2023). Nevertheless, data integration, resource scalability, and data governance have become critical. A clearly defined data integration strategy is necessary to ensure that data from various streams is successfully integrated and evaluated (Rajagopal et al., 2023). Real-time data processing from numerous streams may require a lot of computational power and scalability techniques. Data governance rules must be in place to preserve data security, privacy, and quality across various data sources.

1.4 Applications

FL is used in many domains. It is comprehensively used because it is a cooperative modeling method that allows for efficient ML while retaining data or information privacy and legal conformity across numerous compute nodes. Examples include virtual keyboard prediction, smart shopping, banking, and drone-to-drone communications. We focus on securing DHC applications in the HC sector, but critical principles can also be extended to other industries.

1.5 Healthcare

EHR data using ML face challenges since hospitals have systemic and random biases, impeding generalizability. Standard HC enrollment algorithms abused some stakeholders by delegating identical risks to healthier Caucasians (Guo et al., 2023). Inadequately calibrated algorithms can result from unequal admission to HC or an absence of training data representation. The use of generic EHR data decreases algorithm bias. Data will never be linked in a single database for learning due to data methodologies and privacy concerns. Standardized shared data models (Behera et al., 2023), for instance, observational medical outcome partnerships (Curchoe, 2023), enable more replication analyses while restricting cooperative data access. As a result, EHR data learning must progress beyond discovery-replication. FL has the potential to enable ML on a large-scale using EHR data.

It has been experimented further that FL can attach data in EHR from HC facilities and permit them to reveal their encounters, not information, with confidentiality (Dobischok et al., 2023; Hongbin & Zhi, 2023; Jin et al., 2023; Lu, 2023; Zhang, 2023). FL from extensive and diverse medical data sets will significantly increase ML model operation in these instances. Patient resemblance (Hu et al., 2023), patient representation, phenotyping and learning, and extrapolative modeling have been explored in FL settings in HC. Privacy-preserving federated framework, the model can detect patients across HC facilities without revealing patient data. Tensor factorization prototypes aim to transform huge EHRs into suitable phenotypes for FL data analysis (Chen et al., 2023a).

Hu et al. (2023) present a combination of patient representation learning and obesity phenotyping with good scores. Shenaj et al. (2023) have applied SplitNN to assist health entities in developing DL models without exposing sensitive data or model specifics. Lohmeyer et al. (2023) discuss clinical cohort brain structural correlations to determine their FL architecture, such as clustering nonindependent intensive care unit (ICU) patients into clinically substantial communities with geological locations and comparable diagnoses and training one model per HC facility.

FL allows analytical modeling from multiple sources, helping medics assess the risks and benefits of early treatment. The sparse 1 standardized SVM classifier in FL is used to predict cardiac patient hospitalizations utilizing EHR data from several data sources or agents. FL helps to forecast pharmacological, therapeutic, and infection survival. An uncertainty-aware FL model to analyze the birth data from scattered EHR reduces the high-uncertainty model contribution to the aggregation model, estimated hospital stays, and deaths for 31 institutions in the electronic ICU collaborative research database. Shin et al. (2023) demonstrate a privacy-protected framework for forecasting in-hospital mortality in ICU patients. FL's trained model performs more like primary learning, as summarized in Table 1.1.

1.6 Other applications

Google proposed FL in 2016 utilizing Gboard, a mobile touchscreen keyboard that supports more than 600 languages (Stragapede et al., 2023). As the number

Table 1.1 Overview of recent work on federated learning (FL) for healthcare.

FL concerns	AI method	# of user	Dataset
Mortality estimation (Dubowski et al., 2023)	LRR, MLP-LASSO	5	Mount Sinai COVID-19
Data analysis on images (Singh et al., 2023a)	VAE	4	MNIST, Brain imaging
Disease estimation (Singh et al., 2023a)	NN	10, 5	Cleveland Heart Disease DatabasePima Indians Diabetes Dataset
Arrhythmia recognition (Narayan et al., 2023)	NN	64, 32,16	PhysioNet
Prediction on adverse drug reactions (Melnikov et al., 2023)	LR, MLP, SVM	10	MIMIC, LCED
Activity detection (Singh et al., 2023b)	CNN	5	UCI Smartphones
Mortality estimation (Wang et al., 2023c)	MLP, LR	2	MIMIC-III
Mortality estimation (Yang et al., 2023)	NN, LR	31	e-ICU CollaborativeResearch Database
Preterm-birth estimation (Cozzi et al., 2023)	RNN	50	Cerner Health Facts
Hospitalization estimation (DiLorenzo et al., 2023)	SVM	10, 5	Boston Medical Center
Mortality estimation (Zadorozhny et al., 2023)	Autoencoder	5–50	e-ICU Collaborative Research Database
Representation learning (Krishna et al., 2023)	PCA	10–100	PPMI, MIRIAD, UK Biobank, ADNI
Phenotyping (Tanwar et al., 2023)	NLP	10	MIMIC-III
Phenotyping (Li et al., 2023b)	TF	1–5	UCSD, MIMIC-III
Similarity learning of patient (Ghosh & Ghosh, 2023)	Hashing	3, 20	MIMIC-III, MIMIC-III

AI, *Artificial intelligence*; ICU, *intensive care unit.*

of mobile device users continues to rise, cellular response techniques must involve autonomous rectification, word accomplishment, and next-word extrapolation. For specific natural language processing tasks involving next-phrase estimations, typed text in smart gadgets is more favorable than data from visualized manuscripts or speeches to texts. Nonetheless, this linguistic data frequently presents passwords, search queries, and private text messages. FL can anticipate virtual keyboards in natural language processing (Shafik et al., 2022).

Table 1.2 Identified popular tools for federated learning (FL) literature.

Name of the ML project	Developer	Description
Tensor/Input or Output (Doc.ai: declarative, 2023)	Dow and others	Tensor/Input or Output brings TensorFlow and TensorFlow Lite to mobile operating systems, Android, Windows for smartphones, reactive, and native apps
FATE (Shafik, 2023c)	Webank	FATE supports the Federated AI system's ciphertext encryption-based MPC-secure computation protocol
TFF (Google: Tensorflow federated, 2023)	Google	TFF lets TensorFlow operators mimic distributed computation locally
PySyft (Campagner et al., 2023)	OpenMined	PyTorch's FL, DP, and MPC disconnect private data from model training. TensorFlow extensions

AI, *Artificial intelligence;* DP, *differential privacy;* ML, *machine learning.*

Smart retail and finance are other applications (Oualid et al., 2023) of FL. The intelligent retail environment employs ML to recommend and sell products based on consumer purchasing power and product characteristics. Some banks operate FL technology for credit risk supervision, allowing many institutions to verify a user's credit scores without exchanging information (Goscinski et al., 2023). As FL has expanded, several industries and research groups have established scientific research and product development technologies. Table 1.2 summarizes the other applications of FL.

1.7 Federated learning in healthcare ecosystem

In this subsection, FL is illustrated in the context of the HC sector.

1.7.1 Privacy preservation

Due to the sensitive nature of patient information and the stringent regulatory environment (e.g., HIPAA in the United States), data privacy is of uttermost significance in the HC ecosystem. FL tackles this issue using privacy-preservation methods, including safe aggregation, DP, and homomorphic encryption (Melnikov et al., 2023). These methods guarantee the confidentiality of unprocessed patient data. FL enables medical facilities to train regional models using their data, sending only model updates (gradients) to a centralized server. Sensitive information is decentralized and secured due to the encryption used in these upgrades (Yang et al., 2023). This strategy promotes cooperative model improvement while ensuring compliance with data privacy laws.

1.7.2 Healthcare data fragmentation and silos

The majority of the hospitals, clinics, and research facilities keep HC data in discrete silos that are infamously dispersed and fragmented. FL facilitates decentralized model training to eliminate these data silos (Zadorozhny et al., 2023). Only model updates are exchanged, and each institution trains a local model on its data. The central server can integrate these updates to cooperatively enhance the global model using federated aggregation techniques, such as FedAvg (Li et al., 2023b). To address the problems caused by data fragmentation, FL promotes data exchange and research collaboration throughout the HC ecosystem.

1.7.3 Data heterogeneity

The HC ecosystem works with various data types, such as structured EHRs, genomics data, sensor data from IoT devices, and unstructured medical images. FL has demonstrated that it can allow local models to be tailored to handle particular data types to accommodate this heterogeneity (Ghosh & Ghosh, 2023). Artificial networks, for instance, recurrent neural networks, can be used for time-series data, whereas convolutional neural networks can be used for medical image data. These numerous model updates are combined via the global model's aggregation method, creating a flexible and adaptable AI model that can successfully process various HC data types (Stragapede et al., 2023).

1.7.4 Regulatory compliance

Globally, robust regulatory frameworks are in place for the HC industry, and noncompliance can have serious legal repercussions, as seen in developed countries. FL complies with these rules by implementing encryption, access controls, and auditing systems. This is demonstrated when only authorized personnel can partake in FL procedures due to access constraints (Oualid et al., 2023). All model updates and access are tracked in in-depth audit logs for compliance and auditing purposes. Homomorphic encryption and safe multiparty computing are two encryption methods that protect patient data both during transmission and processing (de Miras et al., 2023). These procedural safeguards make FL comply with all applicable data protection requirements, such as the California Consumer Privacy Act in California, USA, and the Data Protection Directive in European Union nations.

1.7.5 Healthcare resource efficiency

Large-scale AI model training can be computationally demanding and require a lot of processing and storage power. FL reduces the data centralization requirement, which helps decrease this problem. As an alternative, local data from each institution is kept on-site, lessening the computational load on any organization.

Methods such as model compression and federated optimization, which cut communication costs and computing demands, further improve resource efficiency (Nagy et al., 2023; Yang et al., 2023). Because of this, FL encourages equal access to cutting-edge AI capabilities by allowing even tiny HC providers to participate.

1.7.6 Bias mitigation

Unfair or erroneous medical AI-assisted models can produce HC data bias, which may instigate demographic imbalances or other factors. To address bias issues, FL technically uses weighted federated learning (Wang et al., 2023c). According to the quantity and quality of their datasets, local models are given weights. A balanced representation of data sources is ensured by trained models on more extensive and representative datasets (Calbimonte et al., 2023). More still, local applications of data preparation methods, such as oversampling underrepresented groups, are possible with FL.

1.7.7 Collaborative learning and healthcare research

Collaborative research and learning are essential for advancing medical knowledge, smart HC services, medical AI-assisted research, and enhancement of patient care in the HC industry (Behera et al., 2023). By enabling numerous stakeholders, for instance, physicians, medical professionals, public health departments, HC researchers, and HC organizations, to cooperatively train and develop AI models, FL promotes collaboration. Predictive analytics, therapy recommendations, and diagnostic precision are all improved by this combined effort. With the advancement of FL, researchers collaborate to build models, share knowledge, and jointly address HC concerns (Yang et al., 2023). FL's collaborative nature accelerates HC research and development, leading to more effective and innovative solutions.

1.7.8 Real-time medical monitoring and edge computing

Real-time monitoring and decision-making are essential in HC scenarios such as remote patient care and emergency response. To provide real-time model updates and decision support, FL can be customized for edge computing (since it brings computation close to the edge of the network) and IoT devices (due to increased resource pooling) (Cozzi et al., 2023). Wearable health monitors and IoT sensors are edge devices that may process data locally and make predictions using locally trained algorithms. For updates and model upgrades, these devices periodically synchronize with a central server (Krishna et al., 2023). Even in situations with limited resources, FL ensures that HC professionals have quick access to reliable information for patient care and actions at any time of the medical process.

1.8 Federated learning open research questions

This section adds to the previously summarized general issues that arise in an FL environment. Therefore the open research questions depend on the surveyed literature.

1.8.1 Model precision

FL urges remote institutes or devices to distribute their data (information), consequently enhancing ML model performance (Huang et al., 2023; Jahromi et al., 2023). Currently, prediction is restricted and simplistic. Medical treatment is reasonably competent, and wearable medical gadgets are inferior to hospital-based technologies. Based on this fact, Doc.ai processes might redefine data phenotypes, for example, age, sex, weight, and height. Ways through which prediction models can be improved to forecast future health concerns are worth an investigation.

1.8.2 Personalization

Wearables are focused on public health, assisting healthy individuals in exercising, meditating, and sleeping better. It is significant to assist patients in executing scientifically created individualized HC, correcting functional pathology by evaluating indicators, and halting pathological change (Hongbin & Zhi, 2023; Li et al., 2023b). Reasonable chronic illness treatment can lessen the need for hospitalization and emergency room visits. For HC informatics, there is some generic work on FL personalization (Campagner et al., 2023; Hongbin & Zhi, 2023). What are the available merging measures for the medical information (data) domain and equipment utilization for each medical facility or wearable device?

1.8.3 Incentive mechanisms

With technological advances and increased device connectivity to the IoT and third-party gateways, the number of smart FC applications is on the rise in response to their compatibility with other wearables (Shafik, 2023c; Wang, Yi, et al., 2023b). Like hospital data, wearable device data is relevant to researchers and device owners. With increased latency, clients suffer substantial communication and computation costs during federated model training. Another concern is building a practical incentive approach to entice high-level-excellence data devices to FL.

1.8.4 Expert knowledge incorporation

In early 2016, IBM used a natural language processing engine to summarize patients' EHRs and searched its colossal database to advise clinicians on therapies

(Obermeyer et al., 2019). Regrettably, some oncologists trust their judgment above Watson's advice. Thus, doctors should be trained because every data acquired here cannot be of high quality. Introducing evidence-based machine standards would help doctors use AI diagnostic criteria. If inaccurate, doctors have to guide AI to increase ML model accuracy throughout training, which is challenging.

1.8.5 Data quality

FL could bring together independent medical groups, hospitals, and pieces of equipment so that they could share their private experiences. Most health systems have too much data and do not use it well. The data quality of multiple sources is usually not standardized (Alicherif, 2023; Curchoe, 2023). Using dirty data as samples leads to useless results. Medical data requires strategy; therefore, if an individual wants to enhance data quality, data clearing, correcting, and fine-tuning are vital, whether using the ML or FL model. This leaves out the question of which AI model can ensure proper data quality without disturbing the medics on daily operations.

1.9 Federated learning challenges in smart healthcare

FL, which is a distributed ML approach where data from multiple sources is used to train a shared model while the raw data stays on the local devices, is gaining popularity in DHC. However, its implementation in the HC domain is associated with a few challenges:

1.9.1 Data privacy and security

Data privacy and security are noteworthy challenges in FL for digital health systems. FL involves sharing data across multiple institutions, increasing the risk of data breaches and privacy violations (Jahromi et al., 2023). Robust data encryption and access control mechanisms must be implemented to protect sensitive health information. Additionally, legal and regulatory frameworks must be developed to govern data sharing and ensure compliance with data protection laws, ethical guidelines, and industry best practices.

1.9.2 Heterogeneity of data

Data heterogeneity is a momentous challenge in FL for digital health systems. HC data is collected from multiple sources using different protocols and standards, leading to variations in quality, format, and structure (Chen et al., 2023a; Pei et al., 2023). This heterogeneity can lead to inconsistent model performance, making it challenging to develop accurate and reliable models that can be generalized across different institutions and patient populations. Standardization of data formats, metadata, and ontologies is necessary to ensure interoperability and harmonization across different sources, enabling more effective health data sharing and analysis.

1.9.3 Data quality

Data quality is another significant challenge in FL for digital health systems. HC data is often incomplete, inaccurate, and subject to errors and biases, leading to poor model performance and unreliable predictions (Chen et al., 2023a; Hongbin & Zhi, 2023). Data quality assessment and cleaning techniques must be applied to ensure that the data used for training and testing models are accurate, reliable, and free from errors and biases. Furthermore, data governance and stewardship frameworks must be developed to ensure data is collected, processed, and shared in a standardized, transparent, and ethical manner, ensuring stakeholder trust and collaboration.

1.9.4 Communication overhead

FL requires devices and institutions to send and receive many data and model updates. This causes an increase in communication costs and delays (Behera et al., 2023). To solve this problem, we need to use efficient communication protocols and compression techniques to reduce the amount of data that needs to be sent back and forth. This will cut down on latency and bandwidth needs. Moreover, edge computing and other distributed computing approaches can perform local computations on devices, reducing the amount of data that needs to be sent to a central server.

1.9.5 Regulatory compliance

Regulatory compliance is a substantial challenge in FL for digital health systems. HC data is subject to strict regulations and standards, for example, HIPAA, which imposes legal and ethical obligations on data privacy and security (Hongbin & Zhi, 2023). Compliance with these regulations is necessary to ensure that sensitive patient data is protected and used appropriately for research and HC purposes. HC providers and technology developers must comply with regulatory requirements, such as obtaining patient consent, ensuring data encryption, and implementing data access controls while fostering transparency and accountability in data sharing and use.

1.9.6 Resource constraints

FL requires significant computing resources, such as processing power, memory, and storage, which can be challenging for small institutions or devices with limited resources (Hongbin & Zhi, 2023; Lu, 2023). To address this encounter, resource-efficient ML techniques, such as model compression, pruning, and quantization, can reduce models' size and complexity without sacrificing performance. Additionally, cloud computing and other distributed computing approaches can be used to leverage the computing resources of multiple devices and institutions, enabling more efficient and scalable FL. All these identified challenges shape the future trend of FL, as demonstrated in the next section.

1.10 Future federated learning trends

In this section, future FL trends are presented.

1.10.1 Increased use of wearable technology

It makes it possible to use wearable technology in DHC systems by letting data be collected from multiple devices and analyzed without compromising user privacy (Yang et al., 2023). The method can make health predictions more accurate, reduce the need for centralized data storage, and make people use wearable technology more. FL can be used with wearable devices such as smartwatches and fitness trackers to monitor patient health and give real-time insights.

1.10.2 Personalized medicine

For personalized medicine to work in DHC systems, models need to be trained with patient-specific data from multiple sources to make personalized treatment suggestions (Doc.ai: declarative, 2023). The method can improve treatment results, lower HC costs, and make patients happier by making treatments fit each person's needs. It makes personalized medicine possible by analyzing large sets of patient information and making treatment plans that are just right for each person.

1.10.3 Improved patient outcomes

FL helps HC providers make better-informed decisions, improving patient outcomes (Shin et al., 2023). Its ability to analyze diverse patient data while preserving patient privacy can lead to more precise diagnoses, personalized treatment plans, and improved health predictions, resulting in better patient outcomes in DHC systems.

1.10.4 Increased data privacy

FL can make DHC systems more private by letting data be processed on local devices instead of stored in a central place (Wang et al., 2023c). The method ensures patient information stays private and safe, lowering the risk of data breaches and unauthorized access. FL provides a secure way of sharing patient data while protecting patient privacy.

1.10.5 Remote patient monitoring

DHC systems can let doctors check on patients from afar by analyzing data from various sources, such as wearable devices and remote sensors, without risking patients' privacy (DiLorenzo et al., 2023; Jahromi et al., 2023). The method can help find health problems early, allowing for timely interventions, personalized treatment, and better patient outcomes, even in remote places where HC is hard to get to. Moreover, FL lets doctors and nurses watch patients from afar and give them personalized care. This will improve patient outcomes and lower HC costs.

1.10.6 Collaborative healthcare

FL in DHC systems can make it easier for HC providers, patients, and researchers to collaborate by letting them combine and analyze data from many different sources (Krishna et al., 2023; Narayan et al., 2023). The method can make it easier for people to work together, share data, and learn from each other. This can lead to more accurate diagnoses, personalized treatment, and better patient outcomes while keeping data private and secure. Moreover, FL allows HC providers from different institutions to collaborate and share information, leading to better patient care.

1.10.7 Real-time decision-making

FL in DHC systems can make HC more collaborative by letting doctors, patients, and researchers add and look at data from different sources (Tanwar et al., 2023). The method can make it easier for people to work together, share data, and learn from each other. This can lead to more accurate diagnoses, personalized treatment, and better patient outcomes while keeping data private and secure. FL allows HC providers to make real-time decisions based on the latest data, improving patient outcomes and reducing HC costs.

1.10.8 Improved drug discovery

FL can improve drug discovery in DHC systems by making it possible to analyze large amounts of data from different sources, such as patient data, genomics, and research on drug development (Li et al., 2023b). This can speed up the process of finding new drugs, make them more effective, and help patients feel better. It also ensures data privacy, which can encourage data sharing and collaboration among researchers. It enables faster and more efficient drug discovery, leading to better patient treatment.

1.10.9 **Predictive analytics**

FL can help DHC systems use predictive analytics by analyzing patient data from many sources (Ghosh & Ghosh, 2023). The method can predict and stop health problems before they happen. It can also improve treatment planning and lower the cost of HC. It protects data privacy, making patients more likely to share their data and making predictive models more accurate. It lets doctors and nurses use predictive analytics to find patients likely to get certain diseases and treat them before they do.

1.10.10 **Improved clinical trials**

FL can improve clinical trials in DHC systems by allowing the analysis of different patient data from different sources, such as genomics, demographics, and medical history (Shafik et al., 2022; Stragapede et al., 2023). The method can make finding suitable patients easier, finishing the trials faster, and getting more accurate results. It also ensures data privacy, which can encourage more patients to participate in clinical trials and improve data sharing among researchers. It enables more efficient and cost-effective clinical trials, leading to faster drug approval and improved patient outcomes.

1.10.11 **Enhanced medical imaging**

Fl allows multiple medical institutions to collaborate and collectively improve the accuracy and performance of ML models for medical imaging analysis (Oualid et al., 2023). By leveraging vast and diverse datasets across different institutions while maintaining data privacy and security, FL enables the development of more advanced algorithms, leading to enhanced medical imaging capabilities in digital systems. Indeed, the approach will enable more accurate and efficient medical imaging, leading to better diagnosis and treatment of diseases.

1.10.12 **Continuous learning**

FL enables continuous learning in digital systems by allowing models to be trained on data collected from multiple sources over time (Goscinski et al., 2023; Lohmeyer et al., 2023). As more data is collected, the models are retrained to incorporate new information, improving their accuracy and effectiveness. The approach enables the development of highly adaptable models that can evolve and learn from new information in a timely and efficient manner, making them well-suited for use in dynamic and rapidly changing environments.

1.10.13 **Improved resource allocation**

FL enables improved resource allocation in digital systems by distributing computation across multiple devices and locations (Ghosh & Ghosh, 2023; Google:

Tensorflow Federated, 2023). The approach reduces the need for centralized infrastructure and better uses existing resources, reducing costs and increasing efficiency. Also, by leveraging the computational power of edge devices, FL enables real-time data processing and analysis, enabling more responsive and adequate decision-making in a wide range of applications.

1.10.14 Increased patient engagement

FL empowers increased patient engagement in digital systems by empowering patients to participate in the training and improvement of ML models (Google: Tensorflow Federated, 2023). By allowing patients to contribute their data and insights to the development of algorithms, the technique promotes a more patient-centered approach to HC that is more personalized, transparent, and responsive to individual needs. The approach fosters greater trust and collaboration between patients, HC providers, and technology developers, leading to positive health outcomes and improved patient satisfaction.

1.10.15 Improved public health

FL tolerates improved public health in digital systems by allowing large-scale analysis of diverse health data while maintaining patient privacy (Campagner et al., 2023; Kumar et al., 2023). The approach facilitates the identification of patterns, trends, and correlations that can inform public health policies and interventions, leading to more effective disease prevention, early detection, and treatment. By leveraging the power of ML and distributed computing, FL enables a more proactive and data-driven approach to public health that can help address the complex and evolving health challenges facing communities worldwide (D'silva et al., 2023; Shafik et al., 2019; Shafik et al., 2020b).

1.11 Conclusion

In this chapter, we take a detailed overview of the present state of FL across various industries, including but not limited to the DHC aspect. We aim to give researchers a resource that can be of value to them by summarizing the general answers to the many issues associated with FL. We have described some directions encountered or existing research questions when FL is utilized in the traditional HC and DHC sectors. FL, which involves training these shared data on a generic model using a main server while protecting sensitive information (data) in a local facility, can incorporate disintegrated HC information (data) sources with confidentiality. Therefore this chapter discusses the general idea, terms, possibilities, openings, and challenges of FL in DHC systems. The incorporation of biomedical FL technologies affects the statistical, communication, security, and privacy issues in HC. Additionally, open research questions need to be addressed to ensure the proper application of FL in the medical industry.

Abbreviations

AFL	agnostic federated learning
AI	artificial intelligence
CPU	central processing unit
DHC	digital healthcare
DP	differential privacy
DPR	data processing at rest
EHR	electronic health record
FedAvg	federated averaging
FL	federated learning
HC	healthcare
HIPAA	Health Insurance Portability and Accountability Act
ICU	intensive care unit
ML	machine learning
MSDP	multistream data processing
MUL	multitask learning
P2P	peer-to-peer
q-FFL	q-Fair Federated Learning
SMPC	secure multiparty computation

References

Akter, M., Moustafa, N., Lynar, T., & Razzak, I. (2022). Edge intelligence: federated learning-based privacy protection framework for smart healthcare systems. *IEEE Journal of Biomedical and Health Informatics*, 26(12), 5805−5816. Available from https://doi.org/10.1109/JBHI.2022.3192648.

Alicherif, N. (2023). *Privacy preserving in the medical sector: Techniques and applications. Advanced bioinspiration methods for healthcare standards, policies, and reform* (pp. 221−239). IGI Global. Available from https://doi.org/10.4018/978-1-6684-5656-9.ch012.

Amon, M. (2023). Modeling user characteristics associated with interdependent privacy perceptions on social media. *ACM Transactions on Computer-Human Interaction*. Available from https://doi.org/10.1145/3577014.

Anh, T. T., Luong, N. C., Niyato, D., Kim, D. I., & Wang, L. C. (2019). Efficient training management for mobile crowd-machine learning: A deep reinforcement learning approach. *IEEE Wireless Communications Letters*, 8(5), 1345−1348. Available from https://doi.org/10.1109/LWC.2019.2917133.

Batool, Z., Zhang, K., & Toews, M. (2023). Block-RACS: Towards reputation-aware client selection and monetization mechanism for federated learning. *ACM SIGAPP Applied Computing Review*, 23(3), 49−66. Available from https://doi.org/10.1145/3626307.3626311.

Behera, R. K., Bala, P. K., Rana, N. P., & Irani, Z. (2023). Responsible natural language processing: A principlist framework for social benefits. *Technological Forecasting and Social Change*, *188*, 122306. Available from https://doi.org/10.1016/j.techfore.2022. 122306.

Calbimonte, J. P., Aidonopoulos, O., Dubosson, F., Pocklington, B., Kebets, I., et al. (2023). Decentralized semantic provision of personal health streams. *Journal of Web Semantics*, *76*, 100774. Available from https://doi.org/10.1016/j.websem.2023.100774.

Campagner, A., Ciucci, D., & Cabitza, F. (2023). Aggregation models in ensemble learning: A large-scale comparison. *Information Fusion*, *90*, 241. Available from https://doi.org/10.1016/j.inffus.2022.090.015, 52.

Chen, Q., Wang, Z., Zhang, W., & Lin, X. (2023a). PPT: A privacy-preserving global model training protocol for federated learning in P2P networks. *Computers & Security*, *124*, 102966. Available from https://doi.org/10.1016/j.cose.2022.102966.

Chen, Z., Yi, W., Liu, Y., & Nallanathan, A. (2023b). Knowledge-aided federated learning for energy-limited wireless networks. *IEEE Transactions on Communications*, 3368−3386. Available from https://doi.org/10.1109/TCOMM.2023.3261383.

Cheng, K., Fan, T., Jin, Y., Liu, Y., Chen, T., Papadopoulos, D., & Yang, Q. (2021). Secureboost: A lossless federated learning framework. *IEEE Intelligent Systems*, *36*(6), 87−98. Available from https://doi.org/10.1109/MIS.2021.3082561.

Cozzi, G. D., Blanchard, C. T., Szychowski, J. M., Subramaniam, A., & Battarbee, A. N. (2023). Optimal predelivery hemoglobin to reduce transfusion and adverse perinatal outcomes. *American Journal of Obstetrics & Gynecology MFM*, *5*(2), 100810. Available from https://doi.org/10.1016/j.ajogmf.2022.100810.

Curchoe, C.L. (2023). Proceedings of the first world conference on AI in fertility. *Journal of Assisted Reproduction and Genetics*, pp. 1−8, Available from https://doi.org/10.1109/ACCESS.2023.3310400.

DiLorenzo, M. A., Davis, M. R., Dugas, J. N., Nelson, K. P., Hochberg, N. S., et al. (2023). Performance of three screening tools to predict COVID-19 positivity in emergency department patients. *Emergency Medicine Journal*. Available from https://doi.org/10.1136/emermed-2021-212102.

D'silva, G., Batra, J., Bhoir, A., & Sharma, A. (2023). *On-device emotional intelligent iot-based framework for mental health disorders. Emerging technologies in data mining and information security* (pp. 511−521). Singapore: Springer. Available from https://doi.org/10.1007/978-981-19-4193-1_50.

Dobischok, S., Metcalfe, R., Matzinger, E., Palis, H., Marchand, K., et al. (2023). Measuring the preferences of injectable opioid agonist treatment (IoAT) clients: development of a Person-Centered Scale (Best-Worst Scaling). *International Journal of Drug Policy*, *112*, 103948. Available from https://doi.org/10.1016/j.drugpo.2022.103948.

Doc.ai: declarative, on-device machine learning for ios, android, and react native, 2019. https://github.com/doc-ai/tensorio. Accessed on 09.02.2023.

Dubowski, K., Braganza, G. T., Bozack, A., Colicino, E., DeFelice, N., et al. (2023). COVID-19 subphenotypes at hospital admission are associated with mortality: a cross-sectional study. *Annals of Medicine*, *55*(1), 12−23. Available from https://doi.org/10.1080/07853890.2022.2148733.

Díaz, J. S., & García, Á. L. (2023). Study of the performance and scalability of federated learning for medical imaging with intermittent clients. *Neurocomputing*, *51*, 142. Available from https://doi.org/10.1016/j.neucom.2022.110.011, 54.

Elmi, J., Eftekhari, M., Mehrpooya, A., & Ravari, M. R. (2023). A novel framework based on the multi-label classification for dynamic selection of classifiers. *International Journal of Machine Learning and Cybernetics*, 1−8. Available from https://doi.org/10.1007/s13042-022-01751-z.

Frost, E. D., Donlon, J., Mitwally, A., Magnani, G., Tomlin, S., et al. (2023). Paper Charts: A continued barrier to psychiatric care in the midst of a broken and fragmented mental health system. *The Journal of Nervous and Mental Disease*, *211*(1), 1−4. Available from https://doi.org/10.1097/NMD.0000000000001606.

Gao, H., He, N., & Gao, T. (2023). SVeriFL: Successive verifiable federated learning with privacy-preserving. *Information Sciences*, *622*, 98−114. Available from https://doi.org/10.1016/j.ins.2022.110.124.

Ghosh, S., & Ghosh, S. K. (2023). FEEL: Federated learning framework for elderly healthcare using edge-IoMT. *IEEE Transactions on Computational Social Systems*, *10*(4), 1800−1809. Available from https://doi.org/10.1109/TCSS.2022.3233300.

Google: Tensorflow federated, 2023. https://www.tensorflow.org/federated. Accessed on 08.01.23.

Goscinski, A., Delicato, F. C., Fortino, G., Kobusińska, A., & Srivastava, G. (2023). Special issue on distributed intelligence at the edge for the future Internet of Things. *Journal of Parallel and Distributed Computing*, *171*, 157−162. Available from https://doi.org/10.1016/j.jpdc.2022.090.014.

Guo, P., Wang, P., Zhou, J., Jiang, S., & Patel, V. M. (2023). *Improved MR image reconstruction using federated learning. Meta-learning with medical imaging and health informatics applications* (pp. 351−368). Academic Press. Available from https://doi.org/10.1016/B978-0-32-399851-2.00028-4.

Hongbin, F., & Zhi, Z. (2023). Privacy-preserving data aggregation scheme based on federated learning for iiot. *Mathematics*, *11*(1), 214. Available from https://doi.org/10.1109/JIOT.2022.3229122.

Huang, P. H., Tu, C. H., Chung, S. M., Wu, P. Y., Tsai, T. L., et al. (2023). SecureTVM: A TVM-based compiler framework for selective privacy-preserving neural inference. *ACM Transactions on Design Automation of Electronic Systems*. Available from https://doi.org/10.1145/3579049.

Hu, Y., Zhang, Y., Gao, X., Gong, D., Song, X., et al. (2023). A federated feature selection algorithm based on particle swarm optimization under privacy protection. *Knowledge-Based Systems*, *25*(260), 110122. Available from https://doi.org/10.1016/j.knosys.2022.110122.

Imteaj, A., Mamun Ahmed, K., Thakker, U., Wang, S., Li, J., et al. (2023). Federated learning for resource-constrained IoT devices: panoramas and state of the art. *Federated and Transfer Learning*, 7−27. Available from https://doi.org/10.1007/978-3-031-11748-0_2.

Jahromi, A. N., Karimipour, H., & Dehghantanha, A. (2023). An ensemble deep federated learning cyber-threat hunting model for industrial internet of things. *Computer Communication*, *198*, 108−116. Available from https://doi.org/10.1016/j.comcom.2022.110.009.

Jiang, W., Korolija, D., & Alonso, G. (2023). Data processing with FPGAs on modern architectures. *In Companion of the 2023 International Conference on Management of Data*, 77−82. Available from https://doi.org/10.1145/3555041.3589410.

Jiang, X., Zhang, J., & Zhang, L. (June 2023). FedRadar: Federated multi-task transfer learning for radar-based internet of medical things. *IEEE Transactions on Network and Service Management, 20*(2), 1459−1469. Available from https://doi.org/10.1109/TNSM.2023.3281133.

Jin, Y., Zhu, H., Xu, J., & Chen, Y. (2023). *Evolutionary multi-objective federated learning. Federated learning* (pp. 139−164). Singapore: Springer. Available from https://doi.org/10.1007/978-981-19-7083-2_3.

Jun, Y., Craig, A., Shafik, W., & Sharif, L. (2021). Artificial intelligence application in cybersecurity and cyberdefense. *Wireless Communications and Mobile Computing*. Available from https://doi.org/10.1155/2021/3329581.

Kamila, N. K., Pani, S. K., Das, R. P., Bharti, P. K., Najafabadi, H. E., et al. (2023). A near-optimal & load balanced resilient system design for high-performance computing platform. *Cluster Computing*, 1−6. Available from https://doi.org/10.1007/s10586-022-03913-8.

Krishna, B. V., AP, B., HL, G., Ravi, V., Almeshari, M., & Alzamil, Y. (2023). A novel application of k-means cluster prediction model for diabetes early identification using dimensionality reduction techniques. *The Open Bioinformatics The Journal, 16*(1). Available from https://doi.org/10.2174/18750362-v16-230825-2023-18.

Kumari, J., Kumar, D., & Kumar, E. (2023). *Transitions in machine learning approaches for healthcare-sector applications. Designing intelligent healthcare systems, products, and services using disruptive technologies and health informatics* (pp. 269−276). CRC Press. Available from https://doi.org/10.1201/9781003217107-16.

Kumar, P., Kumar, R., Gupta, G. P., Tripathi, R., Jolfaei, A., et al. (2023). A blockchain-orchestrated deep learning approach for secure data transmission in IoT-enabled healthcare system. *Journal of Parallel and Distributed Computing, 172*, 69−83. Available from https://doi.org/10.1016/j.jpdc.2022.100.002.

Lee, H. S. (2022). Device selection and resource allocation for layerwise federated learning in wireless networks. *IEEE Systems Journal, 16*(4), 6441−6444. Available from https://doi.org/10.1109/JSYST.2022.3169461, https://doi.org/10.1109/JSYST.2022.3169461.

Li, T., Sahu, A.K., Zaheer, M., Sanjabi, M., Talwalkar, A., & Smith, V. (2020). Federated optimization in heterogeneous networks. *Proceedings of machine learning and systems*, vol. 2, pp. 429−50, https://api.semanticscholar.org/CorpusID:59316566.

Li, Z., WU, H., & LU, Y. (2023a). Coalition based utility and efficiency optimization for multi-task federated learning in internet of vehicles. *Future Generation Computer Systems, 140*, 196−208. Available from https://doi.org/10.1016/j.future.2022.100.014.

Li, W., Wen, L., Rathod, B., Gingras, A. C., Ley, K., et al. (2023b). Kindlin2 enables EphB/ephrinB bi-directional signaling to support vascular development. *Life Science Alliance, 6*(3). Available from https://doi.org/10.26508/lsa.202201800.

Lohmeyer, Q., Schiess, C., Garcia, P. D., Petry, H., Strauch, E., et al. (2023). Effects of tall man lettering on the visual behaviour of critical care nurses while identifying syringe drug labels: A randomised in situ simulation. *BMJ Quality & Safety, 32*(1), 26−33. Available from https://doi.org/10.1136/bmjqs-2021-014438.

Lukauskas, M., & Ruzgas, T. (2023). *Review and comparative analysis of unsupervised machine learning application in health care. Data intelligence and cognitive informatics* (pp. 751−759). Singapore: Springer. Available from https://doi.org/10.1007/978-981-19-6004-8_56.

Lu, S. (2023). Top-k sparsification with secure aggregation for privacy-preserving federated learning. *Computers & Security, 124*, 102993. Available from https://doi.org/10.1016/j.cose.2022.102993.

Maskeliūnas, R., Pomarnacki, R., Huynh, K. Van, Damaševičius, R., & Plonis, D. (2023). Power line monitoring through data integrity analysis with Q-learning based data analysis network. *Remote Sensing*, *15*(1), 194. Available from https://doi.org/10.3390/rs15010194.

Melnikov, F., Anger, L. T., & Hasselgren, C. (2023). Toward quantitative models in safety assessment: A case study to show impact of dose—response inference on herg inhibition models. *International Journal of Molecular Sciences*, *24*(1), 635. Available from https://doi.org/10.3390/ijms24010635.

Miao, Y., Liu, Z., Li, H., Choo, K.-K. R., & Deng, R. H. (2022). Privacy-preserving byzantine-robust federated learning via blockchain systems. *IEEE Transactions on Information Forensics and Security*, *17*, 2848—2861. Available from https://doi.org/10.1109/TIFS.2022.3196274.

Min, X., Yu, B., & Wang, F. (2019). Predictive modeling of the hospital readmission risk from patients' claims data using machine learning: A case study on COPD. *Scientific Reports*, *9*(1), 2362. Available from https://doi.org/10.1038/s41598-019-39071-y.

de Miras, J. R., Ibáñez-Molina, A. J., Soriano, M. F., & Iglesias-Parro, S. (2023). Schizophrenia classification using machine learning on resting state EEG signal. *Biomedical Signal Processing and Control*, *79*, 104233. Available from https://doi.org/10.1016/j.bspc.2022.104233.

Nagy, B., Hegedűs, I., Sándor, N., Egedi, B., Mehmood, H., et al. (2023). Privacy-preserving federated learning and its application to natural language processing. *Knowledge-Based Systems*, *268*, 110475. Available from https://doi.org/10.1016/j.knosys.2023.110475.

Nair, A. K., Sahoo, J., & Raj, E. D. (2023). Privacy preserving federated learning framework for IoMT based big data analysis using edge computing. *Computer Standards & Interfaces*, *86*, 103720. Available from https://doi.org/10.1016/j.csi.2023.103720.

Narayan, V., Mall, P. K., Alkhayyat, A., Abhishek, K., Kumar, S., & Pandey, P. (2023). Enhance-net: an approach to boost the performance of deep learning model based on real-time medical images. *Journal of Sensors*. Available from https://doi.org/10.1155/2023/8276738.

Nazir, S., & Kaleem, M. (2023). Federated learning for medical image analysis with deep neural networks. *Diagnostics*, *13*(9), 1532. Available from https://doi.org/10.3390/diagnostics13091532.

Obermeyer, Z., Powers, B., Vogeli, C., & Mullainathan, S. (2019). Dissecting racial bias in an algorithm used to manage the health of populations. *Science (New York, N.Y.)*, *366*(6464), 447—453. Available from https://doi.org/10.1126/science.aax2342.

Oualid, A., Maleh, Y., & Moumoun, L. (2023). Federated learning techniques applied to credit risk management: A systematic literature review. *EDPACS*, *68*(1), 42—56. Available from https://doi.org/10.1080/07366981.2023.2241647.

Pei, M., Pei, Y. A., Zhou, S., Mikaeiliagah, E., Erickson, C., et al. (2023). Matrix from urine stem cells boosts tissue-specific stem cell mediated functional cartilage reconstruction. *Bioactive Materials*, *23*, 353—667. Available from https://doi.org/10.1016/j.bioactmat.2022.110.012.

Peng, L., Luo, G., Walker, A., Zaiman, Z., Jones, E. K., et al. (2023). Evaluation of federated learning variations for COVID-19 diagnosis using chest radiographs from 42 US and European hospitals. *Journal of the American Medical Informatics Association*, *30*(1), 54—63. Available from https://doi.org/10.1093/jamia/ocac188.

Pontillo, G. (2023). *Digital medical design: How new technologies and approaches can empower healthcare for society. International conference on design and digital communication, Barcelos, Portugal* (pp. 255−269). Cham: Springer. Available from https://doi.org/10.1007/978-3-031-20364-0_23.

Qin, Y., Li, M., & Zhu, J. (2023). Privacy-preserving federated learning framework in multimedia courses recommendation. *Wireless Networks, 29*(4), 1535−1544. Available from https://doi.org/10.1007/s11276-021-02854-1.

Rahman, A., Hossain, M. S., Muhammad, G., Kundu, D., Debnath, T., et al. (2023). Federated learning-based AI approaches in smart healthcare: Concepts, taxonomies, challenges and open issues. *Cluster Computing, 26*(4), 2271−2311. Available from https://doi.org/10.1007/s10586-022-03658-4.

Rajagopal, A., Redekop, E., Kemisetti, A., Kulkarni, R., Raman, S., et al. (2023). Federated learning with research prototypes: Application to multi-center MRI-based detection of prostate cancer with diverse histopathology. *Academic Radiology, 30*(4), 644−657. Available from https://doi.org/10.1016/j.acra.2023.020.012.

Šajina, R., Tanković, N., & Ipšić, I. (2023). Peer-to-peer deep learning with non-IID data. *Expert Systems with Applications, 214*, 119159. Available from https://doi.org/10.1016/j.eswa.2022.119159.

Razavi-Far, R. (2023). *An introduction to federated and transfer learning. Federated and transfer learning* (pp. 1−6). Cham: Springer. Available from https://doi.org/10.1007/978-3-031-11748-0_1.

Sattler, F., Müller, K.-R., & Samek, W. (2021). Clustered federated learning: Model-agnostic distributed multitask optimization under privacy constraints. *IEEE Transactions on Neural Networks and Learning Systems, 32*(8), 3710−3722. Available from https://doi.org/10.1109/TNNLS.2020.3015958.

Shafik, W., Matinkhah, S. M., & Ghasemzadeh, M. (2019). A fast machine learning for 5g beam selection for unmanned aerial vehicle applications. *Journal of Information Systems and Telecommunication, 7*(28), 262−278. Available from https://doi.org/10.7508/jist.2019.04.003.

Shafik, W., Matinkhah, S. M., Etemadinejad, P., & Sanda, M. N. (2020). Reinforcement learning rebirth, techniques, challenges, and resolutions. *International Journal on Informatics Visualization, 4*(3), 127−135. Available from https://doi.org/10.30630/joiv.4.30.376.

Shafik, W., Matinkhah, M., Asadi, M., Ahmadi, Z., & Hadiyan, Z. (2020b). A study on internet of things performance evaluation. *Journal of Communications Technology, Electronics and Computer Science*, 1−19. Available from https://doi.org/10.22385/jctecs.v28i0.303.

Shafik, W., Matinkhah, S. M., & Ghasemzadeh, M. (2020). Theoretical understanding of deep learning in UAV biomedical engineering technologies analysis. *SN Computer Science, 1*(6), 1−13. Available from https://doi.org/10.1007/s42979-020-00323-8.

Shafik, W. (2024). Wearable medical electronics in artificial intelligence of medical things. *In Handbook of security and privacy of AI-enabled healthcare systems and internet of medical things*, 21−40. Available from https://doi.org/10.1201/9781003370321-2.

Shafik, W., Matinkhah, S. M., Sanda, M. N., & Shokoor, F. (2021). Internet of things-based energy efficiency optimization model in fog smart cities. *International Journal on Informatics Visualization, 5*(2), 105−112. Available from https://doi.org/10.30630/joiv.5.20.373.

Shafik, W., Matinkhah, S. M., & Shokoor, F. (2023). Cybersecurity in unmanned aerial vehicles: A review. *International Journal on Smart Sensing and Intelligent Systems*, *16*(1). Available from https://doi.org/10.2478/ijssis-2023-0012.

Shafik, W. (2023a). A comprehensive cybersecurity framework for present and future global information technology organizations. In *Effective cybersecurity operations for enterprise-wide systems*, (pp. 56−79). IGI Global. Available from https://doi.org/10.4018/978-1-6684-9018-1.ch002.

Shafik, W. (2023b). *Making cities smarter: IoT and SDN applications, challenges, and future trends. Opportunities and challenges of industrial IoT in 5G and 6G networks* (pp. 73−94). IGI Global. Available from http://doi.org/10.4018/978-1-7998-9266-3.ch004.

Shafik, W. (2023c). *Cyber security perspectives in public spaces: drone case study. Handbook of research on cybersecurity risk in contemporary business systems* (pp. 79−97). IGI Global. Available from https://doi.org/10.4018/978-1-6684-7207-1.ch004.

Shafik, W., Matinkhah, S. M., & Shokoor, F. (2022). Recommendation system comparative analysis: Internet of things aided networks. *EAI Endorsed Transactions on Internet of Things*, *8*(29). Available from https://doi.org/10.4108/eetiot.v8i29.1108.

Shafik, W., Tufail, A., Liyanage, C. D. S., & Apong, R. A. A. H. M. (2024). Medical robotics and AI-assisted diagnostics challenges for smart sustainable healthcare. In AI-Driven Innovations in Digital Healthcare: Emerging Trends, Challenges, and Applications, (pp. 304−323). IGI Global. Available from https://doi.org/10.4018/979-8-3693-3218-4.ch016.

Shanmugarasa, Y., Paik, H. Y., Kanhere, S. S., & Zhu, L. (2023). A systematic review of federated learning from clients' perspective: challenges and solutions. *Artificial Intelligence Review*, 1−55. Available from https://doi.org/10.1007/s10462-023-10563-8.

Shenaj, D., Rizzoli, G., & Zanuttigh, P. (2023). Federated learning in computer vision. *IEEE Access*, *11*, 94863−94884. Available from https://doi.org/10.1109/ACCESS.2023.3310400.

Shetty, P. N. (2023). *Trust based resolving of conflicts for collaborative data sharing in online social networks. Emerging technologies in data mining and information security* (pp. 35−48). Singapore: Springer. Available from https://doi.org/10.1007/978-981-19-4052-1_5.

Shin, J. W., Choi, J., & Tate, J. (2023). Interventions using digital technology to promote family engagement in the adult intensive care unit: An integrative review. *Heart & Lung*, *58*, 166−178. Available from https://doi.org/10.1016/j.hrtlng.2022.120.004.

Singh, S. K., Yang, L. T., & Park, J. H. (2023a). FusionFedBlock: Fusion of blockchain and federated learning to preserve privacy in industry 5.0. *Information Fusion*, *90*, 233−240. Available from https://doi.org/10.1016/j.inffus.2022.090.027.

Singh, R., Kushwaha, A. K., & Srivastava, R. (2023b). Recent trends in human activity recognition—A comparative study. *Cognitive Systems Research*, *77*, 30−44. Available from https://doi.org/10.1016/j.cogsys.2022.100.003.

Sirisha, U., & Chandana, B. S. (2023). Privacy preserving image encryption with optimal deep transfer learning-based accident severity classification model. *Sensors*, *23*(1), 519. Available from https://doi.org/10.3390/s23010519.

Stragapede, G., Vera-Rodriguez, R., Tolosana, R., & Morales, A. (2023). BehavePassDB: Public database for mobile behavioral biometrics and benchmark evaluation. *Pattern Recognition*, *134*, 109089. Available from https://doi.org/10.1016/j.patcog.2022.109089.

Tanwar, A., Zhang, J., Ive, J., Gupta, V., & Guo, Y. (2023). *Unsupervised numerical reasoning to extract phenotypes from clinical text by leveraging external knowledge.*

Multimodal AI in healthcare (pp. 11−28). Cham: Springer. Available from https://doi.org/10.1007/978-3-031-14771-5_2.

Tian, X., Jiang, Y., & Tianfield, H. (2023). *Swarm meta learning. Federated and transfer learning* (pp. 167−183). Cham: Springer. Available from https://doi.org/10.1007/978-3-031-11748-0_8.

Tominaga, R., & Masataka, S. E. O. (2023). *Image generation from text using stackgan with consistency regularization. International symposium on distributed computing and artificial intelligence* (pp. 76−85). Cham: Springer. Available from https://doi.org/10.1007/978-3-031-20859-1_9.

Valente, R., Senna, C., Rito, P., & Sargento, S. (2023). *Federated learning framework to decentralize mobility forecasting in smart cities. NOMS 2023-2023 IEEE/IFIP network operations and management symposium* (pp. 1−5). FL, USA: Miami. Available from https://doi.org/10.1109/NOMS56928.2023.10154456.

Wang, W., Li, X., Qiu, X., Zhang, X., Zhao, J., et al. (2023a). A privacy preserving framework for federated learning in smart healthcare systems. *Information Processing & Management*, *60*(1), 103167. Available from https://doi.org/10.1016/j.ipm.2022.103167.

Wang, D., Yi, Y., Yan, S., Wan, N., & Zhao, J. (2023b). A node trust evaluation method of vehicle-road-cloud collaborative system based on federated learning. *Ad Hoc Networks*, *138*, 103013. Available from https://doi.org/10.1016/j.adhoc.2022.103013.

Wang, Z., Zhang, L., Chao, Y., Xu, M., Geng, X., et al. (2023c). Development of a machine learning model for predicting 28-day mortality of septic patients with atrial fibrillation. *Shock (Augusta, Ga.)*. Available from https://doi.org/10.1097/shk.0000000000002078.

Wojtowytsch, S. (2023). Stochastic gradient descent with noise of machine learning type Part I: Discrete time analysis. *Journal of Nonlinear Science*, *33*(3), 45. Available from https://doi.org/10.1007/s00332-023-09903-3.

Xu, J., Glicksberg, B. S., Su, C., Walker, P., Bian, J., & Wang, F. (2021). Federated learning for healthcare informatics. *Journal of Healthcare Informatics Research*, *5*, 1−9. Available from https://doi.org/10.1007/s41666-020-00082-4.

Yang, Q., Liu, Y., Chen, T., & Tong, Y. (2019). Federated machine learning: concept and applications. *ACM Transactions on Intelligent Systems and Technology*, *10*(2), 1−19. Available from https://doi.org/10.1145/3298981.

Yang, Z., Xie, X., Zhang, X., Li, L., Bai, R., et al. (2023). Circadian rhythms of vital signs are associated with in-hospital mortality in critically ill patients: A retrospective observational study. *Chronobiology International*. Available from https://doi.org/10.1080/07420528.2022.2163656.

Zadorozhny, K., Thoral, P., Elbers, P., & Cinà, G. (2023). *Out-of-distribution detection for medical applications: Guidelines for practical evaluation. Multimodal AI in healthcare* (pp. 137−153). Cham: Springer. Available from http://doi.org/10.1007/978-3-031-14771-5_10.

Zhang, Z. (2023). Communication-efficient federated continual learning for distributed learning system with Non-IID data. *Science China Information Sciences*, *66*(2), 1−20. Available from https://doi.org/10.1007/s11432-020-3419-4.

Zhao, M. (2023). Multi-task learning with graph attention networks for multi-domain task-oriented dialogue systems. *Knowledge-Based Systems*, *259*, 110069. Available from https://doi.org/10.1016/j.knosys.2022.110069.

Zhao, Z., Mao, Y., Liu, Y., Song, L., Ouyang, Y., et al. (2023). Towards efficient communications in federated learning: A contemporary survey. *Journal of the Franklin Institute*. Available from https://doi.org/10.1016/j.jfranklin.2022.120.053.

Zhou, Y., Liu, X., Fu, Y., Wu, D., Wang, J. H., et al. (2023). Optimizing the numbers of queries and replies in convex federated learning with differential privacy. *IEEE Transactions on Dependable and Secure Computing*, 1−15. Available from https://doi.org/10.1109/TDSC.2023.3234599.

Zhuang, W. (2020). Performance optimization of federated person re-identification via benchmark analysis. In *Proceedings of the 28th ACM international conference on multimedia*, New York, U.S.A, pp. 955−963, https://doi.org/10.1145/3394171.3413814.

Zhu, X., Wang, J., Chen, W., & Sato, K. (2023). Model compression and privacy preserving framework for federated learning. *Future Generation Computer Systems*, *140*, 376−389. Available from https://doi.org/10.1016/j.future.2022.100.026.

Further reading

Shafik, W., Matinkhah, S. M., Afolabi, S. S., & Sanda, M. N. (2020a). A 3-dimensional fast machine learning algorithm for mobile unmanned aerial vehicle base stations. *International Journal of Advances in Applied Sciences*, *2252*(8814), 8814. Available from https://doi.org/10.11591/ijaas.v10.i1.pp28-38.

Architecture and design choices for federated learning in modern digital healthcare systems

Konstantinos A. Koutsopoulos[1], Christoph Thümmler[2], Angelica Avila Castillo[2],
Alice Abend[2], Stefan Covaci[3], Benjamin Ertl[3], Giannis Ledakis[4],
Stéphane Lorin[5], Vincent Thouvenot[5], Sahar Haddad[6], Gouenou Coatrieux[6],
Reda Bellafqira[6] and Alessandro Bassi[7]

[1]*Qualtek Sprl., Brussels, Belgium*
[2]*6G Health Institute GmbH, Markkleeberg, Germany*
[3]*Agentscape AG, Berlin, Germany*
[4]*UBITECH, Chalandri Athens, Greece*
[5]*Thales SIX GTS France SAS, Campus Polytechnique, Palaiseau cedex, France*
[6]*Inserm, Cyber Health, LaTIM UMR1101, IMT Atlantique, Brest, France*
[7]*Eurescom GmbH, Heidelberg, Germany*

2.1 Introduction

The recent advancements in artificial intelligence (AI) and the integration of machine learning (ML) models into day-to-day tasks and applications (Davenport & Kalakota, 2019) have heightened aspirations for innovative healthcare solutions (Habehh & Gohel, 2021). This potential, however, depends on the availability of medical data that is both of good quality and sufficient quantity (Abraham, 2023; Riskin, 2023). However, apart from these two requirements concerning data sources, a horizontal requirement, arising from a legal standpoint related to privacy protection (Wieringa et al., 2021), poses a challenge regarding how data can be involved and utilized in training and inference workflows.

With the emergence of dataspaces of controlled and fully interoperable infrastructure for data sharing and exploitation and the continuously enriched legal and governance framework concerning data and digital services (Shaping Europe's Digital Future, https), data availability and usability are becoming more straightforward. Consequently, this development is creating better awareness regarding both the value and sensitivity of digital assets (Adekoya & Ekpo, 2022). Although the federation of data in the context of dataspaces addresses to a large extent the vertical business ecosystem, where data ownership and provision are both subject to the governance decisions of a single stakeholder (Usländer &

Federated Learning for Digital Healthcare Systems. DOI: https://doi.org/10.1016/B978-0-443-13897-3.00002-3

Teuscher, 2022), much more complexity is involved when it comes to medical data. The reason is similar to all cases where the interests of individuals are involved. The data subjects who play the primary role in governance decisions do not have a direct connection with the health domains responsible for storing the data on their behalf (Berlage et al., 2022). As the storage, exposure, and usage of data are strictly related to the well-defined consent decisions made by the data subjects, the parties involved need solutions that may require revisiting current practices to allow them to effectively participate in the various dataspace processes. Such solutions should be subject to trust establishment and continuous attestation, expressed in the form of immutable proofs, ensuring adequate and foreseen system and processing integrity (Koutsopoulos, 2022).

The motivation, challenges, and aspirations outlined above have significantly influenced the objectives of the European project PAROMA-MED. The project aims to develop, validate, and evaluate a hybrid-cloud (central and edge) delivery framework that ensures privacy and security for services and applications in federative cross-border environments. This is achieved by providing technologies, tools, and services to support various aspects, including automatic attestation of federation partners; privacy and security by design; continuous risk assessment; privacy-preservation; and trusted data storage and processing in federative environments; AI/ML by design, managed privacy and security operations for automated policy enforcement; and cyberthreat detection and mitigation. The concepts of PAROMA-MED are discussed in this chapter.

2.1.1 Key contributions

The chapter focuses on the following:

i. Current dataspace landscape and evolution of new practices and patterns that relate to the value of data.
ii. Potential of the availability of medical data for the development of ground-breaking AI-based solutions, as well as the privacy concerns and restrictions
iii. Concepts of candidate solution, currently evolving in PAROMA-MED, embracing privacy protection at its core with utilization potential beyond the healthcare domain.

2.1.2 Chapter organization

Section 2.1.2 presents the current landscape of dataspaces and particularly how it may introduce new practices in the healthcare domain due to the potential of medical data for the evolution of AI-based solutions. Section 1.3 presents the details of a privacy-aware and privacy-preserving technical approach that is aligned with the current needs for dataspace establishment in the healthcare domain. This section considers FAIR (findable, accessible, interoperable, reusable) principles as key enabling features for data exploitation and also analyzes how data sovereignty can be assured, and most importantly, the concrete trust establishment mechanisms that will raise user confidence.

2.2 Dataspaces and health domain

2.2.1 State-of-the-art and current practices

The evolution of dataspaces across Europe has been dictated, among others, by the strategic objective of the European Union (EU Strategy, 2022) to become a world leader in digital and data economy. With interoperability being a key aspect (EU Data Act, 2022) that all initiatives in the field are trying to ensure and to avoid fragmentation due to different interpretations and implementations, the Horizon 2020 OpenDEI (OpenDEI, 2021) project brought together experts from several initiatives and organizations to define a set of common design principles and standards, including both technical and governance aspects. One of the key takeaways of this effort has been the definition of a soft infrastructure that identifies the main sector-agnostic building blocks that in turn identify how the participants have to interact either within sector-specific (the ones defined in Common EU Dataspaces and future ones [EU Data Act, 2022]) dataspaces or in intersector scenarios. This soft architecture organizes the building blocks according to the four main categories, three of which relate to technology (interoperability, trust, data value) and one to business and regulation (governance).

Each of the categories identifies a number of important building blocks facilitating the purpose of the category, whereas additional ones can be optionally deployed to aid interoperability and connectivity with additional systems with the data connector architecture by IDSA and the Trust Framework semantics and procedures by Gaia-X identified in this study as the most important (Siska et al., 2023).

Data connectors provide the basic mechanism for enabling a participant to connect and operate within the context of a dataspace, ensuring the support of exchange services and policy enforcement. This, in turn, facilitates technical interoperability. According to the IDSA Reference Architecture Model (IDS RAM, 2023), a connector is deployed, either on the cloud or on local resources, as a set of containers under the command of an application container management functionality. Among the containers, those identified as core containers take control of tasks related to data exchange, metadata management, remote attestation, logging and monitoring, and policy and contract management.

Gaia-X Trust Framework (GAIA-X Trust, 2022) is based on the exchange of verifiable credentials managed and utilized by functional components at any phase of interaction among dataspace participants. The fundamental element of the Gaia-X Trust Framework regards the verification of the validity of the claims stated in the self-descriptions of the participants, including self descriptions and claims of service and resource offerings. To achieve this purpose the utilization of trust anchors and compliance services is predicted. Presently the trust framework deals with participants identified as legal persons and, on an experimental basis, natural persons. It is expected that the roles of provider, consumer, and federator, as defined by the architecture document, will soon be supported by the trust framework.

2.2.2 Impact on machine learning

Dataspaces have a positive impact on ML and AI as they facilitate data access and allow better coordination and interoperability between the participants (EU, 2023). All AI technologies benefit from these new facilitating concepts and solutions, leading to the development of important applications in the healthcare domain, specifically in areas such as diagnosis, treatment planning, and therapy guidance.

In addition to the advantages of ML/AI technologies, the introduction of dataspaces not only transforms the paradigm of data storage and access but also has an impact on ML models. It also brings forth new requirements for the privacy and security of ML, impacting the field (Kerry, 2020).

In the subsequent sections of this chapter we will focus on AI algorithms that aim to protect data and AI processes. These algorithms are designed to ensure protection even in the event of attacks and/or leaks or to minimize their impact. The privacy and security ML techniques outlined here are designed to prevent unnecessary data sharing and exchanges, introduce noise to the data when referencing is necessary, and incorporate traceability measures for all information used and generated by AI models. A more detailed examination of these techniques will be provided in a later paragraph that outlines the proposed approach.

2.2.3 European digital age and development of secure (health) dataspaces

The European Commission is trying to make Europe fit for the digital age. It is determined to promote in Europe the so-called digital decade whose goal is to strengthen digital sovereignty by setting a new set of standards with a focus on data, technology, and infrastructure (EU Digital Decade, 2019).

The Digital Decade policy program contains targets, objectives, and ambitions for 2030 and will guide Europe's digital transformation. The commission will pursue its digital ambitions through concrete terms such as projected trajectories at the EU and national level, an annual cooperation cycle to monitor and report on progress, and through multicountry projects that combine investments from EU member states and the private sector.

On January 26, 2022, the commission proposed an interinstitutional solemn declaration on digital rights and principles in the digital decade. These new rights include, e.g., prioritizing individuals and their rights in the digital transformation, supporting solidarity and inclusion, ensuring freedom of choice online, promoting participation in the digital public space, enhancing safety and security, and consequently empowering individuals (EU Digital Decade, 2019). These rights and principles will complement the already existing rights reported in the Charter of Fundamental Rights of the European Union, as well as data protection and privacy legislation (GDPR) (Charter of Fundamental Rights of the European Union, 2000; Radley-Gardner et al., 2016).

The European Commission made the proposals in December 2020, and on March 25, 2022, a political agreement was reached on the Digital Markets Act (Digital Markets Act, 2022), and on April 23, 2022, on the Digital Services Act. Together they form a single set of new rules governing digital services in the EU that will be applicable across the whole of the EU. The main goals of the DSA and DMA are to create a safer digital space in which the fundamental rights of all users of digital services will be guaranteed and protected and to establish a level playing field to foster innovation, growth, and competitiveness, both in the European Single Market and globally (Digital Services Act, 2020).

The health domain is composed of four essential contributors: data holders and users, application and service providers, data space governance and operating systems, and cloud service providers (Gaia-X Domain Health Position Paper Version 1.0, 2021). Dataspace is the term that primarily refers to any ecosystem of datasets and data models, including ontologies, data-sharing contracts, and data management services, as well as associated soft competencies such as social interactions, governance, business processes, etc. Such competencies follow a data engineering approach whose goal is to optimize data storage and exchange mechanisms, preserving, generating, and making possible knowledge sharing to others. In contrast, data platforms refer to architectures and repositories of a group of interoperable hardware/software components that follow a software engineering approach (Scerri et al., 2022). The two concepts, data engineering and platforms, are interconnected and need to be considered together, as commercial solutions often do not differentiate between them. Therefore, and due to the special requirements for protecting the privacy of the individual, a distinction was made between technology and infrastructure that stores and processes personal and other data.

The nine European data-sharing spaces outlined by the European Strategy for Data are health, industry, agriculture, finance, mobility, green deals, energy, public administration, and skills. They are essential for the implementation of the European digital market and guide European activities toward the data economy (Scerri et al., 2022).

The BDVA (Big Data Value Association) is a community of experts that has been working on the development of dataspaces for many years. Their vision comprises a data space composed of several individual connected spaces. The dataspace should be able to cut across sectoral, organizational, and geographical boundaries (European Big Data Value Association, 2015).

As previously described, the European strategy for data aims to create a single market for data, which should ensure Europe's global competitiveness and data sovereignty. The strategy essentially aims to ensure the flow of data within and across EU sectors, making high-quality data available for innovation in the economy and society (while keeping data owners, companies, or individuals in control). The strategy is based on ensuring full compliance with European rules, regulations, and values as well as setting rules for fair access and usage of data in accordance with the existing data governance mechanisms (EU Digital Decade, 2019). Common European Dataspaces will be central in enabling new technologies such as AI and supporting the marketplace for cloud and edge-based services.

2.2.4 **Potential**

In order to release the full potential of health data, the European Commission is presenting a regulation to set up the European Health Dataspace (EHS). The EHS is a specific ecosystem composed of rules, standards, infrastructure, and a legal governance framework that aims to empower individuals through increased digital access to and control of their electronic personal health data. It supports the use and free movement of health data across the EU for better healthcare delivery, better research, innovation, and policymaking. It enables the EU to use and reuse the full potential of health data offered through a safe and secure exchange. The EHS supports the fostering of a genuine single market for electronic health record systems, medical devices, new technologies, and high-risk AI systems (EU Health Data Space, 2022). Thus, it is a core component of the European Health Union and builds further on the GDPR, the proposed data Governance Act, the draft Data Act, as well as the Network and Information Systems Directives.

Dataspace initiatives aim to access and share highly sensitive personal data in a secure and confidential manner governed and controlled by each EU member state in a consistent way, complying with relevant European regulations. A EHS could facilitate future pandemic management, including fast data transfer and short reaction times. Pattern recognition of disease outbreaks across state borders would be possible. Moreover, national healthcare systems could be relieved of the burden of some bureaucratic processes. For example, doctor appointments necessary to simply transfer data from one medical office to the other by the patient would become obsolete. Doctors would also be able to treat patients at home, which again reduces the pressure on hospital bed occupancy.

In the case of regulators and policy-makers, they will have easier access to health data and be able to make decisions for the better functioning of healthcare systems, leading to a more evidence-based policy-making. This will lead to better access to healthcare, increase its efficiency, reduce costs, and enable new research and innovation.

Also, the industry can benefit from the better availability of electronic health data sourced in an EU-wide market. This will improve people's health by facilitating the production of medical devices and gadgets, leading to improved personalized healthcare coverage.

2.2.5 **Challenges**

Secure health dataspaces require a complex ecosystem involving many different stakeholders connected by a plethora of regulated processes. The set-up and operation of such dataspaces are associated with many challenges. Various sources of public funding involved in the design of health dataspaces need to be reconciled. EU member states can apply different financing models in accordance with their sovereign legislation. Moreover, highly fragmented and heterogeneous EU markets limit the quick rollout of dataspaces or any other kind of digital framework on a large and transnational scale. Stakeholders may struggle to interpret and map the GDPR rules with the local legislators of the member states. Navigating the complex regulatory landscape and ensuring compliance with data protection laws adds another layer of complexity to the development process.

Data interoperability will be a major challenge for the design of health dataspaces. Medical data is often stored in various formats, collected by different organizations across miscellaneous systems. Data interoperability is a prerequisite for unified dataspaces. Otherwise, practitioners will have difficulty accessing, modifying, and exchanging data. Advances in standardization and the development of data exchange protocols are needed to achieve sufficient data interoperability. A single European or international health data standard such as Fast Healthcare Interoperability Resources (FHIR) should be adopted on a European level, as proposed by a study prepared for the European Parliament's committee on industry, research, and energy (Marcus et al., 2022). FHIR is a healthcare data standard with an application programming interface used to represent and exchange electronic health records (FHIR & Cloud Healthcare API', 2023). The standard enables links between medical data across different systems.

Additionally, special attention needs to be paid to security threats and data breaches. Sensitive health data should be protected from unauthorized access and cyberattacks with security measures such as encryption, intrusion detection systems, and the application of strong access control. Data quality conservation and integrity go hand in hand with data security concerns. Medical data needs to be accurate, reliable, free from errors, and protected from unauthorized tampering. Ensuring long-term data quality is a challenge, especially if data stems from various practitioners across different systems. This, in turn, leads to the establishment of data standards and exchange formats. However, the establishment of data silos is discouraged, even though they might simplify data protection measures. On the other hand, these silos could bring about data stockpiling without any useful function or connectivity. Current solutions are mostly too permissive (e.g., exposing data to the public domain or transferring data usage rights to a single commercial company) or too restrictive (e.g., study-specific point solutions or local-for-local solutions without opportunities for reuse). A balanced middle ground between these two extremes should be found in the development of the modern health dataspaces.

From a global perspective, without a more open European market, innovative companies are forced to focus their strategies on the United States and China.

The European Union and its associated transnational legislation could offer a great opportunity to attract international companies if they were not limited by the national borders and limitations of the member states. It remains a complex task to deploy a health data management system throughout the EU. But it is nevertheless a vital step toward medical research and development in Europe.

2.3 Proposed approach

EHDS will create a common space where natural persons can easily control their electronic health data as far as the fundamental rights of the data subject are concerned. Thus the individual can control to which entities (including humans and services) their data can be made available, as well as the constraints enforced. It will also enable researchers, innovators, and policymakers to use electronic health data on the condition that they fulfill the eligibility criteria for a number of added-value cases and workflows, in a trusted and secure way. The added value cases may range from medical inspection to AI training and evidence traceability.

In accordance with the goals of EHDS, PAROMA-MED aims to resolve the challenges mentioned above and proposes an approach that establishes a data life cycle, empowering individuals in the governance of their data. It allows the sharing of nonidentifiable health data and facilitates the trusted execution of data for researchers and other health professionals.

2.3.1 Data life cycle

The governance of data in PAROMA-MED involves four main phases: (1) secure and trusted addition of health data to the local data lake during data generation; (2) preparation of data for exposure in the federation, including encryption, anonymization, and ensuring data interoperability; (3) allowing the search over nonidentifiable health data available in the data space; and (4) using the data in a secure and trusted way, which may include additional consent from the user. The introduced data life-cycle management complies with any stakeholder's archiving procedures and policies as it retains adequate structuring, indexing, and retrieval. Beyond that, it allows data subject consent (including the right to be forgotten) to be appropriately applied. Furthermore, legacy data is planned to be integrated through the deployment of appropriate adapters with privacy and ownership protection and enforcement mechanisms supported by design. PAROMA-MED plans to provide user support dashboards for managing legacy data inclusion tasks.

2.3.2 Data generation and interoperability

Ideally, data should be generated in direct association with the subject they belong to. Assuming that data is generated following some medical procedures that are performed within the relevant medical infrastructure, the data subject has to be identified in the context of the validation of their prescribed examinations (Fig. 2.1). Upon presentation of the prescription, a trusted medical domain (through components that are continuously attested for integrity and adherence to the foreseen procedures—depicted in the figure in the green box) requires an identity challenge to be sent to the data subject. This step involves verifying both the subject's identity (potentially supported by a digital wallet application) and confirming that the medical domain is verified for its adherence to the proper procedures (Step 1). The resolution of the challenge establishes a time-limited association between the subject and the process (Step 2), concluding with the secure storage of the results under clear governance for future use (Step 3). Adhering to FAIR practices, an FHIR server (as shown in the figure within the secured storage

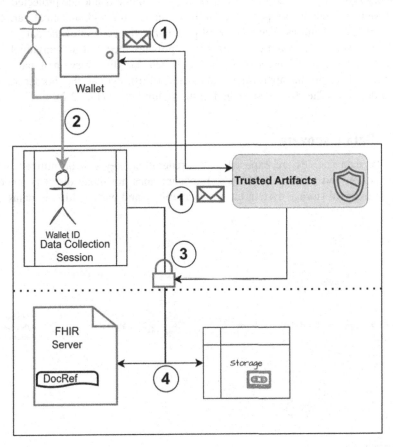

FIGURE 2.1

Data collection steps.

area at the bottom) is used from the moment data is generated. It is protected under the supervision and encryption by trusted domain artifacts of edge nodes for maintaining personal information. Additionally, an object storage solution, also under the same protection measures, is used for storing various medical results. Object references are included within FHIR documents. In the figure below (Fig. 2.1), the data subject is represented through a wallet-based interaction. This aspect has been taken into consideration for further research to ensure compliance with EBSI and eIDAS solutions.

2.3.3 Data exposure

Before the data can be used in the context of specific actions, they have to be discoverable, as far as federation interactions are concerned. For this purpose, a data inventory layer is produced from the data types available inside the protected storage. Inventory updates are performed in batch mode to avoid statistical variation being linked to identities. The process of populating the content of the data inventory layer takes into account constraints from policies specified by individual consents. The outcome is intended to be published for discovery in a dataspace ecosystem through the appropriate connector (illustrated at the border of the trusted domain). The flow is presented in the figure below (Fig. 2.2).

2.3.4 Data discovery

Once the data sources are exposed in the federation they can be utilized in AI model design and training workloads. Exposure does not mean direct population of some external storage system but the availability and participation in dataspace

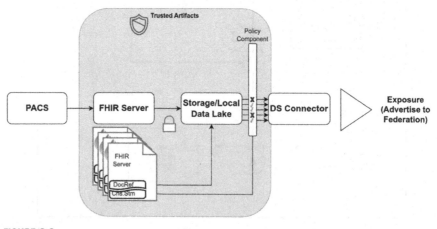

FIGURE 2.2

Dataspace exposure.

interworking sessions for resolutions of queries. Data scientists are able to submit queries to the Dataspace Metadata Brokering subsystem for discovering availability of types and volumes of data according to certain protection levels. The outcome is collected from all the participating domains. Each of the domains contributes to the query resolution by applying internally user policies and resolving the portions of the stored data that can be made available according to the query options.

This process aims at a streamlined and ergonomic approach that relieves the data scientists from the burden of locating data that are highly distributed, but most importantly from the burden of taking all measures to remain with the legal restrictions that private data protection legislations require. This leads to a one stop shop service and enabling mechanism. At this stage data availability is presented under three main categories (assuming local processing in all cases):

- Directly usable data
- Data of application relevance that need additional consent
- Data without known relevance and quantity

If the first category suffices, the ML flow can continue. In the opposite case the consuming side (data scientists on behalf of any organization or by themselves) can suggest rewards for the other two categories in an effort to secure data availability adequate enough to allow proper development of the intended ML model. In such cases the subject is presented with an incoming request containing usage context details to facilitate the creation of a clear decision in the form of an enforceable policy. Subsequently the consent details are updated, and additional usage possibilities are permitted.

2.3.5 Data usage

In the PAROMA-MED approach one of the core concepts is to avoid the transfer of medical data. Instead, with appropriate consent and trust prerequisites, the ability to perform secure computation on health data stored in a node of PAROMA-MED should be possible. As the usage of data is constrained by the intentions and identity of the consumer, based on the options of the producer or the subject whose privacy is to be protected, there is a need to securely enclose the entire flow within the strict borders of an instantiated environment, both in terms of deployed functionality and data with a limited lifespan. The approach is based on the Gaia-X conceptual and composition model, which envisions that resources can be:

- Virtual Resource: It represents static data in any form and necessary information such as dataset, configuration file, license, keypair, an AI model, neural network weights, ...

- Instantiated Virtual Resource: it represents an instance of a virtual resource. It is equivalent to a service instance and is characterized by endpoints and access rights.

According to the envisaged approach, data at rest (stored in local data lake and FHIR server) serves as virtual resources that can be instantiated within a volatile and isolated software enclosure. This enclosure facilitates the application of the intended processing, forming the depicted data usage layer. This step requires that the data is exposed in a uniform manner irrespective of its actual storage format. Additionally, if this step requires any filtering, encryption, anonymization, or watermarking, it is performed during the provisioning phase for the preparation of the data usage layer (Fig. 2.3).

Once adequate data is available for the model training purposes, data is prepared and remains available for the foreseen processing (constrained in terms of usage and time limits) (Fig. 2.4). The preparation phase, as explained earlier, involves adaptation and encryption/crypto-watermarking according to data owner policies and data user requirements.

The negotiation between the designed application and federation resources is not limited to data discovery and usage. It also encompasses the availability of processing resources, which is also subject to discovery and, in several cases, closely related to the resolution of data availability in cases where data cannot be

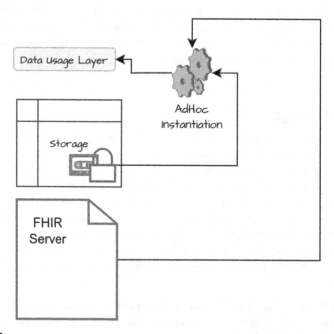

FIGURE 2.3

Data usage layer: provisioning.

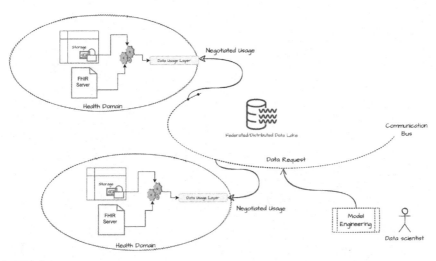

FIGURE 2.4

Model training preparation: data usage negotiation.

transferred outside of the domain borders. Such processes lead to the provisioning of data processing modules that actively participate in the training process. Moreover, processing modules can be deployed on the central cloud for resource-intensive tasks (Fig. 2.5).

The envisaged data usage approach resolves any issues related to the prevention of the use of AI/ML at large that stem from data encryption and protection. This is achieved by bringing training close to the data without necessitating disclosure and transfer among storage systems.

2.4 Dataspaces and participation in ecosystems

2.4.1 Identity governance

The PAROMA-MED approach for identity governance plays a crucial role in ensuring secure and trustworthy data sharing and collaboration. Identity governance within the PAROMA-MED framework focuses on establishing a robust and reliable mechanism for managing identities, access controls, and privacy considerations in federated learning (FL) scenarios.

PAROMA-MED emphasizes the need for a centralized identity management system that governs the identities of all participants involved in the FL ecosystem. This includes healthcare providers, researchers, data custodians, and other stakeholders. The identity management component ensures that each entity is uniquely identified and authenticated prior to their involvement in any data sharing or analysis activities.

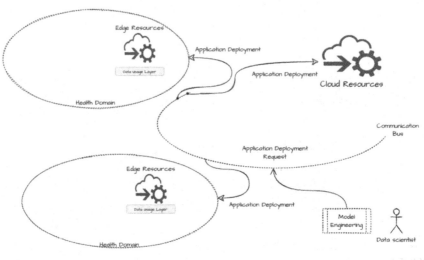

FIGURE 2.5

Model training preparation: deployment of compute modules.

By incorporating RBAC principles to define and enforce access controls within FL systems, different entities can be assigned roles based on their responsibilities and privileges. This approach ensures that only authorized individuals or entities can access specific datasets, participate in collaborative analysis, or contribute to the FL process.

In addition, PAROMA-MED leverages privacy-preserving technologies, such as differential privacy, FL, and secure multiparty computation, to protect sensitive patient data during the collaborative analysis process. These technologies help in minimizing the risk of data breaches and maintaining patient privacy while enabling effective knowledge sharing and model development.

Clear data governance policies that outline the permissible use, access, and sharing of healthcare data and consent management mechanisms ensure that patients have control over their data and can provide informed consent for its use in FL research.

2.4.2 Consent management

The focus on maintaining privacy and control over patient data while enabling its usage for research and learning purposes is one of the key aspects of the consent management approach of PAROMA-MED. Consent management is designed for the seamless integration of data into the architecture and design choices of the system.

The consent management process begins with capturing and recording patient consent for data usage. Various mechanisms, such as consent forms and digital consent processes, are employed to ensure that individuals can make informed decisions about how their data will be used. The aim is to provide transparency and clarity regarding the purpose and scope of data usage.

Consent storage and access control mechanisms are implemented to securely store the consent choices made by individuals and associate them with their data. This ensures that only authorized entities can access the data based on the provided consent, safeguarding patient privacy and ensuring compliance with consent preferences.

To respect the dynamic nature of consent, the PAROMA-MED approach allows individuals to easily revoke or modify their consent preferences. They are provided with interfaces and tools to manage their consent settings, empowering them to have control over the usage of their data and exercise their right to withdraw consent if desired. Moreover, maintaining an audit trail, ensures accountability, transparency, and compliance with privacy regulations. These auditing capabilities include recording the details of consent, such as when it was obtained, the specific terms of consent, any modifications or revocations, and the actions taken based on the provided consent.

In the context of FL, PAROMA-MED's consent management approach is seamlessly integrated into the process. Only data for which explicit consent has been given is included in the FL models. Privacy-preserving techniques, such as data anonymization or encryption, are applied during the learning process to further protect patient privacy.

By incorporating consent management into the architecture and design choices, the principles of responsible and ethical data usage are upheld. The privacy preferences of individuals are respected, and transparency and control over data usage are provided, fostering trust between patients, healthcare providers, and researchers.

2.4.3 Trust establishment

According to the European Data Strategy fact sheet, data processing moves gradually from centralized cloud computing facilities to smart, connected, and edge resources. This pattern, in combination with dataspaces, leads to the formation of data federations, where business domains attach to ecosystems to perform data-related tasks according to specific roles. According to the Gaia-X conceptual model, participants interact within the federation by providing (producers) and consuming (consumers) resources. The basic assumption before engaging in a transaction is that both interacting parties present verifiable credentials to each other, issued by participants and signed by trust anchors. Aiming to enable the automated attestation of federation participants, following the ideas of Gaia-X and the Compliance Service deployment options (licensed, private decentralized, secure private, and public decentralized models) closely, the consolidation of the trust model, grounded in both the secure private and the public decentralized models, can be expressed as follows:

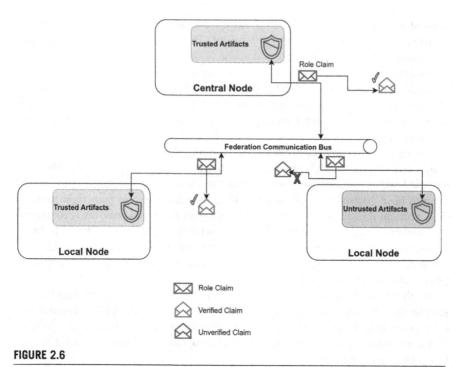

FIGURE 2.6

Trust establishment based on integrity of the participant.

- Elimination of manual process in trust establishment and legal binding (e.g., contracts and/or SLAs)
- Trust is established on proofs related to the integrity of certified components and their exclusive involvement in privacy-sensitive operations, based on trusted execution enclaves and/or trusted platform modules
- Adherence to proper operation is a continuous process, and any verification failure leads to immediate and immutable publication of the status change

This approach is visualized in the following figure (Fig. 2.6). The concept of the Federation Communication Bus serves as a placeholder to be further elaborated and clarified according to the dataspace connector protocols.

2.4.4 Protection and assurance

2.4.4.1 Privacy-preserving technologies

2.4.4.1.1. Agent-based approach

The agent-based approach, specifically using authentication and authorization sidecar proxies along with policy enforcement agents, is an innovative concept in the context of architecture and design choices for FL in modern digital healthcare systems. This approach leverages the benefits of microservice architecture to enhance security, privacy, and consent management.

In this concept, microservices within the FL system are equipped with sidecar proxies responsible for authentication and authorization. These proxies act as intermediaries between the microservices and external systems, handling the authentication process and enforcing access control policies. They authenticate users or entities requesting access to the system and verify their credentials against trusted identity providers. By offloading authentication tasks to dedicated proxies, the microservices can focus on their core functionalities, ensuring a modular and scalable architecture.

Policy enforcement agents are introduced to enforce security, privacy, and consent policies within the system. These agents are responsible for evaluating and enforcing policies related to data access, data sharing, privacy protection, and consent management. They operate in conjunction with the sidecar proxies to enforce fine-grained policies based on user roles, permissions, and other contextual attributes. This enables dynamic and context-aware policy enforcement, ensuring that sensitive data is accessed and shared appropriately and that privacy and consent requirements are adhered to.

By incorporating authentication and authorization sidecar proxies and policy enforcement agents, the agent-based approach enhances the security, privacy, and consent management capabilities of the FL system. It enables centralized policy management and enforcement, ensuring the consistent application of security and privacy controls across microservices. The use of dedicated agents allows for flexibility and agility in adapting to changing policies and regulatory requirements.

Furthermore, this approach promotes interoperability and compatibility with existing authentication and authorization frameworks, enabling seamless integration with external identity providers and policy management systems. It provides a unified and standardized approach to authentication, authorization, and policy enforcement across the FL system, facilitating secure and privacy-preserving data exchange and collaboration. The approach strengthens security, privacy, and consent management capabilities, ensuring the protection of sensitive data and compliance with regulatory requirements while maintaining the flexibility and scalability offered by microservices.

2.4.4.1.2 Machine learning-based approach

FL is the most commonly used ML approach that allows multiple data owners to collaboratively train an ML model without sharing their own training data. Some other approaches are possible, such as model fusion. Here, all participants train a local ML model. When the local trainings have converged, the model weights are smartly aggregated to obtain a model that generalizes well on the data of all participants. Teacher aggregation ensemble is another potential approach. Here, each local data owner trains a local ML model, which is called the teacher model. The teacher models are employed to label a new dataset that is used to train a student ML model that is deployed.

While FL is flexible and resolves data governance and ownership issues, it does not itself guarantee security and privacy unless combined with other methods. Indeed, when using an ML model, information can leak on the learning data, even if the ML objective is to generalize as much information as possible. Many recent works have shown that ML models themselves can be used to derive personal information. In particular, two kinds of attacks have been described:

1. Membership inference attacks: The ability to identify whether a data record was included in the training dataset of the target ML model
2. Attribute inference attacks: The ability to infer missing attributes of a partially known record used in the training dataset by accessing the ML model.

These new needs in terms of security and privacy encourage the use of approaches such as secure multiparty computation (SMPC) or differential privacy (DP) to secure FL processes.

2.4.4.2 Secure multiparty computation

Working on encrypted data is one of the best ways to guarantee security. However, enabling efficient processing of such encrypted data is one of the biggest challenges in the security field. Although fully homomorphic encryption allows one to perform calculations over encrypted data without decrypting it first (Gentry, 2009), it is often judged too slow, complex to use, and impractical. SMPC (Lindell, 2020) is an alternative to homomorphic encryption. It allows owners of private datasets to perform operations on their collective data without disclosing anything except the outcome of the computation. As an example of the SMPC method, private set intersection (PSI) has garnered much attention due to its capability to facilitate efficient comparisons and certain analytics on encrypted data sets. PSI could be used, for example, to determine shared patients between two hospitals without disclosing the specific patient lists held by hospitals.

2.4.4.3 Differential privacy

Intuitively, differential privacy (Dwork, 2014) corresponds to ensuring that the output distribution of a randomized algorithm will not be significantly different considering the presence or absence of one particular individual. An adversary with access to the algorithm will not be able to learn about individuals but will only have access to the global knowledge of the algorithm among them, ensuring the protection of privacy.

2.4.4.4 Data protection and traceability

Data traceability is another major concern for FL, as sensitive data has to be shared between different users. In some cases, data samples may be remotely requested to understand incorrect model behavior, or externalized for annotation when the expertise for annotation is not on site. The risk of information leakage is not negligible. Over 50% of data breaches originate internally. There is a need to be able to hold accountable the entity responsible for the leak and identify it as quickly as possible. Today, this requires a complex and lengthy investigation, generally lasting more than 2 months. Similar issues arise when it comes to models' parameters. Building a model is costly, as it requires expertise in data science and medicine as well as huge computing resources. Herein it is important to protect model ownership.

Data watermarking and model watermarking are technologies that can address such threats in the FL environment. When applied to images, watermarking is defined as the invisible embedding of a message into a host image by imperceptibly modifying its gray values. Watermarking leaves access to the data while keeping it protected by the message (Boenisch, 2021). Depending on its content, the embedded message can fulfill various security services, such as ensuring data authenticity, maintaining data integrity, and enabling data traceability. This may involve embedding proof of ownership or a message tracker to counteract information leaks and identify malicious users. There has been a growing interest in combining watermarking with encryption to achieve both a priori and a posteriori protection simultaneously (Haddad et al., 2021). The integration, known as the crypto-watermarking technique, is designed to provide watermarking-based security services from encrypted data.

2.5 Lessons learned: conclusions and future scope

PAROMA-MED has worked so far on an extended set of functional and nonfunctional requirements that are trying to cover several perspectives from different stakeholders' (data subjects, data scientists, medical experts, medical centers and organizations, application providers, etc.) point of view. The process revealed several aspects regarding interoperability, feasibility, value protection, and adequacy of the technology. More specifically, the role of FHIR has been identified as the most prominent solution with respect to data structuring, management, and storage. Furthermore, involvement in dataspace practices appears to present a clear pathway toward maximizing the utilization of project outcomes, including proposing/contributing to a concrete model that explores the feasibility of close-to-data processing. Furthermore, a clear challenge for utilizing trusted computing practices for the purpose of zero trust and attestation to deliver the trusted components has been made evident. Finally, the interaction with external players, such as medical experts, has significantly clarified the importance of data value protection not only for primary medical data but also for secondary data products that demand domain knowledge.

After fulfilling the identified requirements, the project is currently advancing in the development of key components, including medical imaging adaptation, FHIR server, object storage, FL framework, dataspace connectors, watermarking solutions, and identity and privacy awareness. At the same time technical workshops are being conducted, with the gradual expansion of integration and experimentation scenarios focusing on the overall flow of operations, including data ingestion, data consent management, data advertisement, data discovery, usage negotiation, and ML training.

The current study introduces the main concepts evolving within the context of the European Project PAROMA-MED. With the challenges and potential of dataspaces and federated ML well-identified, the project is soon to enter an experimentation phase. This phase will lead to the realization of the identified concepts to be evaluated in a concrete use case related to the qualitative assessment of cardiac anatomy. Specifically, the project is based on addressing the characterization of myocardial wall thinning using cardiac computed tomography images.

Acknowledgment

This work is funded by the European Union under Grant Agreement 101070222. The views and opinions expressed in this work are solely those of the author(s) and do not necessarily represent the views of the European Union or the European Commission (granting authority). The European Union and the granting authority cannot be held responsible for this work.

References

Abraham, I. Jr., (2023). Importance of data quality in artificial intelligence for healthcare. Accessed June 11, 2023, from https://www.linkedin.com/pulse/importance-data-quality-artificial-intelligence-abraham-ibrahim-jr-/.

Adekoya, O., & Ekpo, E. (2022). Digital assets — An emerging trend in capital markets. Available: https://www.pwc.com/ng/en/assets/pdf/digital-assets.pdf

Berlage, T., Claussen, C., Geisler, S., Velasco, C. A., & Decker, S. (2022). Chapter 18 Medical data spaces in healthcare data ecosystems. *Designing Data Spaces*. Available from https://doi.org/10.1007/978-3-030-93975-5_18.

Boenisch, F. (2021). A systematic review on model watermarking for neural networks. *Frontiers in Big Data*, 4729663.

Charter of Fundamental Rights of the European Union', no. C 364/3, Dec. 2000, [Online]. Available: https://www.europarl.europa.eu/charter/pdf/text_en.pdf

Davenport, T., & Kalakota, R. (2019). The potential for artificial intelligence in healthcare. *Future Healthcare Journal*, 6(2), 94—98. Available from https://doi.org/10.7861/future-hosp.6-2-94.

Dwork, C. (2014). *The algorithmic foundations of differential privacy*.

EU Data Act, EU COMMISSION, On Common European Data Spaces, 23.02.2022.

EU Strategy (2022). *A European strategy for data*. https://digital-strategy.ec.europa.eu/en/policies/strategy-data (June 25, 2023).

EU (May 05, 2023). A European approach to Artificial Intelligence and the role of open data. Accessed May 15, 2023 from https://data.europa.eu/en/news-events/news/european-approach-artificial-intelligence-and-role-open-data.

European Big Data Value Association' (Apr. 08, 2015). BDVA. Accessed May 23, 2023 from https://www.bdva.eu/about.

Europe's Digital Decade: digital targets for 2030'. Accessed May 23, 2023 from https://commission.europa.eu/strategy-and-policy/priorities-2019-2024/europe-fit-digital-age/europes-digital-decade-digital-targets-2030_en.

FHIR | Cloud Healthcare API'. Google Cloud. Accessed Jun. 02, 2023 from https://cloud.google.com/healthcare-api/docs/concepts/fhir.

Gaia-X Domain Health Position Paper Version 1.0 2021'. Accessed: May 24, 2023. [Online]. Available: https://www.bmwk.de/Redaktion/DE/Publikationen/Digitale-Welt/211116-pp-health.pdf?__blob = publicationFile&v = 1

Gaia-X Trust Framework - 22.10 Release.

Gentry, C. (2009). Fully homomorphic encryption using ideal lattices. *STOC*, *9*, 169−178.

Habehh, H., & Gohel, S. (2021). Machine learning in healthcare. *Current Genomics*, *22*(4), 291−300. Available from https://doi.org/10.2174/1389202922666210705124359.

Haddad, S., Coatrieux, G., Moreau-Gaudry, A., & Cozic, M. (2021). Joint watermarking-encryption-JPEG-LS for medical image reliability control in encrypted and compressed domains. *IEEE Transactions on Information Forensics and Security*, *15*, 2556−2569.

IDS RAM (2023). Accessed May 31, 2023 from https://github.com/International-Data-Spaces-Association/IDS-RAM_4_0/.

Kerry, C.F. (2020). Protecting privacy in an AI-driven world. Accessed June 15, 2023 from https://www.brookings.edu/articles/protecting-privacy-in-an-ai-driven-world/.

Koutsopoulos, K., et al. (2022). Federated machine learning through edge ready architectures with privacy preservation as a service. In: *2022 IEEE Future Networks World Forum (FNWF), Montreal, QC, Canada* (pp. 347−350), doi: 10.1109/FNWF55208.2022.00067.

Lindell, Y. (2020). Secure multi-party computation[Online]. Accessed 25 June 2020 from, Available at: https://eprint.iacr.org/2020/300.pdf.

Marcus, J. S., Martens, B., Carugati, C., Bucher, A., & Godlovitch, I. (2022). The European health data space. *SSRN Journal*. Available from https://doi.org/10.2139/ssrn.4300393.

OpenDEI (April 2021). Design Principles for Data Spaces, Position Paper.

Proposal for a Regulation of the European parliament and of the council on a Single Market For Digital Services (Digital Services Act) and amending Directive 2000/31/EC'. Accessed: May 23, 2023. [Online]. Available: https://eur-lex.europa.eu/legal-content/EN/TXT/PDF/?uri = CELEX:52020PC0825

Proposal for a Regulation of the European Parliament and of the council on the European Health Data Space'. Accessed: May 24, 2023. [Online]. Available: https://eur-lex.europa.eu/resource.html?uri = cellar:dbfd8974-cb79-11ec-b6f4-01aa75ed71a1.0001.02/DOC_1&format = PDF

Radley-Gardner, O., Beale, H., & Zimmermann, R. (Eds.), (2016). *REGULATION (EU) 2016/679 OF The European Parliament and of the council of 27 April 2016 on the protection of natural persons with regard to the processing of personal data and on the free movement of such data, and repealing Directive 95/46/EC (General Data Protection Regulation)*. Hart Publishing. Available from 10.5040/9781782258674.

Regulation of the European parliament and of the council of 14 September 2022 on contestable and fair markets in the digital sector and amending Directives (EU) 2019/1937 and (EU) 2020/1828 (Digital Markets Act)'. Accessed: May 23, 2023. [Online]. Available: https://eur-lex.europa.eu/legal-content/EN/TXT/PDF/?uri = CELEX:32022R1925

Riskin,D. (2023). Why healthcare data quality matters in the age of AI. Accessed October 23, 2023 from https://www.forbes.com/sites/forbestechcouncil/2023/09/05/why-health-care-data-quality-matters-in-the-age-of-ai/?sh = 2f4d48883bdd.

Scerri, S., Tuikka, T., de Vallejo, I. L., & Curry, E. (2022). 'Common European data spaces: Challenges and opportunities'. In E. Curry, S. Scerri, & T. Tuikka (Eds.), *Data spaces : Design, deployment and future directions* (pp. 337−357). Cham: Springer International Publishing. Available from 10.1007/978-3-030-98636-0_16.

Shaping Europe's digital future, https://digital-strategy.ec.europa.eu/en, (accessed June 25, 2023).

Siska, V., Karagiannis, V., & Drobics, M. (2023). Building a Dataspace: Technical overview, Gaia-X Hub Austria.

Usländer, T., & Teuscher, A. (2022). Industrial data spaces, Chapter 19. *Designing Data Spaces*. Available from https://doi.org/10.1007/978-3-030-93975-5_19.

Wieringa, J., Kannan, P. K., Ma, X., Reutterer, T., Risselada, H., & Skiera, B. (2021). Data analytics in a privacy-concerned world. *Journal of Business Research, 122*, 915−925. Available from https://doi.org/10.1016/j.jbusres.2019.05.005.

Curation of federated patient data: a proposed landscape for the African Health Data Space

Mirjam van Reisen[1,2], Samson Yohannes Amare[1,2], Ruduan Plug[1],
Getu Tadele[2], Tesfit Gebremeskel[1], Abdullahi Abubakar Kawu[3,4], Kai Smits[2],
Liya Mamo Woldu[2], Joëlle Stocker[5], Femke Heddema[1],
Sakinat Oluwabukonla Folorunso[6], Rens Kievit[7] and Araya Abrha Medhanyie[2]

[1]*Leiden University Medical Center (LUMC), Leiden University, Leiden, the Netherlands*
[2]*School of Humanities and Digital Sciences, Tilburg University, Tilburg, the Netherlands*
[3]*Technological University Dublin, Dublin, Ireland*
[4]*Department of Computer Science, Ibrahim Badamasi Babangida University, Lapai, Nigeria*
[5]*Research Advisers and Experts Europe (RAEE), Brussels, Belgium*
[6]*Artificial Intelligent Systems Research Group (ArISRG), Department of Mathematical Sciences,*
Olabisi Onabanjo University, Ago-Iwoye, Nigeria
[7]*Leiden Observatory, Leiden University, Leiden, the Netherlands*

3.1 Introduction

The COVID-19 pandemic has demonstrated the importance of access to and the shareability of health data so that it can be used optimally for research, policy-making, public health, and innovation (Daniel et al., 2022; Green et al., 2023). The European Union (EU), which is recognized as the frontrunner in data protection, has attempted to deal with this new "gold" with a number of policies and pieces of legislation, including the Cross-Border Healthcare Directive, the General Data Protection Regulation, and the Data Governance Act (Marcus et al., 2022; Testa, 2022). In 2022, it was announced that a European Health Data Space (EHDS) would be created, building upon existing legislation. The main purpose of the EHDS is to unify the governance system (including that of secondary data) for all EU member states and ensure semantic interoperability between the health data infrastructure in each state (Molnár-Gábor et al., 2022). The EHDS would address the four aspects of data curated as FAIR data, i.e., the data should be findable, accessible (under well-defined conditions), interoperable, and reusable (FAIR). So far, interoperability has remained difficult due to the different systems

Federated Learning for Digital Healthcare Systems. DOI: https://doi.org/10.1016/B978-0-443-13897-3.00013-8

and structures for health data in each country, which are operating under different jurisdictions (Stellmach et al., 2022), and the reusability of clinical data has been poorly addressed (Marcus et al., 2022).

Africa has seen a number of initiatives toward the creation of a common data space. In East Africa, it was assessed that a regional approach to storing and distributing health data would boost economic efficiency, improve health systems, and increase the speed and quality of digital health implementation. In 2017, the digital Regional East African Community Health (REACH) initiative was introduced to advance FAIR-based health data interoperability in the region (EAHRC, 2017). Another development in Africa is the adoption of the African Union's Malabo Convention on cyber security and personal data protection, which is the first step toward African data governance standardization (Ball, 2017). This convention establishes a number of principles for processing personal data, including:

- Personal data processing procedures must be lawful, fair, and transparent.
- Objective of data collection, as well as the data to be collected, should be defined, specific, and legitimate.
- Personal information gathered must be correct and up to date.
- Personal data processing and storage must take data security and privacy into account.

Kenya, South Africa, Nigeria, and Tunisia were among the first nations to enact legislation to protect citizens' privacy and the security of sensitive data. To allow for cross-border data flows while addressing privacy and security concerns, a data governance framework needs to be established that answers concerns about the sovereignty of Africa's data gathering, storage, and mining capability (Ndemo & Thegeya, 2022).

This chapter presents a landscape of the use cases developed in Africa by the virus outbreak data network (VODAN) based on the deployment of a federated data infrastructure across 9 African countries connecting clinical and research data from 88 health facilities, as well as data from population groups without access to health clinics. Using the concept of "curation," new aspects of data stewardship in federated, machine-actionable, and semantic format are analyzed for consideration of what parameters are needed for an African Health Data Space.

3.2 Background

3.2.1 Inventory on open science and findable, accessible, interoperable, and reusable data initiatives in Africa

Among the main initiatives promoting open science in Africa are the African Open Science Platform (AOSP), the East African Health Commission's initiative for an East African Open Science Cloud for Health, and the Africa Implementation Network developed by GO FAIR (n.d.), IN-Africa (n.d.), AOSP (n.d.). Some universities have implemented FAIR-data management techniques (Ministry of Education, 2019; University of Cape Town, 2023), and the Africa University Network on FAIR Open Science has embraced FAIRification (Africa University Network on FAIR Open Science, 2022). Similarly, the Consultative Group on International Agricultural Research (CGIAR), a global alliance for food security research, is dedicated to sharing its research data in accordance with FAIR-data principles (CGIAR, n.d.). In addition, the Committee on Data for Science and Technology (CODATA) works on data accessibility for research, making it FAIR, and promoting international collaboration to advance Open Science (CODATA, 2022). Finally, FAIR Forward: Artificial Intelligence for All is working with a number of partner countries to create an inclusive and sustainable approach to AI (BMZ Digital. Global, 2023; PCAST, 2010).

3.2.2 Relevance of federated learning approaches for digital healthcare

Marcus et al. (2022) highlighted the limitations that the EU has struggled with regarding health data, including (1) uncertain demand for health services across borders by patients; (2) the need to maintain the privacy of health data; (3) EU member states not taking up incentives to participate in data-pooling arrangements; (4) lack of a strong mandate for EU-level management; and (5) interference with other EU and national legislation. These issues have made it difficult to respond to crises, such as the COVID-19 pandemic, on an EU or member state level.

These issues are also relevant in Africa, which is transitioning from paper-based to digital patient records. The variability of contextual factors is high. Receiving granular data from scarcely connected and remote areas is challenging. To enable an efficient outbreak response and to improve overall healthcare while also increasing research capacity and quality, a robust, interoperable health data space could offer a solution. Federated learning makes healthcare data accessible because it is privacy preserving and makes learning from the data possible without losing its provenance. This chapter thus explores the relevance of a federated learning approach to such a space and its ability to tackle issues relating to privacy and data quality in healthcare.

3.3 Conceptual framework

In Jean-Paul Martinon's book, *The Curatorial: A Philosophy of Curating* (2013), Stefan Nowotny links the origins of curation to the Latin fable of *Cura* ("care" or "concern") by Hyginus. The fable describes how the first human was made from clay from the banks of a river. Once finished, Cura asks Jupiter to bring this figure to life; Jupiter agrees to do so, but on the condition that it is named after him. They start arguing, which brings the Earth-goddess Tellus into the picture; Tellus feels that the figure should be named after her, as it was molded from her body (the Earth). They call Saturn as a judge, who decides that when the creature dies, the spirit will return to Jupiter and the body will be retrieved by Tellus, while Cura will possess it for the entirety of its life. Thus, if we see humans as the first things to be curated, the curation thereof is both a measure of care and possession.

Yet, curation is a broad term for which the meaning varies extensively. Shott (1996) links the traditional meaning of curation, namely, housing collections to allow use in research and education, with the "contemporary keeping of old things for future use" (p. 260). He describes curation as the "degree of use or utility extracted, expressed as a relationship between how much utility a tool starts with—its maximum utility—and how much of that utility is realized before discard. (Shott, 1989)" Data curation is also centered around utility and future usability. A report by the National Science Foundation (2003) on cyberinfrastructure states that "acquisition, curation, and ready access to vast and varied types of digital content provide the raw ingredients for discovery and dissemination of knowledge" (p. 44). The importance of utility in the act of curation becomes clear when looking at what distinguishes curation from archiving and preservation, which ensures that data or objects are available for discovery and reuse (Yakel, 2007).

However, what these definitions are missing, which is a focal point in the fable of Cura, is the aspect of care. The Latin word *cura*, which means "taking care" or "worrying about," also signifies "to treat and supervise" (Sposito, 2017). Hence, caring for data has to be synonymous with data governance and management. As the value of data is perceived primarily in its usability, caring about data is deeply linked to the use and reuse of data. This includes ensuring the quality of the data, thereby optimizing its usability (Sposito, 2017). Caring for data leads to the creation and enhancement of its value. The curation of health data lies in ensuring that it is accessible but also that it has value, that is, that it is reusable, so that researchers, health workers, and policymakers can benefit from its existence. For instance, data used in research should lead to new insights, increased accuracy and validity of the findings, decreased time and effort spent looking for data, and more efficiency in cleaning and manipulating data (Hedstrom, 2012).

Martinon (2013) also introduces the reader to a work of art that Stéphen Mallarmé started writing in 1866, but never finished. It was to be a 2-hour multisensory event constituting the final Orphic explanation of life on Earth, which he named "*This is.*" All that remains of this work now are some notes, including that it displays the work of others and is, thus, explanatory of the essence, that it brings the past and present together, that it enables a constellation of meaning that is not achievable through any other artform, and that it combines human agency and the absolute, that it is centered on the viewer, that it is multisited (there is no center of significance), and that it does not have predetermined rules, grammar, or syntax, but rather needs to find its own language (cited in Martinon, 2013). Martinon argues that this work of art could be seen as the ultimate act of curation, as it tries to gather all art and philosophy together in a decentralized space that is accessible to the public in their own context, much like a federated health data space could be.

3.4 Methodology

The curation of health data requires adherence to ethical and legal standards, which becomes a complicated topic when research is carried out across national borders (Plug et al., 2022a). Care has to be taken with regard to data access control, privacy, and local regulations (Jati et al., 2022), as well as the interoperability of data (Basajja et al., 2022), to be able to ethically generate value from localized health data. To mitigate bias in the data, as well as to enhance inclusivity, data from population groups that do not have access to health facilities needs to be analyzed (Ghardallou et al., 2022).

3.4.1 Research approach

The paradigm underlying adherence to restrictive and privacy-sensitive data spaces is federated analysis (Plug et al., 2022b). This technique embodies the concept of data ownership to the fullest extent, as localized data is aggregated and only aggregated data may be analyzed by those authorized. Ensuring strong principles in regard to data ownership and digital sovereignty is key to promoting international collaboration within the African Health Data Space (Van Reisen et al., 2022). Without direct access to data, the accuracy, verbosity, and richness of metadata become crucial for machine readability and actionability. Applying FAIR-data principles ensures that such data can be used within the federated data framework, involving both horizontal and vertical data analysis across distinct data repositories.

3.4.2 Virus outbreak data network: establishing a quality data production pipeline in residence

VODAN deployed a federated architecture of machine-actionable linked data in a real-life situation (Van Reisen et al., 2021, 2022). The network curated the data and created a data pipeline to study the interoperability and reusability of the data in a real-life natural experiment. The simple knowledge organization system vocabulary was used to define concepts with a uniform resource identifier, standardize and describe relationships between concepts, and match concepts to those already used in the domain. By publishing this vocabulary on a web-based source, the metadata is stored in an openly accessible repository. While the FAIRification of the metadata was not complete according to VODAN standards, the use cases showed that even when applied partially, the increased linkability of the data greatly enhances opportunities for future research. The objective of deployment in a natural environment was to understand the potential for the adoption of a federated data architecture across different regulatory frameworks with respect to data sovereignty. Use cases were investigated to explore the potential of the data pipeline.

3.4.3 Installation of a federated data network

This study was performed through the development of the VODAN FAIR-based data platform, which is an infrastructure for a data pipeline across borders in different jurisdictions. The data is fully owned by the data producer, analyzed in the health facility, and stored locally. The study was carried out in Africa, where a lack of data control and data loss is perceived to be a problem in the creation of integrated healthcare centered around the interests of the patient. It was concluded that the architecture developed by VODAN has the capability to conduct cross-silo federated learning on data that is standardized and stored as RDF triples in a decentralized manner.

Recognizing the value of federated data and the importance of keeping data in residence, VODAN's implementation is certified as FAIR with ownership, localization, and regulatory (OLR) compliance features (Van Reisen et al., 2023). The OLR features reveal data curation as a process that aims for ethical awareness, which means that recognition is awarded to the original producer and data is situationally located. The OLR facets help to closely define the qualities that should be considered in the data curation process. These facets help to distinguish who cares *about* the data and who cares *for* the data, as well as to investigate whether the various aspects of data care can be bridged in a meaningful way in the new paradigm of a health data space.

3.5 Results: use cases

The value of the federated data infrastructure was tested by performing a number of use cases. The findings are presented here as distinct cases, all of which are based on the implementation of the VODAN federated data curation structure.

3.5.1 Pandemic early warning

Central to this use case is the adaptation of federated data infrastructures, through which stakeholders can contribute and analyze their data locally while preserving data privacy and security. This distributed approach allows organizations and researchers to collaboratively analyze large-scale datasets without the need for centralized data sharing. With FAIR-data principles guiding the process, data is made findable and accessible, enabling timely and efficient information sharing across different domains and geographic regions, which is essential for early warning on disease spread.

The use of federated data for early warning in global pandemics offers several advantages. It facilitates the horizontal and vertical aggregation of diverse datasets, ranging from clinical records and epidemiological data to specialized clinical measurements and environmental variables. This amalgamation of heterogeneous data sources enhances the comprehensiveness and richness of the analysis, providing a more holistic understanding of the factors influencing the spread and impact of pandemics (Bertozzi et al., 2020) Furthermore, federated data places strong safeguards on data privacy and protection, a critical consideration when dealing with sensitive health information. By locally storing and controlling data, both the risk and scope of unauthorized access or data breaches are minimized. This approach promotes trust and encourages participation by stakeholders, such as healthcare providers, researchers, and public health agencies, across national borders (Fig. 3.1).

To realize the full potential of federated learning and analytics for early warning systems, the advancement of distributed and aggregate analytical techniques for federated data is essential. These techniques enable collaborative analysis across distributed datasets without the need to share data, allowing information to be extracted in aggregate form. This places strong requirements on the quality of the underlying data, which requires careful curation (Nguyen et al., 2022). Federated analysis techniques also facilitate the standardization and interoperability of analysis methods, ensuring consistent and comparable results across different data sources and time horizons. This results in more replication and repeatable science, allowing researchers to make maximum use of past and future data. Altogether, these techniques advance the development of robust predictive

FAIR Federated Analysis

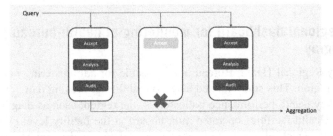

FIGURE 3.1

FAIR federated analysis of aggregation based on permissions per locale where data is produced for queries. *FAIR*, Findable, accessible, interoperable, and reusable.

models and early warning systems that can effectively detect and monitor the emergence and spread of infectious diseases, supporting timely interventions and mitigation strategies on a global scale.

3.5.2 Integration of data on incidence of COVID-19

This use case on the FAIRification of data on the impact of COVID-19 on migrants, refugees, and asylum seekers relates to a three-phase process to create FAIR data out of a database compiled from in-depth interviews, press articles, and research reports (Ghardallou et al., 2022). The process started with the observation that COVID-19 data from Africa was limited and that vulnerable communities were not represented. As refugees and migrants are highly mobile, to enable better access to services for them, it was deemed critical to include them in COVID-19 prevalence data. These population groups are mostly without access to health facilities and are, thus, left out of datasets. By making the data FAIR-compliant, access to the data for cross-analysis under well-defined conditions was guaranteed throughout the life cycle of the data.

This use case included 118 interviews and 565 reports on the prevalence of COVID-19, utilizing data from migrants, refugees, and asylum seekers residing in Tunisia, Libya, and Niger. The curation of the data as FAIR, based on the VODAN parameters, included the development of new data templates and vocabularies specifically designed for the research data. Analysis of the data across the different sources showed a low incidence of COVID-19 infections among migrants, refugees, and asylum seekers (5%) and a general lack of access to healthcare. The data indicated that although instances of COVID-19 were low, other health outcomes were adversely affected due to their hazardous movement across borders and exclusion from visiting health facilities. Analysis across the datasets showed that despite the expectation that the movement of migrants and refugees would be hindered by COVID-19 border closures, in reality, the smugglers developed new and more dangerous routes for increased prices. This use case shows how the analysis of incidences across different datasets can be facilitated through federated curation using VODAN parameters (Fig. 3.2).

3.5.3 Regional dashboard for monitoring at health-bureau level in Tigray

The Tigray Regional Health Bureau is responsible for administering health facilities in the region. This study looked at the possibility of reusing data for the monitoring of key health performance indicators in the region, such as diagnostics and treatment. Health facilities operated a dashboard at the facility level (Van Reisen et al., 2022), which allowed the Tigray Regional Health Bureau to monitor the aggregated health data in near-to-real time.

The VODAN parameters arrange the curation of the data produced at the source as machine-readable data and metadata, which are kept within the health

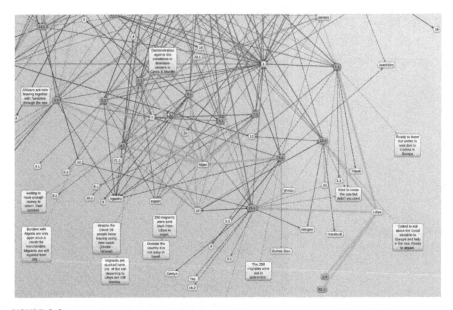

FIGURE 3.2

Sample of a knowledge graph created in AllegroGraph from FAIRified data on COVID-19 among refugees, migrants, and asylum seekers in Tunisia, Libya, and Niger.

facility where it is produced. The data and metadata are pushed from local facilities into regional and national systems based on well-defined ownership and regulation. The benefits reaped by the facilities, as well as the regional administration, from using dashboards include improved decision-making, improved efficiency, and, ultimately, better care for patients (Fig. 3.3).

3.5.4 Syphilis cases in Ayder Referral Hospital

The use of VODAN data was tested in routine syphilis screening and treatment during antenatal care visits to the Ayder Comprehensive and Specialized Hospital in Ethiopia, which is one of the health facilities serviced by VODAN data curation. VODAN provided a software platform that helped data production, data visualization, and the performance of remote queries using the RDF query language SPARQL.

A simple visual representation of syphilis reactive and nonreactive cases and adolescent pregnancy was obtained. Narrowing the visual query, warning signs for congenital syphilis—a disease that occurs when a mother with syphilis passes the infection on to her baby during pregnancy—were filtered (Fig. 3.4).

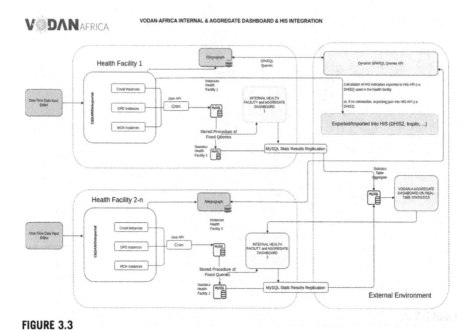

FIGURE 3.3

VODAN architecture for a one-data entry machine-actionable semantic curation for a multiple functionalities architecture based on queries through data visiting of federated local depositories in AllegroGraph. *VODAN*, Virus outbreak data network.

3.5.5 Scaling up health system implementation research: the *Saving Little Lives* project

The maternal, adolescent, reproductive, and child health (MARCH) research center at Mekelle University, Ethiopia, deployed VODAN principles for the data produced in the *Saving Little Lives* project. This project is a scaled-up research project implemented in the four major regions of Ethiopia. The objectives of the research were to facilitate: (1) the inclusion of data from a larger number of health facilities; (2) the inclusion of research data through several VODAN data points; (3) the inclusion of federated locally stored data in the health facilities; and (4) the harmonization of machine-actionable semantic data production across different sites in Ethiopia.

The project uses standardized indicators that are accessible across all sites. This facilitates the adaptation of the tools and formats to VODAN standards. It creates an opportunity to construct a dashboard for the team coordinator to monitor the data and information effectively. Moreover, adopting VODAN principles to make data reusable enables the use of project data for future research and projects long after the project has been completed (Fig. 3.5).

(A)   (B)   (C)

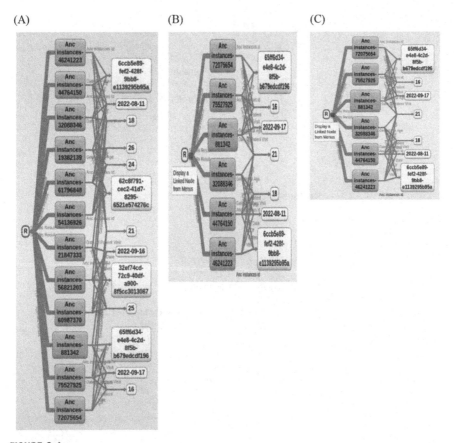

FIGURE 3.4

(A) All syphilis reactive cases. (B) Syphilis reactive adolescent cases. (C) Congenital syphilis at-risk cases.

3.5.6 Retrospective FAIRification to identify drivers of perinatal care outcomes in Kenya and Tanzania (MomCare)

This use case demonstrated how the content and structure of data contribute to actionable insights (value creation) for the MomCare program, a perinatal care intervention. MomCare covers a care trajectory for pregnant women in Kenya and Tanzania and brings vulnerable segments of the population into contact with health services using a mobile health application for accessibility and the registration of patients, as well as billing data (Aksünger et al., 2022).

Although currently aggregated data is disseminated to steer resource allocation, advanced analytics looking at specific use cases, such as the identification of factors contributing to individual risk within the population, or the analysis of

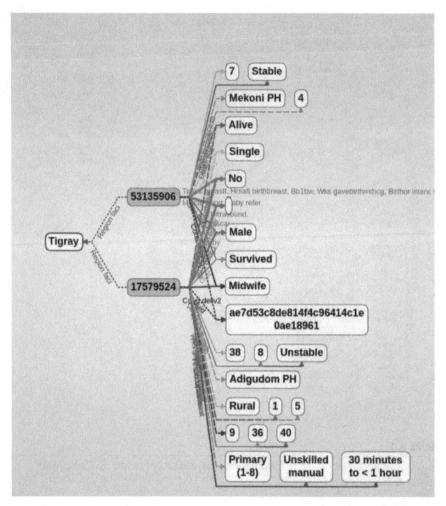

FIGURE 3.5

Sample of a knowledge graph created in AllegroGraph from FAIRified data on labor and delivery from the *Saving Little Lives* project in Tigray.

interdisciplinary factors at a country level, are not yet derived from the data, which is at present not curated as machine-actionable semantic data at the source.

Through an explorative design, this use case points out the limitations of the FAIRification process applied retrospectively. By defining a semantic model and the creation of a triple-store graph database, the provenance and quality of semantic data that could be provided at the source was lost. Hence, this use case provides clear recommendations on the relevance of the curation of data as machine-actionable semantic instances at the point of production.

3.5.7 **FAIR data curation of clinical research data on vaccines**

This use case concerns the creation of FAIR metadata in the female-only controlled human *Schistosoma mansoni* infection (CoHSI2) model. The study's metadata relates to clinical research that looked at a controlled human infection with female *S. mansoni cercariae*. The purpose of the study was to evaluate the effectiveness and safety of infection in healthy volunteers using primary and secondary outcome measures and any associated adverse events. This metadata holds valuable information for monitoring and evaluating the impact of interventions, as well as enhancing the management of adverse events associated with schistosomiasis infection.

The curated metadata was published on a FAIR data point. The FAIR data point was set up on an Azure instance located and governed in the Netherlands to make the (meta)data available in a machine-readable semantic format, using widely used ontologies in the field to improve the quality of the (meta)data. The metadata includes essential fields such as patient registration visit, event description, ICD-10 code, onset and end date and time, severity, relationship to the CoHSI, and whether the event is classified as a serious adverse event or not. It also provides information about the given treatment related to the adverse event.

3.5.8 **Integration of cardio-related patient-generated health data**

For patients with cardiac-related ailments such as hypertension, the home monitoring of key vital signs is an important activity in managing their health. A person's vital signs measure some of the body's basic functions, such as pulse rate, blood pressure (systolic and diastolic), body temperature, and respiration rate. While these are usually checked routinely in health facilities during episodic visits, some of these vitals are monitored at home too, especially when a patient has chronic diseases such as hypertension and diabetes. This data can be produced at home and, if interoperable with the records of the patient at the health facility, can facilitate individual patient monitoring.

Taking the VODAN data pipeline as the basis for the production of data at home for health facilities, a low-end technology was identified for measuring the vital signs of hypertensive patients, which was parsed and curated using VODAN parameters.[1] An interactive voice response system was tailored to individuals and localized to languages within a region to serve indigenous patients in Africa in relation to their healthcare (Demena et al., 2020). The data produced is then parsed to the VODAN-based digital patient health data system in the health facility, where it is stored in a safe and secure store. The data is made interoperable with the unique identifier of the patient and can be further processed to identify spikes and unusual readings. This helps health workers identify patients with

---

[1] This use case is part of the outcome of the LIACS DSIP module undertaken by Kievit and colleagues.

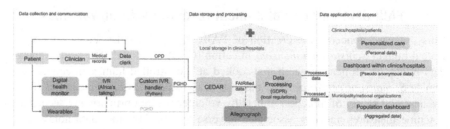

FIGURE 3.6

Proposed architecture for federated PGHD interoperability with patient data curated at a health facility. *PGHD*, Patient-generated health data.

worrying signs, triggering faster response times. A real-life implementation of this architecture will be tested (Kawu et al., 2023), following the architecture presented in Fig. 3.6.[2]

3.5.9 Interoperability across sectors: data on human trafficking

The use case on human trafficking research data relates to creating FAIR tools to share data in a way that protects the identity of victims and survivors, while allowing others to access and reuse the data under clearly defined conditions. Typically, human trafficking data, in relation to health outcomes, has not yet been coordinated toward FAIRification. Although some efforts have been made to develop ontologies and tools to find victims of human trafficking on the web (Hultgren et al., 2016), these tools have not led to the development of the tools necessary to FAIRify human trafficking research data, especially concerning refugees. The VODAN standards were used for the data curation of health outcomes. The researchers identified the ontologies to be constructed and the tools to convert the data into a FAIR VODAN-compliant format and investigated the potential to make knowledge graphs from the data and test their interoperability with other datasets.

Making the data FAIR not only helps with the reuse of data but also assists in enabling additional analysis. Developing community-relevant tools to record metadata is key to this process. One of the challenges in developing FAIR tools for human trafficking data is the differences in terms used and nuances in their

---

[2] Fig. 3.6 gives an overview of the data pipeline used for the patient-generated health data collection described. The pipeline is divided into three sections, each containing the steps for data collection, processing, and visualization. The procedures drawn with black arrows correspond to procedures present in the current VODAN implementations. The blue arrows correspond to the new pipeline proposed to be integrated and developed. The gray arrow leading from "wearables" to semantic vocabularies represents a direct connection between the digital wearable and semantic vocabularies.

meaning. As the reused data is based on ethnographic interviews, the researchers found it key to ensure that nuances are not lost in the ontology.

In relation to health outcomes, the data shows adverse health outcomes along the human trafficking trajectories, which have so far been insufficiently understood and documented. These outcomes include severe physical and mental trauma due to trafficking events and also tuberculosis (TB) within detention facilities. Cases of TB during the COVID-19 pandemic raise the question of whether some cases of TB could have been misdiagnosed as COVID-19, or whether the prevalence of TB may have increased vulnerability to COVID-19 for detainees. These hypotheses require further investigation across datasets.

3.6 Landscape for an African Health Data Space

Various stakeholders engage in the process of data curation, which is based on an understanding of the world as a place where different situations come together. The VODAN approach informs the agentic engagement of different actors in the data curation process. The approach views the caring for data and the curation of data in a social situation as a dynamic process. The VODAN data infrastructure demonstrates the emergence of an African Health Data Space based on federated data curated as machine-actionable semantic instances in respective jurisdictions (Fig. 3.7).

The VODAN qualities of data curation for a work format enable different interoperability and reusability functions. To allow this the VODAN life cycle of

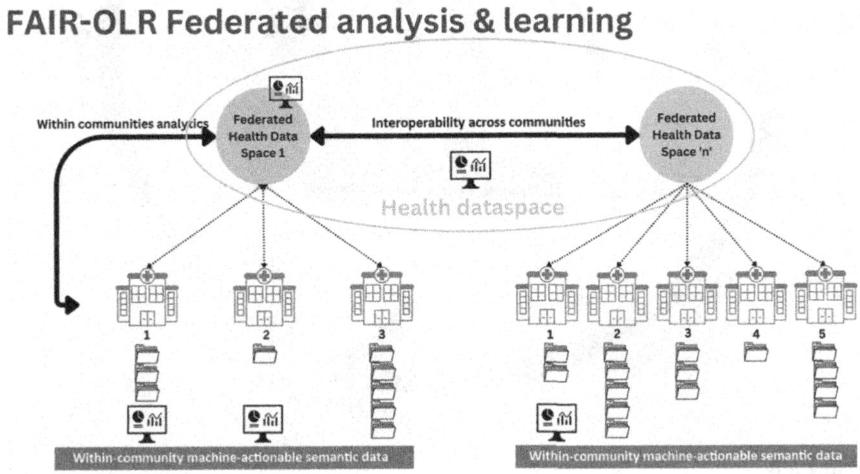

FAIR-OLR Federated analysis & learning

FIGURE 3.7

Architecture of FAIR-OLR federated analysis and learning. *FAIR-OLR*, Findable, accessible, interoperable, and reusable with ownership, localization, and regulatory.

the pipeline can be distinguished into three phases: (1) the creation of machine-actionable semantic data at the source where the data is produced; (2) the application of services determining the data capacities and storage; and (3) the facilitation of intelligence and user-experience for use and reuse of the data. These layers are held together by the seven qualities that guide the VODAN curation process: data is curated as findable, accessible (under well-defined conditions), interoperable, and reusable, with the localization of data production and ownership in a locale with regulatory compliance where the data is produced (Fig. 3.8).

The proposed landscape that emerges for a federated health data space is built along the three layers: (1) curation of the data at the sources as machine-actionable and semantic; (2) infrastructure services and services determining the generic data capabilities; and (3) the availability of standardized application programming interfaces across the ecosystem of the federated data spaces, which

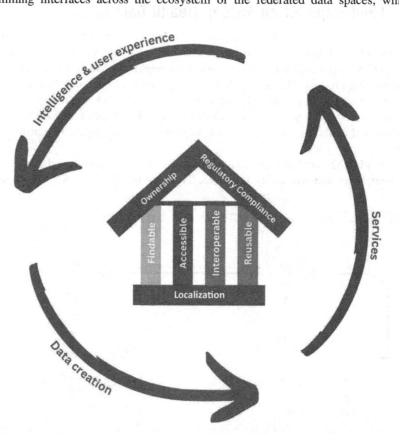

FIGURE 3.8

Schema of workbenches for FAIR-OLR federated analysis and learning. *FAIR-OLR*, Findable, accessible, interoperable, and reusable with ownership, localization, and regulatory.

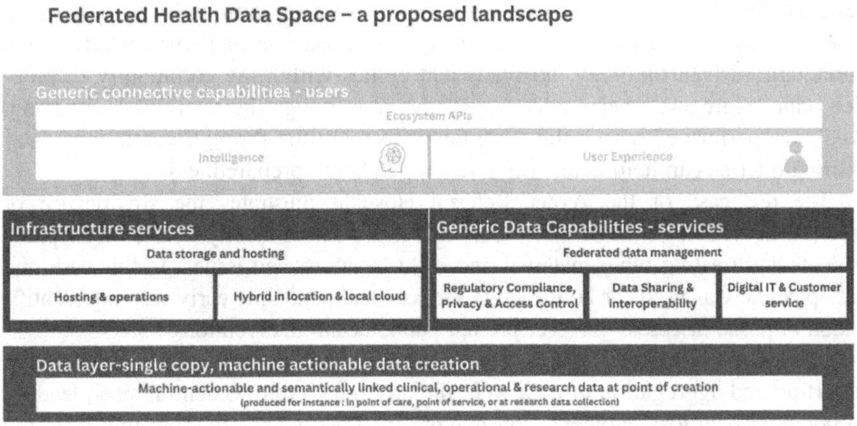

Federated Health Data Space – a proposed landscape

FIGURE 3.9

Proposed landscape for a federated health data space in Africa.

provide intelligence through algorithmic procedures and user experience (see Fig. 3.9).

The Federated Data Space is an extension of the Federated Health Data Space, allowing data from other domains to be used for the production of information and knowledge by visiting data in different disciplines. Preferably, the data is curated in a locale so that it generates value for the African continent. In this way the data space emerges as a visiting federation of data curated with linked semantic and machine-operable qualities that relate to the care awarded to the data at its source. Enhancing the value of the data through curation, services, and intelligence layers will require a value proposition that reflects the return on such added value.

3.7 Discussion

The use of federated data has emerged as a promising approach to the creation of early warning systems for global pandemics. The transfer of medical data across national borders is a restrictive and often infeasible process (Dove & Phillips, 2015). Federated learning and analytics provide a pathway to detect emergent diseases and their spread at an aggregate level at early onset (Naz et al., 2021). This enables researchers and policymakers to make informed decisions at an early stage and gives initial direction for more resource-intensive and granular studies. Such analyses may also provide evidence for health experts to cooperate across borders to thwart future pandemics, which provides value to all stakeholders.

The VODAN initiative focuses on the curation of federated data that allows verified queries to be performed through data visiting based on clinical inpatient data. This enables federated analytics within a medical community of practice in a secure way (Jati *et al.*, 2021). By incorporating aspects of FAIR metadata as a community-driven ontology, hospitals and clinics within the community of practice can aggregate their analyses while preserving their original meaning. Analytics performed across the community in real-time through continuous data-visiting queries can help detect outliers for pandemic preparedness.

The use case of the Ayder Referral Hospital illustrates the significance of machine-actionable semantic data curation within a health facility in assisting syphilis patients and treating other maternal and child health-related issues. In this case, the data pipeline was used for both routine services and pandemic early warning/identification of prone diseases to trigger prompt public health interventions.

FAIR-data principles address the issue of data ownership. Previously, the reporting and aggregation of data were directly transferred to central sites, leading to potential data loss. However, by making the data generated from the federated *Saving Little Lives* project VODAN compliant, health facilities have data readily available for both clinical and research purposes.

The use cases performed on scientific data demonstrate that the curation of research data at the source improves data quality and assists in the evaluation process of data produced at different sites. One of the challenges faced is the frequent changes in indicators and data elements. The adoption of a defined ontology would address such challenges, and new sites joining would follow a similar pattern of vocabulary and interact seamlessly with existing sites.

Use cases from across different disciplines demonstrate the power of imagining the health data space as federated. The queries performed using different types of data through data visiting in a safe and secure way resulted in the development of new hypotheses among populations that often remain out of reach.

Creating data interoperability from data produced at home and in health facilities creates a strong use case for the African Health Data Space. This data can be accessed for other uses, such as population health studies and drug and consumable planning, using federated learning. The use cases showed that data collected by patients at their convenience can be integrated into a digital health system that uses VODAN parameters. Through the FAIR-OLR format, different jurisdictions can be linked without any problem, given that the data is federated and visited on a permission-based process. The use cases show that existing tools with VODAN parameters can be used and adapted to expand the data space to different domains.

3.8 Conclusion

The curation of patient data in health facilities in Africa as federated machine-actionable semantically linked data at the source demonstrates the potential for the creation of an African Health Data Space. VODAN's exploration of use cases involving the application of federated learning to a federated data pipeline produced promising results. The value added at the source, ensuring that the data is FAIR and complies with OLR in the jurisdiction, establishes an important framework for the qualities for data curation. The data curation process has quality assurance in place during data entry at the point of care/service. Federated learning also allows individual data owners to define their governance processes and associated privacy policies, establish data access control, and enable data revocation. The curation of data as federated data that can be visited with clearly defined permission allows multiple data producers to develop a shared, robust machine learning model based on data visiting, thereby tackling serious issues such as data privacy, security, and access rights while gaining access to both granular and heterogeneous data. Creating value through the curation of the data, the services providing the capacity to use and reuse the data, and the intelligence layer derived from it can create a return on quality data production. Ultimately, the creation of an African Health Data Space will enable the better curation of digital data to improve patient care.

3.9 Ethical clearance

Tilburg University, Research Ethics and Data Management Committee of Tilburg School of Humanities and Digital Sciences REDC#2020/013, June 1, 2020-May 31, 2024 on Social Dynamics of Digital Innovation in remote non-Western communities

Uganda National Council for Science and Technology, Reference IS18ES, July 23, 2019-July 23, 2023

Acknowledgments

The researchers are grateful to all partners who helped facilitate the research presented in this study.

References

Africa University Network on FAIR Open Science. (2022). *Data is the new gold and the oil of the new economy*. Kampala International University (KIU) Available at. Available from https://dish-portal.kiu.ac.ug/aun/.

Aksünger, N., Sanctis, T. D., Waiyaiya, E., Van Doeveren, R., Van der Graaf, M., & Janssens, W. (2022). What prevents pregnant women from adhering to the continuum of maternal care? Evidence on interrelated mechanisms from a cohort study in Kenya. *BMJ Open*, *1;12*(1)e050670. Available from https://doi.org/10.1136/bmjopen-2021-050670.

AOSP. (n.d.) *Welcome to the African Open Science Platform*. African Open Science Platform (AOSP). Available at: https://aosp.org.za/.

Ball, K. M. (2017). African union convention on cyber security and personal data protection'. *International Legal Materials*, *56*(1), 164—192. Available from https://doi.org/10.1017/ilm.2016.3.

Basajja, M., Suchánek, M., Taye, G. T., Amare, S., Nambobi, M., Folorunso, S., Plug, R., Oladipo, F. O., & Van Reisen, M. (2022). Proof of concept and horizons on deployment of FAIR data points in the COVID-19 pandemic. *Data Intelligence*, *4*, 917—937. Available from https://doi.org/10.1162/dint_a_00179.

Bertozzi, A., Franco, E., Mohler, G. O., Short, M. B., & Sledge, D. (2020). The challenges of modeling and forecasting the spread of COVID-19. *Proceedings of the National Academy of Sciences of the United States of America*, *117*, 16732—16738. Available from https://doi.org/10.1073/pnas.2006520117.

BMZ Digital. Global. (2023). *FAIR forward — Open data for AI*. Available at: https://www.bmz-digital.global/en/overview-of-initiatives/fair-forward/.

CGIAR. (n.d.) *Open and FAIR Data Assets*. CGIAR. Available at: https://www.cgiar.org/annual-report/performance-report-2021/open-and-fair-data-assets/.

CODATA. (2022). *CODATA's mission*. Committee on Data for Science and Technology (CODATA). Available at: https://codata.org/about-codata/our-mission/.

Daniel, C., Paris, N., Pierre, O., Griffon, N., Breant, S., Orlova, N., Serre, P., Leprovost, D., Denglos, S., Mouchet, A., Dubiel, J., Gozlan, R., Chatellier, G., Bey, R., Frank, M., Hassen-Khodja, C., Mamzer, M.-F., & Hilka, M. (2022). AP-HP health data space (AHDS) to the test of the COVID-19 pandemic. *Studies in Health Technology and Informatics*, 28—32. Available from https://doi.org/10.3233/SHTI220390.

Demena, B. A., Artavia-Mora, L., Ouedraogo, D., Thiombiano, B. A., & Wagner, N. (2020). A systematic review of mobile phone interventions (SMS/IVR/Calls) to improve adherence and retention to antiretroviral treatment in low- and middle-income countries. *AIDS Patient Care and STDS*, *34*(2), 59—71. Available from https://doi.org/10.1089/apc.2019.0181.

Dove, E.S., & Phillips, M. (2015). *Privacy law, data sharing policies, and medical data: A comparative perspective medical data privacy handbook*. https://doi.org/10.1007/978-3-319-23633-9_24.

EAHRC (2017). *Digital REACH initiative roadmap*. East African Health Research Commission (EAHRC). Available at: https://vitalwave.com/wp-content/uploads/2018/05/digital-reach-initiative-roadmap.pdf.

Ghardallou, M., Wirtz, M., Folorunso, S., Touati, Z., Ogundepo, E., Smits, K., Mtiraoui, A., & Van Reisen, M. (2022). Expanding non-patient COVID-19 data: towards the FAIRification of migrants' data in Tunisia, Libya and Niger. *Data Intelligence*, *4*(4). Available from https://doi.org/10.1162/dint_a_00181.

GO FAIR. (n.d.) *Go FAIR*. Available at: https://www.go-fair.org/.

Green, S., Hillersdal, L., Holt, J., Hoeyer, K., & Wadmann, S. (2023). The practical ethics of repurposing health data: How to acknowledge invisible data work and the need for prioritization. *Medicine, Health Care, and Philosophy*, *26*(1), 119—132. Available from https://doi.org/10.1007/s11019-022-10128-6.

Hedstrom, M.L. (2012) *Digital data curation — Examining needs for digital data curators*. Paper presented at Cultural Heritage online: Trusted Digital Repositories, International Conference, Fondazione Rinascimento Digitale, Florence, Italy. Available at: https://hdl.handle.net/2027.42/135735.

Hultgren, M., Persano, J., Jennex, M., & Ornatowski, C. (2016) Using knowledge management to assist in identifying human sex. In: *49th Hawaii International Conference on System Science*. Available at: https://doi.org/10.1109/HICSS.2016.539.

IN-Africa. (n.d.). *GO FAIR*. Available at: https://www.go-fair.org/implementation-networks/overview/implementation-networks-archive/in-africa/.

Jati, P. H., Reisen, M. V., Flikkenschild, E., Oladipo, F., Meerman, B., Plug, R., & Nodehi, S. (2022). Data access, control, and privacy protection in the VODAN-Africa architecture. *Data Intelligence*, *4*, 938–954. Available from https://doi.org/10.1162/dint_a_00180.

Kawu, A. A., O'Sullivan, D., Hederman, L., & Van Reisen, M. (2023). FAIR4PGHD: A framework for FAIR implementation over PGHD. *FAIR Connect*, *1*(1), 35–40. Available from https://doi.org/10.3233/FC-230500, IOS Press. Available at:.

Marcus, J. S., Martens, B., & Carugati, C. (2022). The European Health Data Space. *European Parliament Policy Department Studies*. Available from https://doi.org/10.2139/ssrn.4300393.

Martinon, J.-P. (ed.). (2013) *The curatorial: A philosophy of curating*. London: Bloomsbury. Available at: https://www.bloomsbury.com/uk/curatorial-9781472525604/.

Ministry of Education. (2019). *National open access policy of ethiopia for higher education*. Ministry of Education, Ethiopia. Available at: https://nadre.ethernet.edu.et/record/4193/preview/Ethiopia%20National%20OA%20Policy.pdf.

Molnár-Gábor, F., Sellner, J., Pagil, S., Slokenberga, S., Tzortzatou-Nanopoulou, O., & Nyström, K. (2022). Harmonization after the GDPR? Divergences in the rules for genetic and health data sharing in four member states and ways to overcome them by EU measures: Insights from Germany, Greece, Latvia and Sweden. *Seminars in Cancer Biology*, *84*, 271–283. Available from https://doi.org/10.1016/j.semcancer.2021.12.001.

National Science Foundation. (2003). *Revolutionizing science and engineering through cyberinfrastructure: Report of the National Science Foundation Blue-ribbon Advisory Panel on Cyberinfrastructure*, Arlington, VA: National Science Foundation. Available at: books.google.com.

Naz, S., Phan, K. T., & Chen, Y. (2021). A comprehensive review of federated learning for COVID-19 detection. *International Journal of Intelligent Systems*, *37*, 2371–2392. Available from https://doi.org/10.1002/int.22777.

Ndemo, B., & Thegeya, A. (2022) *A Data Governance Framework for Africa*. African Economic Research Consortium. Available at: https://africaportal.org/wp-content/uploads/2023/06/DG001.pdf.

Nguyen, T. V., Dakka, M. A., Diakiw, S. M., VerMilyea, M., Perugini, M., Hall, J. M., & Perugini, D. (2022). A novel decentralized federated learning approach to train on globally distributed, poor quality, and protected private medical data. *Scientific Reports*, *12*. Available from https://doi.org/10.1038/s41598-022-12833-x.

PCAST (2010). *Report to the President realizing the full potential of health information technology to improve healthcare for Americans: The path forward*. Available at: https://obamawhitehouse.archives.gov/sites/default/files/microsites/ostp/pcast-health-it-report.pdf.

Plug, R., Liang, Y., Basajja, M., Aktau, A., Jati, P.H., Amare, S., Taye, G.T., Mpezamihigo, M., Oladipo, F.O., and Reisen, M.V. (2022a). 'FAIR and GDPR compliant population health data generation, processing and analytics', SWAT4HCLS. Available at: https://ceur-ws.org/Vol-3127/paper-7.pdf.

Plug, R., Liang, Y., Aktau, A., Basajja, M., Oladipo, F., & Van Reisen, M. (2022b). 'Terminology for a FAIR framework for the virus outbreak data network-Africa'. *Data Intelligence*, *4*(4). Available from https://doi.org/10.1162/dint_a_00167.

Shott, J. T. (1989). On tool-class use lives and the formation of archaeological assemblages. *American Antiquity*, *54*(1), 1−30.

Shott, M. J. (1996). An exegesis of the curation concept. *Journal of Anthropological Research*, *52*(3), 259−280. Available from https://doi.org/10.1086/jar.52.3.3630085.

Sposito, F.A. (2017). *What do data curators care about? Data quality, user trust, and the data reuse plan.* Paper presented at IFLA WLIC 2017, Wrocław, Poland, Libraries, Solidarity, Society, in Session S06 − Satellite Meeting: Library Theory and Research Section joint with Preservation and Conservation Section and Information Technology Section. Available at: library.ifla.org.

Stellmach, C., Muzoora, M. R., & Thun, S. (2022). Digitalization of health data: Interoperability of the proposed European Health Data Space. *Studies in Health Technology and Informatics*, *298*, 132−136. Available from https://doi.org/10.3233/SHTI220922.

Testa, S. (2022). An overview of the secondary use of health data within the European Union: EU-driven possibilities and civil society initiatives. In *Privacy symposium 2022: Data protection law international convergence and compliance with innovative technologies (DPLICIT)* (pp. 3-20). Cham: Springer International Publishing. Available at: https://doi.org/10.1007/978-3-031-09901-4_1.

University of Cape Town. (2023). *FAIR Data: As open as possible, as closed as necessary.* University of Cape Town Research Support Hub. https://uct.ac.za/research-support-hub/research-data-managing-research-data/fair-data-open-possible-closed-necessary.

Van Reisen, M., Amare, S. Y., Nalugala, R., Taye, G. T., Gebreselassie, T. G., Medhanyie, A. A., Schultes, E., & Mpezamihigo, M. (2023). Federated FAIR principles: Ownership, localisation and regulatory compliance (OLR). *FAIR Connect*, *1*(1), 1−7. Available from https://doi.org/10.3233/FC-230506, IOS Press.

Van Reisen, M., Oladipo, F., Mpezamihigo, M., Plug, R., Basajja, M., Aktau, A., Purnama Jati, P. H., Nalugala, R., Folorunso, S., Amare, Y. S., Abdulahi, I., Afolabi, O. O., Mwesigwa, E., Taye, G. T., Kawu, A., Ghardallou, M., Liang, Y., Osigwe, O., Medhanyie, A. A., & Mawere, M. (2022). Incomplete COVID-19 data: The curation of medical health data by the virus outbreak data network-Africa. *Data Intelligence*, *4*(4). Available from https://doi.org/10.1162/dint_e_00166.

Van Reisen, M., Oladipo, F., Stokmans, M., Mpezamihigo, M., Folorunso, S., Schultes, E., Basajja, M., Aktau, A., Yohannes Amare, S., Tadele Taye, G., Hadi Purnama Jati, P., Chindoza, K., Wirtz, M., Ghardallou, M., van Stam, G., Ayele, W., Nalugala, R., Abdullahi, I., Osigwe, O., … Musen, M. A. (2021). Design of a FAIR digital data health infrastructure in Africa for COVID-19 reporting and research. *Advanced Genetics*, *2*(2). Available from https://doi.org/10.1002/ggn2.10050.

Yakel, E. (2007). Digital curation. *OCLC Systems and Services: International Digital Library Perspectives*, *23*(4), 335−340. Available from https://doi.org/10.1108/10650750710831466.

Recent advances in federated learning for digital healthcare systems

Pooja Mohnani[1], Christoph Thümmler[2], Angelica Avila Castillo[2], Rasha Tolba[2], Alessandro Bassi[1], Antoine Simon[3], Anastasius Gavras[1], Orazio Toscano[4] and Pascal Haigron[3]

[1]*Eurescom GmbH, Heidelberg, Germany*
[2]*6G Health Institute GmbH, Markkleeberg, Germany*
[3]*Univ Rennes, Inserm, LTSI - UMR 1099, Rennes, France*
[4]*Ericsson, Genova, Italy*

4.1 Introduction

In traditional healthcare systems, sharing and analyzing patient data for research, diagnosis, and treatment is hampered by privacy concerns and regulatory restrictions. Federated healthcare platforms enable the deployment of federated learning (FL), a novel framework to overcome these barriers by allowing healthcare organizations to collaborate without sharing sensitive patient information (Li, Wang, et al., 2020; McMahan, Ramage, et al., 2017; Yang et al., 2019). Instead, each organization retains control of its data, and only aggregated statistics or model updates are shared between organizations. This distributed approach ensures that patient privacy is protected, while still allowing for collaborative sharing, analysis, and model training.

The concept of federated healthcare platforms addresses the growing recognition of the value of healthcare data and the need for data-driven insights to improve patient care and outcomes. As shown in Fig. 4.1, by bringing together diverse datasets from multiple organizations, federated healthcare platforms enable the development of more accurate diagnostic models, personalized treatment recommendations, and real-time predictive analytics (Rieke et al., 2020). These platforms facilitate knowledge sharing and collaboration, ultimately improving the effectiveness and efficiency of healthcare delivery.

Thus this opens many research challenges and possibilities that our society can and must address to prepare for current and future innovation in healthcare. Through the implementation of FL and the adoption of secure, distributed healthcare data management practices, can unlock the full potential of digital healthcare systems, leading to improved patient outcomes and a more efficient healthcare ecosystem.

Federated Learning for Digital Healthcare Systems. DOI: https://doi.org/10.1016/B978-0-443-13897-3.00005-9

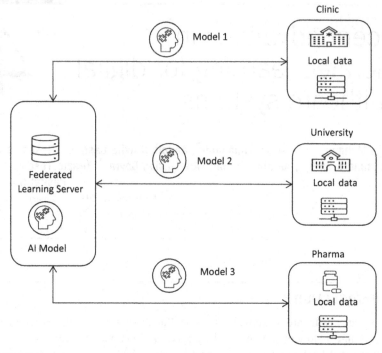

FIGURE 4.1

The concept of federated healthcare platforms.

4.1.1 **Key contributions of the chapter**

The following are the significant contributions of this chapter:

1. Identification of key FL application areas.
2. Highlighting crucial considerations essential for the development of FL systems, with a specific focus on ensuring scalable and reliable privacy preservation.
3. Addressing significant challenges and exploring the prospects of FL in healthcare, examining the anticipated evolution of these systems in the coming times.

4.1.2 **Chapter organization**

Section 4.1.2 presents the related works on federated learning and how it is evolving. Section 4.2 presents the perceptions of federated digital healthcare platforms in medical decision systems. Section 4.3 elaborates on the need and importance of privacy preservation, security, and ethics in healthcare data. Section 4.4 discusses the challenges, future research directions, trending technologies, and recent advances. Section 4.5 presents the view on FL and its integration with future 6G mobile communications networks. Finally, Section 4.6 concludes the chapter.

4.2 Related works

FL enables healthcare records that are located across different institutions to be connected without revealing personal information. Thus researchers, doctors, and data scientists can harness these extensive datasets from multiple hospitals without centralizing the data in one place. This approach effectively resolves crucial challenges related to the access rights of heterogeneous data (Dasaradharami Reddy & Gadekallu, 2023). This is advantageous as it reduces data security and privacy concerns by maintaining local data stores, in comparison to centralized machine learning (ML) techniques that require datasets to reside on one server (Song et al., 2022).

Rieke et al. (2020) discuss the current FL efforts for digital health and their impact on stakeholders, clinicians, patients, hospitals and practices, researchers and artificial intelligence (AI) developers, healthcare providers, and manufacturers. FL enables healthcare and related professionals to tackle the challenges of building unbiased models from datasets with optimized utilization of time, effort, and cost. By training algorithms within a hospital's secure firewall and sharing only the models, FL effectively addresses data governance concerns and ensures the maintenance of data security (Rieke et al., 2020). FL is equipped to capture a wide range of data variables, facilitating the analysis of patients based on their age, sex, and demographic characteristics. For instance, by accessing electronic medical records, patients with similar characteristics (cardiac arrest, mortality, ICU stay, etc.) could be known, and the need for their hospitalization could be predicted (Huang et al., 2019).

FL provides AI developers with access to larger and more diverse datasets that better represent current patients. As a result, AI-based healthcare solutions will have the capability to scale globally on an unprecedented level (Dasaradharami Reddy & Gadekallu, 2023). FL can have a significant impact on a wide range of stakeholders, including clinicians, patients, hospitals, medical researchers, and healthcare providers (Dasaradharami Reddy & Gadekallu, 2023).

4.3 Perceptions of federated digital platforms and their use in healthcare

Perceptions of federated systems are influenced by privacy concerns, interoperability, and governance. While benefits such as increased privacy and community participation attract users, challenges such as fragmentation, scalability, and trustworthiness may favor centralized systems. Addressing these perceptions can support the development and adoption of federated systems, creating a more decentralized and user-centric landscape.

4.3.1 Use of federated learning in medical image processing

FL has gained significant attention in medical image processing due to its ability to leverage decentralized data while preserving patient privacy, as shown in Fig. 4.2. In the real world, numerous sources of medical data, including magnetic resonance imaging (MRI), X-ray, positron emission tomography (PET), and computerized tomography (CT), provide doctors with vast volumes of information in many different medical applications (Rehman et al., 2020; Shen et al., 2017). ML, especially deep learning, has revolutionized the automatic analysis of these medical images with different applications, from exam or object classification (e.g., normal versus abnormal mammogram, benign or malignant tumor) to image registration (alignment of multiple images) or image segmentation (e.g., brain tumor delineation). However, as the models are typically trained on data from a single center, they face challenges in generalization, leading to a decline in performance when applied to data from another center.

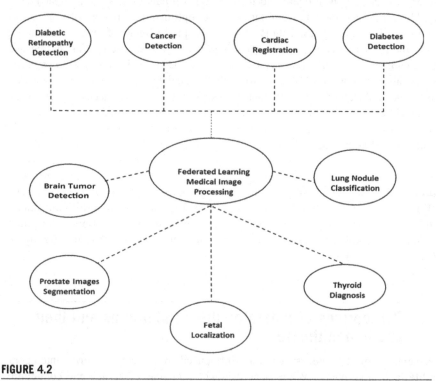

FIGURE 4.2

Applications of FL in medical image processing. *FL,* Federated learning.

FL holds immense potential in medical image analysis by facilitating collaborative training of ML models across healthcare institutions, ensuring adherence to privacy compliance. Li et al. (2020) demonstrated its feasibility in pneumonia detection using chest X-ray images from different hospitals. Zhang et al. (2021) demonstrated federated deep learning's ability to achieve COVID-19 detection accuracy while preserving patient data privacy. Different studies have demonstrated the effectiveness of FL in analyzing brain images. Abadi et al. (2016), and Silva et al. (2019) introduced this technique to aggregate encrypted updates without revealing sensitive information.

FL holds significant potential for medical image processing by enabling collaboration among healthcare institutions while ensuring data privacy. The referenced studies highlight the feasibility and effectiveness of FL in these domains, paving the way for future research and applications in healthcare.

4.3.2 Use of federated learning in Internet of Things-based smart healthcare applications

Recently, FL has emerged as a distributed collaborative AI approach, facilitating a range of intelligent Internet of Things (IoT) applications by enabling AI training on distributed IoT devices without the need for data exchange (Nguyen et al., 2021).

Fig. 4.3 illustrates a typical smart healthcare application based on FL. In this scenario, clinical data is collected from patients using onboard sensors. Multiple edge devices collaboratively execute the FL algorithm, and the resulting ML models assess the physical health of the patients. In urgent situations, the system

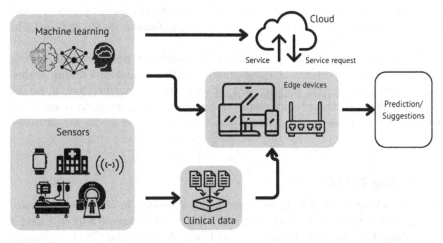

FIGURE 4.3

A typical FL-based smart healthcare application (Chang et al., 2021). *FL,* Federated learning.

can even request cloud-based emergency services. However, a limitation of traditional FL is its reliance on a reliable central server to aggregate model parameters uploaded by devices and distribute the global model to all participating devices (Chang et al., 2021).

In the domain of disease diagnosis and medical image processing, FL can leverage advancements in edge computing for enhanced benefits. Edge devices, such as smartphones and wearable devices, can participate in the FL process while preserving data privacy. Li et al. (2020) proposed a FL framework that leverages edge devices for the classification of electrocardiogram (ECG) signals. Their study demonstrated the feasibility of real-time disease diagnosis using FL at the edge.

4.4 Privacy preservation, security, and ethical needs

Federated healthcare systems must prioritize interoperability, adherence to standards, and implementing robust security measures to ensure seamless collaboration.

4.5 Privacy and security Needs

FL reduces data security and privacy concerns by maintaining local data stores, as opposed to centralized ML techniques, which require datasets to reside on one server (Song et al., 2022). FL is a potential concept for safe, reliable, and impartial models of data. It makes it possible for several parties to work together without exchanging or centralizing datasets (Dasaradharami Reddy & Gadekallu, 2023).

In healthcare, FL involves training ML models on multiple data sources, emphasizing the importance of safeguarding sensitive patient data during processing. It presents several security and privacy challenges; therefore a careful implementation, combined with other privacy-enhancing techniques, is necessary to effectively mitigate the associated risks. Concerns regarding this are discussed below.

4.5.1 Data privacy

FL aims to keep the data localized, restricting its transfer. In FL, instead of transferring data to the central servers, the ML model itself is deployed to each device to be trained on the data (Dasaradharami Reddy & Gadekallu, 2023). However, there is a risk of data leakage, while an adversary could attempt to reconstruct sensitive information by analyzing the model updates during the learning process (Bonawitz, Ivanov, et al., 2019).

4.5.2 **Model poisoning attacks**

Model poisoning exploits the fact that FL gives malicious participants direct influence over the joint model, enabling much more powerful attacks compared with training-data poisoning. For example, in a healthcare setting, an attacker might introduce biased medical records, leading to harmful decisions and incorrect predictions (Bagdasaryan et al., 2020).

4.5.3 **Differential privacy**

FL can employ differential privacy techniques to protect individual data privacy. The addition of noise to safeguard privacy may impact the quality and accuracy of predictions, especially in healthcare applications where precision is of utmost importance (Abadi et al., 2016).

4.5.4 **Secure model aggregation**

In FL, models aggregate the updates from different participants. The aggregation process should be resistant to attacks attempting to extract information from these updates. Safeguarding against collusion attacks poses a significant challenge (McMahan, Moore, et al., 2017).

4.5.5 **Data minimization**

Another way of minimizing risks is to transfer a minimal amount of data and potentially limit privacy breaches (Yang et al., 2019).

4.5.6 **Secure and encrypted communication**

Secure communication protocols can be used to transmit data between participants and prevent unauthorized access. Encryption techniques further add in protecting sensitive healthcare data (Bonawitz, Eichner, et al., 2019).

4.6 **Ethical needs**

FL, with its promising opportunities, is garnering a lot of attention in healthcare. Regulations, such as the Health Insurance Portability and Accountability Act (HIPAA) in the United States or the General Data Protection Regulation (GDPR) in the European Union, ensure the protection of patient rights and privacy.

Nevertheless, several requirements need to be considered to ensure responsible and ethical implementation. This includes the need for unbiased and FAIR (findable, accessible, interoperable, and reusable) data, minimizing the risk of reidentification of the individuals within the data, and enforcing strict control of what data is used and for what purpose (Voigt & von dem Bussche, 2017).

When complying with GDPR, the federated approach introduces the complexity of identifying and sharing responsibilities with multiple data controllers, conducting data protection impact assessments, and auditing the environments to ensure that the ML models function as expected.

To ensure privacy, techniques such as encryption, anonymization, and pseudonymization must be used, as they help to protect sensitive information during the model training process. It is important to ensure that patients (data subjects) provide explicit consent, and that they are well-informed about the purpose, use, and processing of their data, while retaining the right to opt out at any given time. This brings transparency to data usage and sharing and ensures the incorporation of privacy protection mechanisms (Alysa et al., 2022).

There should be clear guidelines and agreements on data ownership, control, and access to ensure accountability. Robust security measures must be implemented to safeguard data transfer and storage. Policy-based access controls and other cybersecurity protocols must be in place to protect against unauthorized access or data breaches.

In the event of a data breach data processors and controllers must be prepared to minimize the impact and establish planned steps for promptly informing authorities and patients (data subjects).

Establishing ethical guidelines for FL initiatives is crucial. Addressing these ethical needs can help foster trust, protect patient privacy, ensure fairness, and maximize the benefits of FL in healthcare while minimizing the associated risks.

4.7 Role of federated learning in future digital Healthcare 5.0

Healthcare 5.0 is a new era of healthcare that focuses on advanced technologies, personalized care, and patient empowerment. It represents a shift toward a patient-centric approach where individuals are actively involved in managing their own health. The concept leverages cutting-edge technologies to create a seamless healthcare ecosystem that empowers individuals, promotes preventative care, and provides personalized treatment options (patient-centric care) (Rieke et al., 2020).

By training algorithms locally and sharing only model updates rather than raw data, FL addresses data governance challenges while ensuring patient privacy. It enables healthcare organizations to comply with regulations, such as HIPAA or GDPR. FL increases trust and encourages data sharing between healthcare organizations, fostering large-scale collaboration and knowledge sharing in the healthcare ecosystem and enhancing the overall quality of experience/service in healthcare.

One of the key benefits of FL in Healthcare 5.0 is that it enables learning from diverse datasets, enabling knowledge sharing among healthcare organizations with unique patient populations, demographics, and expertise. This allows the creation of robust, generalizable models that adapt to various patient contexts. However, FL is limited by the need for a trusted central server to aggregate model parameters and distribute the global model. To overcome this limitation, researchers are exploring advancements such as secure multiparty computation, cryptographic techniques, and blockchain-based solutions. These aim to improve the security and decentralization of FL in healthcare (Chang et al., 2021).

4.8 Federated learning and blockchain for healthcare

FL and blockchain are two powerful promising technologies that could be combined to revolutionize healthcare. FL enables collaborative model training while maintaining privacy, while blockchain provides a decentralized and secure framework for storing and sharing data. FL combined with blockchain technology offers several benefits in healthcare:

1. FL ensures privacy and security by keeping sensitive patient data localized, minimizing the risk of data breaches. The integration of FL with blockchain further enhances security, providing a tamper-proof and transparent system for data storage and access.
2. The combination enables data integrity and trust through immutable and auditable records on the blockchain, fostering accountability among healthcare stakeholders.
3. Blockchain facilitates interoperability and data sharing, enabling secure FL across institutions and promoting collaboration in healthcare research (Rehman et al., 2022).

Moreover, FL and blockchain empower patients by giving them control over their health data, allowing secure management of permissions and consent. Finally, this integration drives research advancements by aggregating data from multiple sources, creating larger and more diverse datasets that can lead to breakthroughs in disease prediction, treatment development, and precision medicine.

4.9 Federated learning for collaborative robotics in healthcare

FL offers significant potential for collaborative robotics in healthcare, enabling intelligent and adaptive systems while protecting patient privacy. It enables real-time model updates and decentralized patient data sharing, allowing robots to make informed decisions while maintaining confidentiality. Research focuses on optimizing FL architectures, communication protocols, and privacy-enhancing techniques for collaborative healthcare robotics.

Collaborative robot (cobot) technology has gained significant adoption in the healthcare and medical device industries, serving as a valuable tool for increasing workforce efficiency, streamlining safety procedures, and facilitating improved workflows. Collaborative healthcare robots are automated systems deployed across the medical sector, capable of performing a range of tasks, including administrative tasks, laboratory testing, patient care, and surgical assistance. By bridging the gap caused by labor shortages, these cobots are helping to ease the burden on the medical industry. Demand for automation in healthcare has been driven by the need to reduce the risk of infection to frontline workers and advancements in inpatient care (Dasaradharami Reddy & Gadekallu, 2023).

In particular, cobots in the healthcare sector demonstrate exceptional efficiency in laboratory testing tasks, offering high precision, fast turnaround times, and reduced reliance on manual processes. In addition, cobots are playing a key role in patient care, performing tasks such as medication dispensing, specimen collection, temperature and blood pressure monitoring, and various tests. These capabilities have freed healthcare workers from tedious tasks. They can prioritize urgent matters and effectively optimize their time.

4.10 Federated learning for integration with 6G in healthcare

By 2030, the sixth generation (6G) of mobile technology is expected to be ubiquitous, as it can be integrated into most sectors of the industry, improving the performance of communications standards and enhancing the current communications network infrastructure. Additionally, 6G is expected to leverage more spectrum, providing even lower latency and higher bandwidth transmission capabilities compared with 5G. Also it is anticipated to extend its reach to rural or remote areas currently lacking cellular signals.

Future wireless systems are expected to significantly improve existing wireless capabilities in terms of network throughput, IoT connectivity, latency (from 1 to 10 ms), reliability, availability, energy efficiency, and security.

Moreover, 6G is expected to deliver a 1000-fold improvement in network throughput when compared with 5G technology. This advancement will enable seamless communication and intelligent connectivity for millions of smart devices, as shown in Fig. 4.4. The increased processing power of 6G wireless networks and devices supporting AI will facilitate the proliferation of augmented reality, more advanced imaging and telepresence technology, and more autonomous robots that can communicate with other devices to perform complex tasks.

The technology of 6G can enable high-quality and immersive telemedicine experiences. By integrating FL with 6G, healthcare providers can use distributed

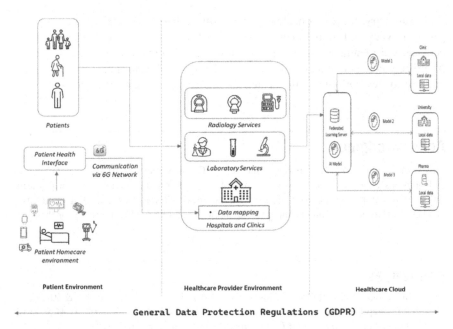

Patients

Patient Health Interface

6G

Communication via 6G Network

Radiology Services

Laboratory Services

• Data mapping

Hospitals and Clinics

Patient Homecare environment

Federated Learning Server

AI Model

Model 1

Model 2

Model 3

Clinic

Local data

University

Local data

Pharma

Local data

Patient Environment   Healthcare Provider Environment   Healthcare Cloud

General Data Protection Regulations (GDPR)

FIGURE 4.4

FL for integration with 6G in healthcare. *FL*, Federated learning.

learning to analyze patient data collected from different remote devices and deliver personalized healthcare services. This combination can support remote monitoring, diagnosis, and treatment, bringing healthcare services closer to patients regardless of their location. It has great potential to revolutionize healthcare delivery, improve patient outcomes, and advance medical research. As 6G technology continues to evolve, further research and innovation will be required to explore the specific applications, challenges, and benefits of integrating FL with 6G in healthcare settings.

4.11 Conclusion

Federated digital health platforms have emerged as an innovative approach to addressing the challenges of privacy and collaboration in healthcare. These systems use the principles of FL, a decentralized ML technique, to enable collaborative analysis and model training on distributed healthcare data.

By enabling collaborative model training, personalized care, and privacy, FL has significant potential for Healthcare 5.0. Its ability to leverage distributed datasets while maintaining security and privacy makes it an attractive approach for advancing healthcare systems toward patient-centric care in the era of Healthcare 5.0.

In summary, federated healthcare platforms utilize distributed data to improve patient privacy and unlock new insights. By utilizing FL, these platforms enable personalized treatments, improved diagnoses, and improved healthcare outcomes. As privacy concerns remain paramount, federated healthcare platforms hold great promise for revolutionizing the healthcare landscape and driving innovation.

Acknowledgment

This work is funded by the European Union under Grant Agreement 101070222. The views and opinions expressed are, however, those of the author(s) only and do not necessarily reflect those of the European Union or the European Commission (granting authority). Neither the European Union nor the granting authority can be held responsible for them.

References

Abadi, M., Chu, A., Goodfellow, I., McMahan, H.B., Mironov, I., Talwar, K., & Zhang, L. (2016). Deep learning with differential privacy. *Proceedings of the 2016 ACM SIGSAC Conference on Computer and Communications Security* (pp. 308–318).

Alysa, Z.T., Han, Y., Lizhen, C., & Qiang, Y., Fellow, IEEE. (2022). *Towards personalized federated learning IEEE transactions on neural networks and learning systems.*

Bagdasaryan, E., Veit, A., Hua, Y., Estrin, D., & Shmatikov, V. (2020). How to backdoor federated learning. In *Proceedings of the 1st conference on machine learning and systems* (pp. 265–278).

Bonawitz, K., Eichner, H., Grieskamp, W., Huba, D., Ingerman, A., Ivanov, V., & Wang, S. (2019). Towards federated learning at scale: System design. *arXiv preprint arXiv, 1902,* 01046.

Bonawitz, K., Ivanov, V., Kreuter, B., Marcedone, A., McMahan, H.B., Patel, & Zhang, Z. (2019). Practical secure aggregation for privacy-preserving machine learning. In *Proceedings of the 2017 ACM SIGSAC conference on computer and communications security* (pp. 1175–1191).

Chang, Y., Fang, C., & Sun, W. (2021). A blockchain-based federated learning method for smart healthcare. *Computational Intelligence and Neuroscience, 2021.*

Dasaradharami Reddy, K., & Gadekallu, T. R. (2023). A comprehensive survey on federated learning techniques for healthcare informatics. *Computational Intelligence and Neuroscience, 2023.*

Huang, L., Shea, A. L., Qian, H., Masurkar, A., Deng, H., & Liu, D. (2019). Patient clustering improves efficiency of federated machine learning to predict mortality and hospital stay time using distributed electronic medical records. *Journal of Biomedical Informatics, 99,* 103291.

Li, T., Sahu, A. K., Talwalkar, A., & Smith, V. (2020). Federated learning: Challenges, methods, and future directions. *IEEE Signal Processing Magazine, 37*(3), 50–60.

Li, X., Wang, Z., Wu, T., Jiang, H., & Pang, C. (2020). Privacy-preserving chest X-ray classification with split learning and federated learning. *IEEE Access, 8,* 147937–147945.

McMahan, B., Moore, E., Ramage, D., Hampson, S., & y Arcas, B. A. (2017). *Communication-efficient learning of deep networks from decentralized data. Artificial intelligence and statistics* (pp. 1273−1282). PMLR.

McMahan, H.B., Ramage, D., Talwar, K., & Zhang, L. (2017). Learning differentially private recurrent language models. In *Proceedings of the 2017 conference on empirical methods in natural language processing* (pp. 2792−2802).

Nguyen, D. C., Ding, M., Pathirana, P. N., Seneviratne, A., Li, J., & Poor, H. V. (2021). Federated learning for internet of things: A comprehensive survey. *IEEE Communications Surveys & Tutorials, 23*(3), 1622−1658.

Rehman, A., Abbas, S., Khan, M. A., Ghazal, T. M., Adnan, K. M., & Mosavi, A. (2022). A secure healthcare 5.0 system based on blockchain technology entangled with federated learning technique. *Computers in Biology and Medicine, 150*, 106019.

Rehman, Z. U., Zia, M. S., Bojja, G. R., Yaqub, M., Jinchao, F., & Arshid, K. (2020a). Texture based localization of a brain tumor from MR-images by using a machine learning approach. *Medical Hypotheses*, 109705.

Rieke, N., Hancox, J., Li, W., Milletari, F., Roth, H. R., Albarqouni, S., & Cardoso, M. J. (2020). The future of digital health with federated learning. *NPJ Digital Medicine, 3*(1), 119.

Shen, D., Wu, G., & Suk, H.-I. (2017). Deep learning in medical image analysis. *Annual Review of Biomedical Engineering, 19*, 221−248.

Silva, S., Gutman, B. A., Romero, E., Thompson, P. M., Altmann, A., & Lorenzi, M. (2019). *Federated learning in distributed medical databases: Meta-analysis of large-scale subcortical brain data. 2019 IEEE 16th international symposium on biomedical imaging (ISBI 2019)* (pp. 270−274). IEEE.

Song, J., Wang, W., Gadekallu, T. R., Cao, J., & Liu, Y. (2022). EPPDA: An efficient privacy-preserving data aggregation federated learning scheme. *IEEE Transactions on Network Science and Engineering.*

Song, J., Wang, W., Gadekallu, T. R., Cao, J., & Liu, Y. (2022). Eppda: An efficient privacy-preserving data aggregation federated learning scheme. *IEEE Transactions on Network Science and Engineering.*

Voigt, P., & von dem Bussche, A. (2017). *The EU General Data Protection Regulation (GDPR)*. Springer International Publishing.

Yang, Q., Liu, Y., Chen, T., & Tong, Y. (2019). Federated machine learning: Concept and applications. *ACM Transactions on Intelligent Systems and Technology (TIST), 10*(2), 1−19.

Zhang, W., Zhou, T., Lu, Q., Wang, X., Zhu, C., Sun, H., & Wang, F. Y. (2021). Dynamic-fusion-based federated learning for COVID-19 detection. *IEEE Internet of Things Journal, 8*(21), 15884−15891. Available from https://www.theodi.org/article/federated-learning-an-introduction-report/.

Performance evaluation of federated learning algorithms using breast cancer dataset

Sakinat Oluwabukonla Folorunso[1], Joseph Bamidele Awotunde[2],
Abdullahi Abubakar Kawu[3,4] and Oluwatobi Banjo[1]

[1]*Artificial Intelligent Systems Research Group (ArISRG), Department of Mathematical Sciences, Olabisi Onabanjo University, Ago-Iwoye, Nigeria*
[2]*Department of Computer Science, Faculty of Information and Communication Sciences, University of Ilorin, Ilorin, Nigeria*
[3]*Technological University Dublin, Dublin, Ireland*
[4]*Department of Computer Science, Ibrahim Badamasi Babangida University, Lapai, Nigeria*

5.1 Introduction

The automated classification of breast cancer is regarded as an efficient diagnostic tool for individuals infected with breast cancer. The approach uses the Internet of Medical Things (IoMT) for data collection. In developing a robust classification task in an IoMT environment, breast cancer images are obtained from different patients affected by cancer from diverse locations and transmitted to a central server. However, keeping patient data on a central server could compromise their privacy, and the central server could also be subjected to malicious malevolent assaults, leading to inefficient alternatives. Unfortunately, patient data is moved to a central location due to the fact that the proposed deep learning (DL) resolutions for breast cancer recognition are trained on a central device. However, a cloud server can be employed to store cancer patient data by taking advantage of the IoMT system, which has a robust storing and computational capability (Awotunde et al., 2022). Data transmission to the cloud incurs resource consumption, communication overhead, and time delays. Also, given the confidential nature of patients' medical information, health institutions may be reluctant to exchange such information even with the cloud.

Thus an excellent option to address these challenges is federated learning (FL), which takes into account confidentiality-related concerns while designing a learning-based solution employing vast, multisourced, and geographically dispersed medical data (Ding et al., 2023). The FL approach involves multiple machine learning (ML) models deployed as local clients. The clients transmit only the localized parameters of the model, which are accumulated at a central ML model. The updated global parameters are calculated before being aired back to the local ML models regarded as clients. A significant focus of AI in the fight against breast cancer is centered on translating radiological visuals, particularly chest CT (Folorunso et al., 2021). These scans have been extensively utilized to identify alterations, provide information for patient care, assess severity, and track the progression of the disease.

Therefore this chapter presents a performance comparison of two FL algorithms, federated averaging (FedAvg) and federated match averaging (FedMA), for breast cancer classification models using breast cancer histopathological BreakHis images. The study uses a three-layered convolutional neural network (CNN) model for both the local and global models for classification tasks deployed to both clients connected to the server. VGG (visual geometry group)-16 and -19 transfer learning models in an FL setting were applied to extract features from the breast cancer images to compare the results of the FL algorithms. To employ both transfer learning methods and fine-tuning on pretrained models, batch normalization and dropout layers are incorporated in both the convolutional and completely interconnected layers to reduce overfitting. Other enhancement methods, such as flip, shift, brightness alteration, and rotation, are used to introduce variations in the attributes of every image.

The major contributions of this study are:

1. Performance evaluation of two FL algorithms, FedAvg and FedMA (federated match averaging), against regular centralized learning.
2. Classification of patient disease by combination of the outcomes from locally trained models of various decentralized and heterogeneous hospital clients.
3. Designing an FL-based model for the classification of patients infected with breast cancer.

The structure of this chapter is organized as follows: Section 5.2 provides a comprehensive overview of the research context and related works. Section 5.3 outlines the methodology employed in this research. Section 5.4 presents the research outcomes. Finally, Section 5.6 draws conclusions from the study and suggests avenues for future research.

5.2 Related works/literature reviews

A review of relevant papers is presented in this section, with a particular emphasis on the utilization of FL in IoMT applications and the efficacy of DL in diagnosing breast cancer. The body of literature already in existence demonstrates contributions made by researchers from around the globe who applied computational techniques to classify breast cancer in patients at various hospitals. One of the most popular approaches in this context is ML, which typically analyzes historical and current data sourced from diverse sources within the health sector with the objective of categorizing diseases (Folorunso, Banjo, et al., 2020; Folorunso, Fashoto, et al., 2020) or predicting future trends (Folorunso et al., 2022a). Several studies have performed a supervised learning task on breast cancer classification with ML models in a centralized learning environment, as summarized in Table 5.1. Chandiramani et al. (2019)

Table 5.1 Summary of related works.

Author/year	Dataset	ML method	FL	Pros	Cons
Chandiramani et al. (2019)	Fashion MNIST	ANN	ML, DML, and FML	Can maintain data privacy; fast; allows deployment at scale	Privacy, data bias
Nilsson et al. (2018)	MNIST	ANN-MLP	FedAvg, CO-OP, and FSVRG	Used both IID and non-IID datasets breakdown of data. comparison between FedAvg and centralized learning	Mainly focuses on classification task and federated optimization
Siddique et al. (2023)	COVID-19 X-rays	VGG-16, ResNet50, CNN	FedAvg, FedMA	Can accommodate variation in client data distributes by local updates, and iteration complexity needed by the clients	Data heterogeneity is to be addressed.
Baghersalimi et al. (2021)	Epilepsy EEG	ResNet, 1D-CNN; MLP	FedAvg	Exploration of DNN convolutional neural network architectures to characterize the trade-off between the detection performance and energy consumption.	Privacy, data bias
Dou et al. (2021)	COVID-19 CT	RetinaNet18, 2D-CNN	FedAvg	Build generalizable, low-cost, and scalable AI tools for image-based disease diagnosis and management, both for research and clinical care	Data heterogeneity is to be addressed.

AI, Artificial intelligence; ANN, artificial neural network; CNN, convolutional neural network; CO-OP, cooperative; DML, distributed machine learning; EEG, electroencephalogram; FedAvg, federated averaging; FedMA; federated match averaging; FL, federated learning; FML, federated meta learning; FSVRG, federated stochastic variance reduced gradient; IID, independent and identically distributed datasets; ML, machine learning; MLP, multilayer perceptron; MNIST, Modified National Institute of Standards and Technology; Non-IID, nonindependent and identically distributed datasets, VGG-16, visual geometry group-16.

propose the performance evaluation of basic ML, distributed ML (DML), and FL by modeling an artificial neural network (ANN) on the Fashion-MNIST (Modified National Institute of Standards and Technology) dataset. The results achieved for the three models are not significantly different from each other, but basic ML gave a superior accuracy value of 87%. Nilsson et al. (2018) conducted a performance apprai-sal of the three FL algorithms: FedAvg (McMahan et al., 2017a), cooperative (CO-OP) ML (Wang, 2017), and federated stochastic variance reduced gradient (FSVRG) (Konečný et al., 2016) and compared the performances to a centralized strategy where MNIST data is stored on the server. The MNIST dataset was employed, using both independent and identically distributed (IID) datasets and nonindependent and identi-cally distributed (non-IID) datasets for data partitioning. The findings revealed that, irrespective of the data partitioning method, the FedAvg algorithm obtained the maxi-mum accuracy value among the three FL algorithms.

Siddique et al. (2023) propose an FL scheme for the efficient classification of COVID-19 cases obtained from the Kaggle database using a decentralized dataset. VGG-16 and residual network 50 (ResNet50) are proposed for feature extraction on a custom three-layered CNN model designed for deployment on the local edge devices and a central server. All these were deployed using FedMA and FedAvg FL algorithms. The results indicated that FedAvg, implemented on the custom CNN master model, achieved the most favorable results in terms of accuracy, recall, precision, and f1_score, with 99.03%, 98.22%, 97.16%, and 98.84% values, respectively. Similarly, Baghersalimi et al. (2021) proposed a personalized FL framework based on the FedAvg algorithm for epileptic seizure detection using four-layered perceptron and 1-dimensional CNN (1D-CNN) ML models. They initially used ResNet to extract features from the electrocardiography of the epileptic seizures of the patient being monitored based on local data before classi-fication. The results, obtained from sensitivity, specificity, and a geometric mean metric, indicate that the proposed scheme utilizing ResNet + 1-DNN achieved the highest values of 90.24%, 91.58%, and 90.90%, respectively, when compared with conventional FL and distributed learning settings. Dou et al. (2021) also pro-posed a study to automate COVID-19 CT scan image classification using the FedAvg algorithm on 2D-CNN. This model was evaluated based on the receiver operating characteristic (ROC) area under the curve (AUC) with a 95% confi-dence interval (CI). The FL method obtained the best ROC_AUC result of 91.99% compared with the results of ensemble and joint models with AUC values of 87.29% and 90.63%, respectively.

5.3 Federated learning

The problem of patient data privacy is one of the main reasons why there aren't many large readily accessible datasets for assessing ML models for medical pro-blems. The conventional way of training data is to access the data that has been gath-ered from various sources, such as health institutions. However, it is not practicable due to the related privacy issues. As a result, our approach replicates a distributed method of model training while maintaining the confidentiality of each institution's data. This method is referred to as FL (Hard et al., 2018; McMahan et al., 2017b). The server aggregates the data by utilizing the averaged-out method, i.e., FedAvg receives local parameters from all of the institutions involved that were collected after numerous training epochs (McMahan et al., 2017b). The averaged parameters are subsequently revised in each client's local models. The model is assessed by the clients using the divided localized testing data. This sequence of communication proceeds through multiple iterations until optimal model efficiency and generalization are achieved, as shown in Fig. 5.1. Every client from N clients has their own BreakHis data, X_i, and its matching labels, Y_i, where i indicates the institution. This feature set $\{X_1, X_2, \cdots, X_N\}$ denotes the data belonging to the institutions, while the target class set is $\{Y_1, Y_2, \cdots, Y_N\}$. The institutions first agree on one standard model whose parameter θ are initialized with θ^0. Every client then duplicates, localizes, and initializes the model θ_k with the global parameters $\theta_k \leftarrow \theta$. After each iteration, the parameters, $\{\theta_k, \cdots, \theta_N\}$, are aggregated with the averaging method, FedAvg. The resultant mean is updated as the parameter θ of the recent server, as illustrated by Eq. (5.1). The ML models of the partaking institutions will be updated using the new parameters for the next training cycle.

$$\theta \leftarrow \frac{1}{|C|} \sum_{k=1}^{K=|C|} \theta_k \qquad (5.1)$$

Here, $|C|$ signifies the size of the partaking institutions.

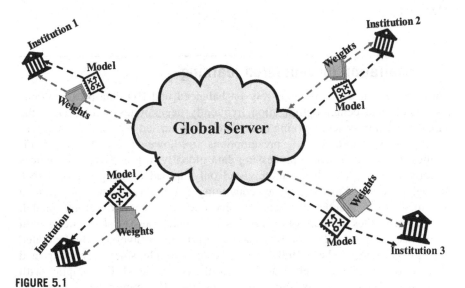

FIGURE 5.1

FL approach for training the dataset on clients. *FL*, Federated learning.

Algorithm 5.1 for federated averaging (FedAvg)

Requirement: The K clients are indexed by k; B is the local minibatch size; E is the number of local epochs; and η is the learning rate.

1. Server executes:
2. Initialize w_0
3. **for** each cycle $t = 0, 1, \ldots, M$ **do**
4. $\quad m \leftarrow \max(\lfloor C \cdot K \rfloor, 1)$
5. $\quad S_t = random\ set\ of\ m\ clients$
6. **for** each client $k \in S_t$ **in parallel do**
7. $\quad w_{t+1}^k = ClientUpadte(k, w_t)$
8. $\quad w_{(t+1)}^k \leftarrow \sum_{k=1}^{K} \frac{n_k}{n} w_{t+1}^k$
9. **end for**
10. **end for**
11. *ClientUpadte*(k, w)//Run on client k
12. initialize $w \leftarrow w_t$
13. $B \leftarrow$ (split P_k into batches of size B)
14. **for** each local epoch i from 1 to E **do**
15. **for** batch $b \in B$ **do**
16. $\quad w \leftarrow w - \eta \nabla l\ell(w; b)$
17. **end for**
18. **end for**
19. return w to server

The aim is to optimize θ over every client, such that every client achieves optimal model efficiency with θ. The deployed method is expressed by Algorithm 5.1.

5.4 Inspiration for federated learning

Considering that these methods rely on balanced and IID data presumptions, distributed data center optimization frequently necessitates regulation of the data distribution. When learning duties are being carried out on a diverse ecosystem of edge devices, these presumptions are, however, overly general. FL implies that, as opposed to processing data globally, a group of edge devices perform learning activities remotely and only upload a changed model to a managing server. Particularly, by combining locally trained models from a potentially huge number of clients, a server learns a common global model. Additionally, these clients often possess poor data transfer skills along with imbalanced and non-IID data. But when data privacy is involved, the federated strategy outperforms the centralized one. As an example, since only the trained model and not the data itself is analyzed at the central level, FL complies with the General Data Protection Regulation's (GDPR) data minimization principle (European Commission, 2018). In accordance with the GDPR's goal and storage restriction rules, communicated models are also transient, as they are promptly deleted after being integrated into the global model.

5.5 Federated learning algorithm

This segment will briefly explain the two algorithms adopted in this study.

5.5.1 Federated averaging

The learning strategy for FedAvg is a repetitive method comprising both the local and global phases. Individual data proprietors train an ML model on a local dataset using local iterations after receiving it from a global server (Darzidehkalani et al., 2022). The global server aggregates the updated local models to update the global model. Then it returns it to clients for the subsequent round. The optimization problem for FedAvg can be expressed by Eq. (5.2):

$$w^{t+1} = \sum_{i=1}^{N} p_i w_i^t; w_i^t = arg \min_{w_i} \left(\mathcal{L}\left(\mathcal{D}_i; w^t\right)\right) \qquad (5.2)$$

where N is the count of data proprietors; $\mathcal{L}(\mathcal{D}_i; w^t)$ is the loss function showing the global model parameters w^t of resident datasets; and p_i is the likelihood of choosing client i. Local optimization is expressed as $w_i^{t+1} \leftarrow w^t - \eta \cdot \nabla \mathcal{L}(w^t; \mathcal{D}_i)$; where η is the learning rate. The global model is updated according to the local models w_i and is distributed for aggregation, as expressed by Eq. (5.3):

$$w^{t+1} = \sum_{i=1}^{N} p_i w_i^{t+1} \qquad (5.3)$$

5.5.2 Federated match averaging

The FedMA (Wang, 2020) algorithm is a more contemporary model aggregation algorithm. Its guiding principle is a layer-wise learning strategy that includes matching and integrating nodes that share comparable weights. The server is informed of each layer's separate training. The FedMA algorithm, in particular, alters the neural model architecture by integrating a layer-wise accumulation process that allows related neurons to be joined and new ones to be included. As a revision to the work of Yurochkin et al. (2019) on CNN and RNN, this method addresses the size of nodes in a layer as a subproblem to be fixed, rather than a hyperparameter to be defined. FedMA believes that neurons in a neural network layer must be permutation resilient to accomplish intelligent accumulation. The core principle of the method is that all clients might include neurons that are identical and should be fused together. Each of these neurons can be grouped nonparametrically by averaging all neurons in that cluster to form a comprehensive neuron. The approach uses a 2D permutation vector estimated repeatedly from improving rank layers to determine which of the neurons can be merged. This vector is computed using the Beta-Bernoulli Process—Maximum a Posteriori (Thibaux & Jordan, 2007). Subsequently, the Hungarian method (Kuhn, 1955) is applied to the resulting vector to determine which specific neurons can be merged and which other neurons should be included. Clients receive updated layers in stages. It is worth noting that the SoftMax layer, such as FedAvg, employs weighted averaging.

5.6 Materials and methods/methodology/design

5.6.1 Study dataset

This study used the BreaKHis 1.0 dataset (Spanhol et al., 2016), consisting of microscopic images of breast cancer histopathology. The images were 700×460 pixels in size and presented in three-channel RGB of 8-bit depth. The dataset includes images from the cancer medium of 82 distinct patients. The breast cancer visuals were curated for investigation purposes for a period of one year spanning from January 2014 through to December 2014. It was obtained from the breast cancer tissue samples through the process of surgical (open) biopsy. These samples are tainted by hematoxylin and eosin and formed by a typical paraffin procedure in which specimen infiltration and embedment are done in paraffin. The dataset has 7783 breast cancer binary instances of benign and malignant tumors in.png format.[1,2] Table 5.2 shows the sample size and class distribution for the dataset.

Out of 7783 visuals, 5448 were utilized to train the model (70% of the total), while 2335 were utilized for testing the model (30% of all images). Since data in FL is dispersed and cannot be associated across various sets, the data has been divided into two groups, each of which is entirely distinctive and has no correlation with another set. All instances, including the augmented images, were split into two groups and distributed to their respective client nodes. The process of augmentation of the dataset adopted in this study was achieved by rotating the images by 90 degrees, flipping by 180 degrees, and an additional rotation of 270 degrees. The benign and malignant image dataset distribution, split into train-tests of 70%−30%, is displayed in Table 5.3. The distribution of the dataset, when split into the two respective client nodes and into their train-test split of 70%−30% is displayed in Table 5.4. The varying sample sizes for the different datasets across clients for the two nodes demonstrate the capability of supporting several clients with datasets of various sizes.

---

[1] http://web.inf.ufpr.br/vri/breast-cancer-database
[2] https://www.kaggle.com/datasets/anaselmasry/breast-cancer-dataset

Table 5.2 BreakHis cancer data.

S/N	Cancer image cases	Image sample	Sample size
1	Benign		2479
2	Malignant		5304

Table 5.3 Augmented breast cancer dataset image distribution for training and testing.

Image class	Training	Testing	Total
Benign	3000	1000	4000
Malignant	3000	1000	4000
Total	6000	2000	8000

Table 5.4 Distribution of datasets to clients for the federated learning scheme.

Image class	Client 1		Client 2		Total
	Training	Testing	Training	Testing	
Benign	2000	600	1000	400	4000
Malignant	2000	600	1000	400	4000
Total	4000	1200	2000	800	8000

5.7 Transfer learning

The extraction of representative numerical information in the identification and classification of breast cancer using transfer learning is illustrated in Fig. 5.2. The goal was to categorize the breast cancer images into dual classes: benign and malignant. The preprocessing and classification are the two phases deployed for this chapter. The preprocessing approach involves the batch normalization and data augmentation procedure while the classification phase involves the application of the pretrained model for feature extraction (Folorunso et al., 2023; Kumar et al., 2021). For the study, the images were rescaled, and every pixel was reproduced by a ratio of 1/255. The display of transitional activations in a CNN framework during training enhances the understanding of the segmentation and feature mining processes, especially on an image dataset (Zhang et al., 2021; Folorunso et al., 2022b). The feature map is formed from the results of the many pooling and convolutional

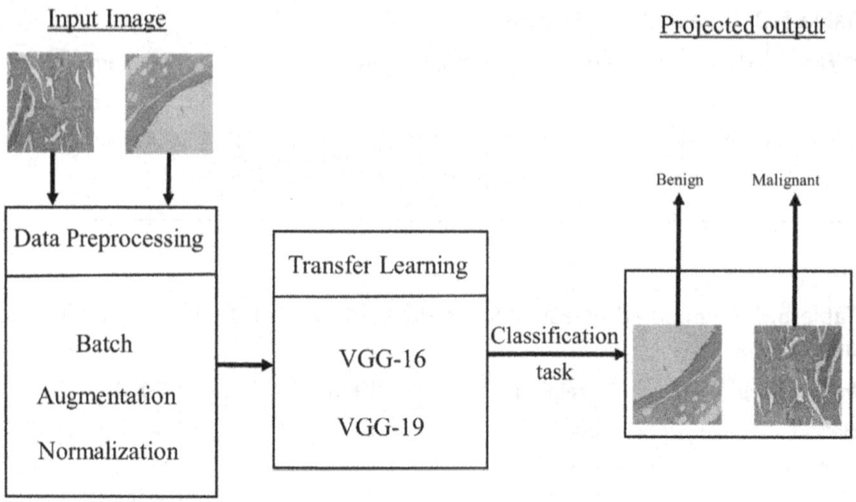

FIGURE 5.2

Transfer learning strategy for BC diagnosis. *BC*, Breast cancer.

layers. Hence, the purpose of exhibiting the mid-activations was to highlight these feature vectors to better understand the steps of the network deconstructing an input visual into various filters learned. These filters represent a variety of image-based characteristics to help in identifying major and distinct elements of the input visuals. The classifier uses these features to accurately project its outcome.

5.8 Pretrained classifiers: a short overview

A standard CNN model is used as the basic model for this research. The server makes this model available to the linked clients upon request. For the recommended application, its performance is contrasted with two renowned pretrained models: VGG-16 and VGG-19.

5.9 Visual geometry group network

VGG network (VGGNet) is a deep CNN architecture with multiple layers, based on VGG-16 or VGG-19 convolutional layers. It is a popular image recognition architecture and surpasses baselines beyond ImageNet.

5.9.1 Visual geometry group-16

The VGG-16 architecture was pretrained with the ImageNet dataset (Russakovsky et al., 2015) and is made up of 16 weighted layers with about 138

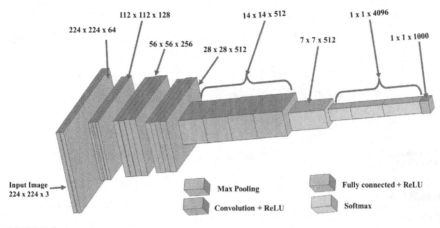

FIGURE 5.3

Basic architecture of VGG-16 model. *VGG-16*, Visual geometry group-16.

million parameters that comprise 13 convolutional and 3 completely connected dense layers. It emphasizes taking convolutional layers of a 3×3 filter with stride 1 rather than an enormous amount of hyperparameters and continually utilizes the matching padding and max pool layer of a 2×2 filter with stride 2. This blend of convolutional and max pool layer is constant around the network. Following the conclusive Max pooling layer, there are three layers that are completely interconnected. Then there are three totally connected layers. It employs the SoftMax classifier as the last layer. All hidden layers are activated by ReLU (rectified linear unit). The basic features of the architecture of the VGG-16 model are shown in Fig. 5.3

5.9.2 **Visual geometry group network-19**

VGG-19 (Simonyan & Zisserman, 2014) is made up of 19 weighted layers with about 138 million parameters, comprising 16 convolutional layers and 3 fully connected dense layers, 5 MaxPool layers, and 1 SoftMax layer as output. It focuses on having 3×3 filters of convolutional layers with stride 1 instead of an unprecedented amount of hyperparameters and constantly utilizes the matching paddings and MaxPool layer of 2×2 filters with stride 2. This convolutional and max pool layer combo is utilized all over the network. The basic features of the architecture of the VGG-16 model are shown in Fig. 5.4. This model was developed using a portion of the ImageNet database, which was utilized in the ImageNet large-scale visual recognition challenge (Russakovsky et al., 2015). The VGG-19 has been trained on more than one million visuals and can categorize visuals into 1000 categories of objects, such as keyboard, mouse, pencil, and a variety of animals. Consequently, the model has acquired intricate feature vectors for a diverse set of visuals.

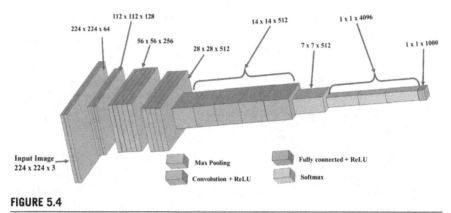

FIGURE 5.4

Basic structure of VGG-19 model. *VGG-19*, Visual geometry group-19.

5.10 Convolutional neural network model for federated learning

The adopted CNN is a three-layered model that will be the starting point for subsequent models provided by the server to its allied clients when responding to a request. Each of the three convolutional layers comprises a total of 128, 32, and 64 filters on every one of the three convolutional layers, collectively making up the algorithm with ReLU as an activation function. The server starts confirming and certifying the clients for the relevant dataset as soon as they are connected. The CNN model is trained using the local data after the data has been verified. The suggested model specifies the use of a 224×224 pixel grayscale image.

5.11 Model evaluation

The performance of an ML model can be appraised based on the cells of a confusion matrix (CM), as displayed in Table 5.5. The true positive (tp) instances are the sum of benign class examples correctly classed as the benign class, while the true negative (tn) instances are the sum of malignant class examples correctly classed as the malignant class (Folorunso et al., 2023; Folorunso, Banjo, et al., 2020). The right diagonal of the CM indicates the correctly classed values of the target classes. A substantial rate for the "tp" and "tn" instances in the CM points to a good model performance. The false positives (fp) are the sum of benign label examples erroneously classed as malignant, while the false negatives (fn) are the sum of malignant label examples erroneously classed as benign. A small or null rate for fn and fp instances by the CM table specifies that the ML model is doing well. The derived

Table 5.5 Confusion matrix.

		Predicted	
		Benign	Malignant
Real	Benign	*tp*	*fn*
	Malignant	*fp*	*tn*

performance indicators utilized by this research are from the CM presented in Table 5.5 and are stated in Eqs. (5.4) to (5.8).

$$Accuracy = \frac{tp + tn}{tp + tn + fp + fn} \tag{5.4}$$

$$Precision = \frac{tp}{tp + fp} \tag{5.5}$$

$$FPR = \frac{fp}{fp + tn} \tag{5.6}$$

$$TPR/Recall = \frac{tp}{tp + fn} \tag{5.7}$$

$$F1\_Score = 2 \times \frac{Recall + Precision}{Recall + Precision} \tag{5.8}$$

5.12 Results and analysis

This segment discusses the specifics of the experiments performed, the outcome of the performance evaluation of the two FL algorithms proposed in this study, and the related discussions. The experiment was performed using Python 3.8.0 programming language Scientific Python research library named SciPy (Virtanen et al., 2020). The Tensorflow open-source tool version 2.3.0 and Keras version 2.4.3 were used to develop CNN models. All investigations were run on a x64-bit Windows 11 Pro with an Intel (R) Core i5−6200U CPU @ 250GB, 2208 Mhz, 6 Core(s), accelerated with two NVIDIA GeForce RTX GPUs with 4GB DRAM.

The full hyperparameters of the FL are displayed in Table 5.6, while Table 5.7 shows the parameters for the data augmentation of the breast cancer images. In our experiment, 5 localized epochs are utilized to train client models (this is common with FL approaches); for FedMa, this means 25 localized epochs for every communication cycle (5 for every first and second layers, and an additional 15 for the SoftMax layer). Given the significant number of repetitions across the train sets (in comparison to other algorithms), we chose to limit our FedMA investigation to 100 communication loops. Considering the FedMA

Table 5.6 Hyperparameters of federated learning.

Name of hyperparameters	Value of hyperparameters
# Epochs	50
# Batch Size	64
# Drop out	0.3
Initialization	LeCun Uniform
Learning Rate	(0−500) 0.001
500−1000	0.0003
1000 +	0.0007
Momentum	0.9
Optimizer	Adamw
#Communication Rounds	200
# Clients	2

Table 5.7 Parameters for data augmentation.

rotation_range = 40,
width_shift_range = 0.2,
height_shift_range = 0.2,
shear_range = 0.2,
zoom_range = 0.2,
horizontal_flip = True,
vertical_flip = True

method for the three-layered CNN in this study, two intermediary communications with the server are conducted per communication cycle with the intent of transmitting layer weights for the successful synchronization of the model across the network.

For training and testing purposes, the diagnosis of breast cancer was conducted utilizing the binary class BreakHis breast cancer image dataset, containing a total of 7783 instances specifically chosen for this research. The entire set of 8000 images was split into 70%−30% train-tests to be used for training and testing the deployed model. So, of the 8000 images, 6000 were utilized for training the model, covering 70% of all images, and 2000 were utilized for testing the model, covering 30%. Since the known fact about the data in FL is that the data is geographically dispersed, and there is no communication between them, we split them in this study into two groups called clients. The dataset of the clients is unique and noncorrelated. The dispersion of testing data is shown based on both the real and projected target variables in a confusion matrix in Fig. 5.5 to demonstrate the effectiveness of the CNN model being trained while employing distributed data on clients or institutions.

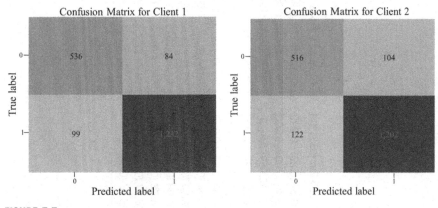

FIGURE 5.5

Confusion matrix of both clients.

The trained model's parameters for performance are listed in Table 5.7. According to Table 5.8, the deployed CNN model performs better than the two prior-trained models, VGG-16 and -19, because it uses the same decentralized dataset at every node. As illustrated in Table 5.8, two typical techniques for FedAvg are used to obtain the federated model. When compared with the model produced by using FedMA, FedAvg delivers a greater degree of accuracy for the server model.

In Fig. 5.6, the accuracy and loss for the client 1 classification model are presented graphically. Despite being chosen at random from the test dataset, all images predicted by the model were entirely accurate.

Fig. 5.7 demonstrates the client 2 model training and testing accuracy as well as their losses.

In real-world data, multimodality, such as geodiversity, existing within each class is normal. Classification algorithms trained on this type of data "learn" such biases and produce bad outcomes on marginalized domains since the associated overwhelming domain and class associations may prevent them from learning significant associations underlying the attributes and classes. The federated method enables users to train their models without sending their own dataset to the server. In this approach, the user's data from the localized node will remain protected, and even if the dataset is updated later, the client/institution would merely require the server to deploy the model again so that it may train their newly created dataset. To create a new master model, each client synchronizes with the server to help train its localized data and adjust their old weights and biases. Using two FL algorithms, FedMA and FedAvg, Fig. 5.8 shows how the accumulated model functions at the server end. FedMA builds a collaborative global model layer by layer that averages and matches concealed elements with comparable feature mining patterns, such as mediums for convolutional layers, concealed conditions for LSTM (long short-term memory), and neurons for layers that are completely linked. In addition to outperforming current cutting-edge FL algorithms, FedMA also significantly

Table 5.8 Assessment of related and adopted models.

FL algorithms	Accuracy (95% CI)	Recall (95% CI)	Precision (95% CI)	F1_Score (95% CI)
Client 1 (CNN) (Benign or Malignant)	93.81 (0.874, 0.990)	95.65 (0.908, 0.966)	93.57 (0.877, 0.990)	98.95 (0.972, 0.990)
Client 2 (CNN) (Benign or Malignant)	94.02 (0.938, 0.980)	94.89 (0.847, 0.954)	93.45 (0.932, 0.990)	94.81 (0.978, 0.990)
FedAvg deployed master model (CNN)	98.71 (0.952, 0.992)	97.72 (0.930, 1.000)	97.1 (0.988, 0.990)	98.94 (0.967, 0.990)
FedMA deployed master model (CNN)	97.42 (0.967, 0.987)	95.64 (0.925, 0.976)	96.69 (0.923, 0.990)	96.82 (0.997, 0.989)
Client 1 (VGG-16) (Benign or Malignant)	90.75 (0.876, 0.976)	88.85 (0.918, 0.982)	90.11 (0.901, 0.990)	90.12 (0.978, 0.966)
Client 2 (VGG-16) (Benign or Malignant)	87.96 (0.830, 0.980)	85.57 (0.967, 0.979)	84.85 (0.875, 0.990)	86.93 (0.979, 0.990)
FedAvg deployed master model (VGG-16)	95.72 (0.938, 0.990)	94.85 (0.897, 0.990)	95.94 (0.965, 0.990)	96.81 (0.925, 0.990)
FedMA deployed master model (VGG-16)	93.57 (0.874, 0.985)	93.76 (0.928, 0.989)	93.54 (0.897, 0.987)	93.72 (0.892, 0.989)
Client 1 (VGG-19) (Benign or Malignant)	90.83 (0.893, 0.886)	93.61 (0.816, 0.981)	91.96 (0.864, 0.990)	91.83 (0.826, 0.966)
Client 2 (VGG-19) (Benign or Malignant)	83.69 (0.855, 0.870)	85.53 (0.851, 0.983)	84.73 (0.865, 0.990)	84.73 (0.953, 0.990)
FedAvg deployed master model (VGG-19)	96.25 (0.935, 0.978)	95.93 (0.887, 0.994)	96.97 (0.952, 0.990)	93.79 (0.854, 0.990)
FedMA deployed master model (VGG-19)	93.59 (0.862, 0.981)	93.77 (0.874, 0.989)	95.27 (0.929, 0.987)	92.82 (0.892, 0.991)

All outcomes are obtained with a 95% confidence interval (in brackets) by resampling the validation task 100 times.

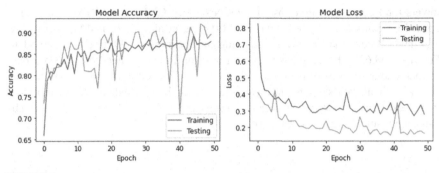

FIGURE 5.6

Accuracy and loss metrics of CNN model for client 1. *CNN*, Convolutional neural network.

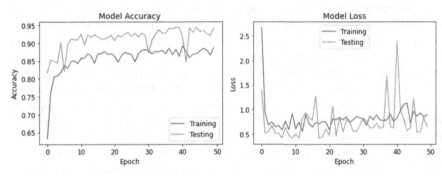

FIGURE 5.7

Accuracy and loss metrics of CNN model for client 2. *CNN*, Convolutional neural network.

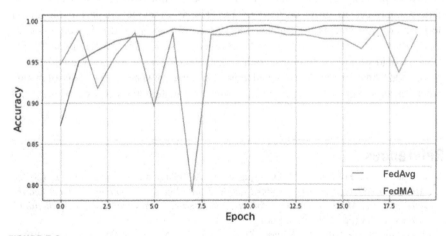

FIGURE 5.8

Accuracy of FedAvg and FedMA compared. *FedAvg*, Federated averaging; *FedMA*, federated match averaging.

lowers total communication costs for DCNN (deep CNN) and LSTM frameworks trained on datasets from the actual world (Qiang et al., 2019). FedAvg performed marginally better than FedMA, achieving an accuracy of 99.03% as opposed to FedMA's 98.63%. FedAvg has been in a position to comprehend the standard data format among the jobs of the customers because of its ability to profit from the variation in client data spreads via local adjustments. If the fundamental shared form is a linear map, the degree of iterative complexity required by the clients to establish such results is clearly defined (Kit et al., 2021).

5.13 Conclusion and future scope

Breast cancer is a fatal disease that requires an early diagnosis and timely and accurate intervention. In developing countries, where a large number of cases are reported, providing timely care for patients becomes an immense challenge for physicians. In such contexts, the adoption of computer-aided diagnosis proves to be an excellent approach to expediting the diagnostic process. The experiments performed in this study have demonstrated the potential of FL in terms of performance, data privacy, and protection—a critical consideration when dealing with sensitive medical data. In the context of FL settings, client datasets are geographically dispersed and operate in isolation, unaware of each other, with only weights of models being transferred, protecting data privacy. The adopted model integrates an FL method that utilizes a three-layered CNN model to train distributed data on clients or localized institutions that are linked to the server, taking into account the fact that the data residing at every client site may be available but that the institution might not be granted permission to share it with the server or data center where it will be trained to provide services. The system will be continuously able to learn as the dataset continues to grow because of FL. The FL method enhanced the model derived from the clients by utilizing the FedAvg algorithm and achieved an accuracy value of 98.71%, while FedMA achieved 97.42%. The evolution of the FL model would provide pathologists/radiologists/oncologists a tool they could use to gather additional image datasets and make an early or speedy diagnosis of breast cancer or any other sickness presented with an adequate dataset of its kind.

References

Awotunde, J. B., Folorunso, S. O., Ajagbe, S. A., Garg, J., & Ajamu, G. J. (2022). *AiIoMT: IoMT-based system-enabled artificial intelligence for enhanced smart healthcare systems. Machine learning for critical internet of medical things: Applications and use cases* (pp. 229—254). Springer.

Baghersalimi, S., Teijeiro, T., Atienza, D., & Aminifar, A. (2021). Personalized real-time federated learning for epileptic seizure detection. *IEEE Journal of Biomedical and Health Informatics, 26*(2), 898—909.

Chandiramani, K., Garg, D., & Maheswari, N., Performance analysis of distributed and federated learning models on private data. In *International conference on recent trends in advanced computing (ICRTAC). Procedia Computer Science*, 2019.

Darzidehkalani, E., Ghasemi-Rad, M., & van Ooijen, P. M. A. (2022). Federated learning in medical imaging: Part i: Toward multicentral health care ecosystems. *Journal of the American College of Radiology: JACR, 19*(8), 969−974.

Ding, W., Abdel-Basset, M., Hawash, H., Abdel-Razek, S., & Liu, C. (2023). Fed-ESD: Federated learning for efficient epileptic seizure detection in the fog-assisted internet of medical things. *Information Sciences, 630*, 403−418.

Dou, Q., So, T. Y., Jiang, M., Liu, Q., Vardhanabhuti, V., Kaissis, G., ... Heng, P. A. (2021). Federated deep learning for detecting COVID-19 lung abnormalities in CT: A privacy-preserving multinational validation study. *NPJ Digital Medicine, 4*(1), 60.

European Commission. (2018). *What data can we process and under which conditions?* https://commission.europa.eu/law/law-topic/data-protection/reform/rules-business-and-organisations/principles-gdpr/overview-principles/what-data-can-we-process-and-under-which," 2018. [Online]. [Accessed 3 6 2023].

Folorunso, S. O., Awotunde, J. B., Rangaiah, Y. P., & Ogundokun, R. O. (2023). EfficientNets transfer learning strategies for histopathological breast cancer image analysis. *International Journal of Modeling, Simulation, and Scientific Computing*2441009.

Folorunso, S. O., Banjo, O. O., Ayo, F. E., Ogunyinka, P. I., Folorunso, T. S., & Folorunso, M. T. (2020). Classification model for COVID-19 and pulmonary (TB) from X-ray images using HOG-PCA-learning algorithms. *African Journal of Science and Nature, 11*, 90−98.

Folorunso, S. O., Fashoto, S. G., Olaomi, J., & Fashoto, O. Y. (2020). A multi-label learning model for psychotic diseases in Nigeria. *Informatics in Medicine Unlocked, 19* (100326), 11.

Folorunso, S. O., Ogundepo, E. A., Awotunde, J. B., Ayo, F. E., Banjo, O. O., & Taiwo, A. I. (2022a). *A multi-step predictive model for COVID-19 cases in Nigeria using machine learning. In decision sciences for COVID-19: Learning through case studies* (pp. 107−136). Cham: Springer International Publishing.

Folorunso, S. O., Awotunde, J. B., Adeboye, N. O., & Matiluko, O. E. (2022b). Data classification model for COVID-19 pandemic. In A. E. Hassanien, S. M. Elghamrawy, & I. Zelinka (Eds.), *Advances in data science and intelligent data communication technologies for COVID-19* (378, pp. 93−118). Springer.

Folorunso, S. O., Ogbuju, E., & Oladipo, F. (2021). Artificial intelligence and the control of covid-19: A review of machine and deep learning approaches. *Artificial intelligence for COVID-19*, 167−185.

Hard, A., Rao, K., Mathews, R., Ramaswamy, S., Beaufays, F., Augustein, S., Eichner, H., Kiddon, C., & Ramage, D. (2018). Federated learning for mobile keyboard prediction," arXiv preprint arXiv:1811.03604, 2018. *arXiv preprint arXiv, 1811*, 03604.

Kit, S., Lu, Q., Wang, C., Paik, H., & Zhu, L. (2021). A systematic literature review on federated machine learning: From a software engineering perspective. *ACM Computing Surveys, 5*(5), 1−39.

Konečný, J., McMahan, H. B., Ramage, D., & Richtárik, P. (2016). Federated optimization: Distributed machine learning for on-device intelligence. *arXiv preprint arXiv, 1610*, 02527.

Kuhn, H. W. (1955). The hungarian method for the assignment problem. *Naval Research Logistics Quarterly, 2*(1−2), 83−97.

Kumar, R., Khan, A. A., Kumar, J., Golilarz, N. A., Zhang, S., et al. (2021). Blockchain-federated-learning and deep learning models for COVID-19 detection using ct imaging. *IEEE Sensors Journal, 12*(14), 16301−16314.

McMahan, B.H., Moore, E., Ramage, D., Hampson, S., & Arcas, B.A. Y. (2017a). Communication-efficient learning of deep networks from decentralized data. In *Proceedings of the 20th international conference on artificial intelligence and statistics, PMLR.*

McMahan, B., Moore, E., Ramage, D., Hampson, S., & Arcas B.A. (2017b). Communication-efficient learning of deep networks from decentralized data. In *Artificial intelligence and statistics, PMLR.*

Nilsson, A., Smith, S., Ulm, G., Gustavsson, E., & Jirstrand, M. (2018). *Performance evaluation of federated learning algorithms. In Proceedings of the second workshop on distributed infrastructures for deep learning (DIDL'2018).* France: Rennes.

Qiang, Y., Liu, Y., Chen, T., & Tong, Y. (2019). Federated machine learning: Concept and applications. *ACM Transactions on Intelligent Systems and Technology (TIST), 10*(2), 1−19.

Russakovsky, O., Deng, J., Su, H., Krause, J., Satheesh, S., Ma, S., . . . Fei-Fei, L. (2015). ImageNet large scale visual recognition challenge. *International Journal of Computer Vision, 115*(3), 211−252.

Siddique, A. A., Talha, S. U., Aamir, M., Algarni, A. D., Soliman, N. F., & El-Shafai, W. (2023). COVID-19 classification from X-ray images: An approach to implement federated learning on decentralized dataset. *Computers, Materials & Continua, 75*(2), 3883−3901.

Simonyan, K., & Zisserman, A. (2014). Very deep convolutional networks for large-scale image recognition. *arXiv technical report.*

Spanhol, F. A., Oliveira, L. S., Petitjean, C., & Heutte, L. (2016). A dataset for breast cancer histopathological image classification. *IEEE Transactions on Biomedical Engineering, 63*(7), 1455−1462. Available from https://doi.org/10.1109/TBME.2015.2496264.

Thibaux, R., & Jordan, M.I. (2007). Hierarchical beta processes and the Indian buffet process. In *Proceedings of the eleventh international conference on artificial intelligence and statistics*, San Juan, Puerto Rico.

Virtanen, P., Gommers, R., Oliphant, T. E., Haberland, M., Reddy, T., Cournapeau, D., Burovski, E., Peterson, P., Weckesser, W., Bright, J., et al. (2020). Scipy 1.0: Fundamental algorithms for scientific computing in python. *Nature Methods, 17*(3), 261−272.

Wang, H. Y. M. S. Y. P. D. & K. Y. (2020). Federated learning with matched averaging. *arXiv preprint arXiv, 2002*, 06440.

Wang, Y. (2017). *Co-op: Cooperative machine learning from mobile devices.* University of Alberta.

Yurochkin, M., Agarwal, M., Ghosh, S., Greenewald, K., Hoang, N., & Khazaeni, Y. Bayesian nonparametric federated learning of neural networks. In *Proceedings of the 36th international conference on machine learning*, Long Beach, California, USA, USA), pp. 7252−7261, 09−15 Jun 2019.

Zhang, Z., Zhou, T., Lu, Q., Wang, X., et al. (2021). Dynamic-fusion-based federated learning for COVID-detection. *IEEE Internet of Things Journal, 8*(21), 15884−15891.

CHAPTER

Taxonomy for federated learning in digital healthcare systems

6

Friday Udeji[1], Samarendra Nath Sur[2], Vinoth Babu Kumaravelu[3] and K.V.N. Kavitha[3]

[1]*Department of Mechanical Engineering, Faculty of Engineering, Ambrose Alli University, Ekpoma, Nigeria*
[2]*Department of Electronics and Communication Engineering, Sikkim Manipal Institute of Technology, Rangpo, Sikkim, India*
[3]*Department of Communication Engineering, School of Electronics Engineering, Vellore Institute of Technology, Vellore, Tamil Nadu, India*

6.1 Introduction

The rapid emergence of the information age has resulted in significant breakthroughs in the domain of machine learning (ML), especially with the advent of deep learning (DL) methodologies (Roscher et al., 2020). DL models, which apply hierarchical feature extraction, have shown outstanding capacity in recognizing patterns across different data sources, such as pictures, audio, and spoken language. In some specialized applications, DL models have even surpassed human performance norms (Jere et al., 2020). The success of DL projects relies largely on getting enough training data and ensuring that the learning target is well-represented within the data. As intelligent devices and applications continue to grow, the decentralized and heterogeneous nature of data produced at the network "edge" presents advantages and risks (Guo et al., 2019). Aggregating and analyzing this huge amount of data has the potential to uncover key discoveries and drive scientific and technological advancement. However, the increased concern for data privacy and ownership necessitates imaginative research and technology to offer secure and privacy-preserving means of gaining insights without compromising sensitive information (Kang et al., 2019).

Federated learning (FL) has emerged as a promising method for tackling the privacy dilemma in ML. It is a distributed learning method that allows a network of client devices to collectively train a DL model while retaining the training data on the devices themselves (Asad et al., 2020; Chai et al., 2021). By conducting model training at the network edge, FL enables aggregate analytics without the requirement for centralizing the data. Achieving the optimal balance between data

Federated Learning for Digital Healthcare Systems. DOI: https://doi.org/10.1016/B978-0-443-13897-3.00008-4

© 2024 Elsevier Inc. All rights are reserved, including those for text and data mining, AI training, and similar technologies.

privacy and model usefulness is a vital challenge that depends on numerous elements such as data type, model architecture, and intended use. Additionally, safeguarding the security and privacy of the FL infrastructure is crucial for generating trust among participating clients (Tanuwidjaja et al., 2020).

Following the COVID-19 pandemic, there have been new pressures and increasing demand on healthcare and automated solutions systems, such as artificial intelligence (AI), to help provide treatment and assist medical doctors or healthcare workers (Allam & Jones, 2020; Awotunde et al., 2023; Imoize et al., 2022; Hagos et al., 2024). This does not come as a surprise considering the modern technological breakthroughs within AI and their application in healthcare to enhance the quality and longevity of life (Amin & Hossain, 2020; Dimitrov, 2016; Imoize et al., 2022; Qadri et al., 2020). These advances have led to disruptive innovations in radiology, pathology, genomics, and other fields. Simply put, health measures and health practices are practically being switched from manual to digitized form, such that ML and especially DL is becoming the de facto knowledge discovery approach in many industries. Modern DL models feature millions of parameters that need to be learned from sufficiently large curated datasets to achieve clinical-grade accuracy while being safe, fair, equitable, and generalizing well to unseen data (Rieke et al., 2020). However, successfully implementing data-driven applications requires large and diverse datasets.

For example, training an AI-based tumor detector requires a large database encompassing the full spectrum of possible anatomies, pathologies, and input data types. Data like this is difficult to obtain because health data is sensitive, and its usage is tightly regulated (Oyebiyi et al., 2023). Data sharing generally requires hospitals to deal with the General Data Protection Regulation (GDPR) restrictions and have approval from the institutional review board. An institutional review board or ethical committee determines to what degree a hospital can share information with other hospitals and ensures that hospitals comply with the GDPR restrictions. Additionally, sharing data in healthcare is not systematic in that collecting, curating, and maintaining a high-quality data set involves considerable time, effort, and expense (Voigt & dem Bussche, 2017).

FL is a novel ML paradigm that allows multiple institutions or individuals to collaborate in building a shared model by training on their local datasets while keeping the data locally stored and without exchanging raw data (Zhang, 2021). FL has shown its potential to improve model accuracy while preserving the privacy of data sources (Li, Sharma, et al., 2020). However, as with any emerging technology, FL poses challenges regarding its adoption in real-world settings, such that despite its potential to revolutionize healthcare research, there remains a critical gap in the literature—a lack of a comprehensive taxonomic view of FL in digital healthcare systems (Abdulrahman, 2021; Massaoudi, 2021).

This chapter aims to bridge this gap by presenting a taxonomy for FL applied to digital healthcare systems. The taxonomy serves as a framework to categorize and classify the various design and use cases of FL in healthcare, providing researchers and practitioners with a comprehensive understanding of its applications and implications. By systematically organizing and categorizing the existing technologies and methods in this field, the taxonomy enables researchers to identify the key dimensions and factors influencing FL in healthcare, facilitating further exploration and advancements in this domain.

Hence, the problem addressed in this chapter is the absence of a comprehensive taxonomic view of FL in digital healthcare systems. The solution method proposed in this chapter is the development and presentation of a comprehensive taxonomy for FL in healthcare. This taxonomy will encompass various dimensions of FL, including its underlying technologies, system architectures, challenges, privacy-preservation methods, and critical aspects such as centralized ML versus FL, qualitative comparison of existing surveys, cloud computing versus edge computing, and security and privacy considerations. By categorizing and classifying these aspects, the taxonomy will provide a clear and organized view of FL in healthcare, enabling researchers and practitioners to navigate the complexities of this field.

6.1.1 Key contributions of the chapter

The key contributions of this chapter can be summarized as follows: Primarily, we have provided a comprehensive taxonomy of FL in healthcare systems. This taxonomy systematically organizes and categorizes the various aspects of FL, serving as a reference point for researchers and practitioners, facilitating a deeper understanding of FL and its applications in healthcare settings. In achieving this objective, we have presented a detailed overview of FL, elucidating its working principles, advantages, and challenges. We have highlighted the significance of data privacy, data heterogeneity, trust and cooperation, and regulatory compliance. Furthermore, we have explored diverse applications and trending use cases of FL in healthcare, showcasing its potential in personalized medicine, disease surveillance, medical imaging analysis, drug discovery, remote monitoring, and clinical decision support systems.

Also, we have conducted an elaborate literature review through which we have identified gaps in the field of FL in healthcare and demonstrated how this chapter fills those gaps. By synthesizing existing research, we have provided insights into the challenges and considerations of implementing FL while offering potential solutions and directions for future research.

We also outlined future directions and challenges in FL within the healthcare system, shedding light on potential research avenues and implications for the field. Our investigation emphasizes the need for enhanced privacy-preserving techniques, federated transfer learning, standardization of data representation, resource optimization, ethical and regulatory frameworks, addressing communication constraints, and validation and interpretability of federated models.

6.1.2 Chapter organization

This chapter is organized in the following order: Section 6.1 provides an introduction to the topic, states the problem being addressed, and presents the solution method proposed in this chapter. Section 6.2 offers an elaborate literature review on FL while focusing on related works published for healthcare systems. Section 6.3 delves into the fundamentals of FL, including its definition, working principles, advantages, and limitations. The section also explores the underlying technologies, system architectures, and challenges in implementing FL. Section 6.4 presents the taxonomy of critical aspects of FL in healthcare systems, covering various dimensions and considerations. Progressively, the section also provides a classification and clustering of the state-of-the-art FL applications in healthcare. Finally, Section 6.5 concludes the chapter by summarizing the key findings and implications for researchers and practitioners, while also highlighting future research directions in FL applied to healthcare.

6.2 Related works

FL is a popular method for training ML models on decentralized data (Li, Sharma, et al., 2020; Zhang & Li, 2022). This technique has several benefits over traditional centralized ML, including increased privacy and security, faster training times, and the ability to include data from multiple sources. In the healthcare industry, FL has significant potential to improve patient outcomes by enabling the development of more accurate and reliable models while ensuring patient privacy (Hao, 2020). In this literature review, we will discuss several recent studies and papers that focus on the use of FL in healthcare and the taxonomy proposed for implementing FL in healthcare.

In an investigation of the feasibility of FL for medical imaging applications, the authors (Cao et al., 2021), conducted a case study by training DL models on radiology images at three different institutions. The results showed that FL systems performed better than traditional centralized ML while maintaining data privacy and security. The study highlights the potential applications of FL for medical image analysis and emphasizes the importance of privacy and security measures.

Another study (Liu et al., 2021) evaluated the impact of FL on clinical natural language processing (NLP). In this study the authors propose a novel FL approach to address the challenges of training DL models for NLP in a distributed environment. The study highlights the potential for FL to improve the efficiency and accuracy of NLP models while providing data privacy guarantees.

To provide a more comprehensive framework for FL implementation, a taxonomy for implementing FL in various fields that consists of four categories (including hardware and device requirements, data privacy and security, model complexity, and model optimization) was proposed by Shaheen et al. (2022). They primarily emphasized the adoption of FL in multiple application areas and the preservation of client privacy. Our study goes beyond these aspects to address additional gaps in the field of FL in healthcare. We achieve this by systematically organizing and sorting the various facets of FL specific to the challenges and requirements of the healthcare domain.

Previously, a study had proposed a secure federated transfer learning (FTL) framework for healthcare applications, which introduced a privacy-preserving technique for FL (Cheng, 2020). The authors aimed to reduce the privacy risks associated with sharing sensitive health data across healthcare institutions while still enabling the generation of large and diverse datasets. Their study addressed privacy-preserving metrics such as differential privacy (DP) and secure multiparty computation (SMC) to enhance the privacy-preserving objectives of FL. The experiment conducted with breast cancer dataset showed an improvement in model accuracy compared with other FL and central learning approaches. Moving further, Chen et al. (2020) introduced a federated transfer learning (FTL) method based on generative adversarial networks (GANs) for privacy-preserving healthcare applications. Their proposed method aimed to overcome the shortcomings of traditional transfer learning by applying a generative approach with GANs.

In another study (Shi et al., 2021) FL was introduced to develop clinical decision support systems for patients with diabetes. The approach utilized federated averaging, whereby a global model was created using data distributed across different institutions. Their experiment results showed that the FL approach saved time and improved personal data privacy, and their proposed method had high accuracy compared with individual models in a private dataset of electronic health records of more than 1.6 million people with one or more chronic diseases in South Korea. These findings indicate the potential impact of FL on personalized healthcare applications.

Similarly, in Pfitzner et al. (2021) a taxonomy of FL algorithms concerning attack-related issues in a medical context were identified. Their research focused on evaluating the challenges of FL algorithms based on criteria such as communication overhead, model accuracy, privacy, and latency. The taxonomy proposed identified the various attacks but did not give a general categorization for use case scenarios. Also, in a recent systemic survey for FL systems (Li et al., 2021), a categorization was presented in which FL was classified based on different distribution patterns of sample space and feature space of data used in its execution. The highlights of the taxonomy thus presented were horizontal FL (HFL), vertical FL (VFL), and federated transfer learning (FTL). Although the survey helps researchers understand the nature of FL data and their applications in healthcare, it focuses on the blockchain-based FL (BCFL) framework. As such, the categorization was limited to the execution of the BCFL framework and not to a general taxonomy of FL in healthcare applications. This chapter addresses the need for a taxonomy that can be used as a general overview of FL applications in digital healthcare systems.

In response to the identified challenges of FL implementation in healthcare systems, a few studies have proposed varying methods for efficient application of FL for clinical purposes. For example, a 2022 research (Darzidehkalani et al., 2022)

Table 6.1 Summary of relevant related works.

References	Scope and Focus	Contributions	Limitations
Li et al. (2020)	FL concepts	Identified fundamental FL concepts and challenges	No taxonomy of FL for applications
Zhang & Li (2022)	Data privacy	Proposed a fault diagnosis method	Not applicable to healthcare
Hao (2020)	Security and privacy of EHR protocols in FL.	Developed and designed PRCL method that ensures privacy, accuracy, and efficiency of FL	No general overview of the various sections and categories of DL and FL technologies.
Cao et al. (2021)	DL in medical image recognition	Improvement of image quality in DL for healthcare systems	Focus not based on FL and there was no categorization of FL
Liu et al. (2021)	NLP in FL models and algorithms for medical applications	Extensive review of common FL tasks associated with NLP	Categorization of FL in healthcare not explicitly documented
Shaheen et al. (2022)	FL in adoption in edge network	Extensive introduction to FL across multiple applications	Broad, not tailored to health industry
Pfitzner et al. (2021)	Data privacy	Review of FL and its applicability for safe healthcare data applications	Categorization only based on data privacy leaving out other possible dimensions of categorization.
Li et al. (2021)	Blockchain in FL systems	Identification of challenges of BCFL implementation.	Taxonomy based on data types and FL architecture only.
Darzidehkalani et al. (2022)	FL in Healthcare systems	Illustration of FL utilization across various image processing-based medical applications	No categorization of the implemented FL algorithms and use cases covered
Rahman et al. (2023)	AI and FL in healthcare systems	Identification of the synergies between FL, AI, and XAI in healthcare systems	Proposed taxonomy for FL only based on four dimensions, which are not exhaustive
Nguyen (2021)	Blockchain and FL chain integration	Survey of FL chain fundamentals and representation of its taxonomies in terms of design, use cases, and edge computing	Taxonomy for healthcare systems was not considered

AI, *artificial intelligence*; BCFL, *blockchain in federated learning*; DL, *deep Learning*; EHR, *electronic health record*; FL, *federated Learning*; NLP, *natural language processing*; PRCL, *privacy aware and resource-saving collaborative learning protocol*; XAI, *explainable artificial intelligence*.

introduced a comprehensive taxonomy of FL in medical imaging. The study included different FL methodologies, such as federated averaging, federated inverse covariance estimation, and federated multitask learning. The authors also addressed critical technical challenges of FL algorithms in healthcare, such as communication efficiency and nonidentically and independently distributed (non-IID) data.

Extensive research has been conducted to ensure the integration and wide adoption of FL technologies into healthcare systems. Rahman et al. (2023) presented a practical integration of AI, FL, and explainable AI (XAI) in healthcare systems. The study also outlined the advantages and limitations of each technology and its applications in terms of security, privacy, stability, and reliability in the healthcare field (Table 6.1).

Summarily, the following can be inferred from the table:

i. Most literature with scope in healthcare systems do not present a taxonomy.
ii. Literature with taxonomies were targeted toward specific criteria, for instance cyber-attacks.
iii. Some literature with taxonomy are not specifically applicable to digital health systems
iv. Literature with FL categorizations did not explicitly show the levels of categorization and examples.

This chapter presents an explicit taxonomy that categorizes the various FL algorithms, methods, and technologies used in digital healthcare systems.

6.3 Fundamentals of federated learning

In recent years, DL has achieved considerable success in several areas, such as computer vision and NLP. Traditionally, DL models are trained via a centralized technique where data is stored and processed on powerful cloud servers (Steinberg, 2021; Wang, 2021). However, with the emergence of mobile devices equipped with sophisticated capabilities, such as a 5 G network, there has been a push toward shifting intelligence from the cloud to the edge, known as the mobile edge computing (MEC) paradigm (Lim et al., 2020). Additionally, considerations regarding data privacy have led to the emergence of FL as a concept. DL plays a crucial role in the understanding and implementation of FL (Zhang et al., 2021). Unlike conventional ML approaches that rely on hand-engineered feature extractors, DL models have the potential to automatically learn and extract relevant features from raw data (Abdulrahman et al., 2021). This feature discovery approach eliminates the necessity for domain expertise and specialized feature selection for each new challenge, making DL models more adaptable and effective.

DL operates within the field of brain-inspired computing, with neural networks being a crucial component. The architecture of a neural network mirrors the structure and functions of a neuron. It includes three major levels: the input layer, the hidden layer(s), and the output layer. In a conventional feedforward neural network, inputs are weighted and adjusted for biases before being passed through nonlinear activation functions, such as ReLU and softmax. The output of the network is produced from these computations. DL models, especially deep neural networks (DNNs), generally have several hidden layers that link inputs to outputs. For example, in image classification tasks, a DNN aims to generate a vector of scores where the index of the biggest score corresponds to the predicted class of the input picture. Training a DNN includes modifying the weights of the network to minimize the loss function, which quantifies the difference between the expected output and the ground truth (Hadzima-Nyarko, 2019; Ramesh & Tasdizen, 2019).

FL involves the collaborative training of DL models on end devices. The FL training technique consists of two main steps: local model training on individual devices and the global aggregation of updated model parameters on a central FL server. While FL may be used to train other ML models, our focus in this context is primarily on DL model training. DL models thrive on accumulating information and outperform typical ML techniques, especially when dealing with enormous datasets. FL in healthcare systems may utilize the processing capability and rich data obtained by scattered end devices in mobile edge networks. This alignment with the rise of DL is impacted by the above reasons (Abdulrahman et al., 2021).

6.3.1 Federated learning

In response to privacy concerns among data owners, FL was presented as a collaborative way to model training while respecting the privacy of personal data. FL enables participants to train a shared model by keeping their data securely on their devices, making it particularly well-suited for ML model training in healthcare applications. Different categorizations of FL settings, such as vertical and horizontal FL, are described in length (Nguyen, 2021).

The FL system consists of two basic entities: the data owners, commonly referred to as participants, and the model owner, represented by the FL server. Let $N = \{1,\ldots,N\}$ be the set of N data owners, each having a private dataset $D = U_i \in N$. Each data owner i employs their individual dataset D_i to train a local model w_i. Instead of transferring the complete dataset, just the local model parameters are transmitted to the FL server (Nguyen, 2021). The FL server combines all the incoming local models to build a global model w_G. This is in contrast to standard centralized training where the data, $D = \cup_i \in N$. Di from individual sources is first aggregated, and then a model w_T is trained centrally. Mathematically, the aggregation process in FL may be expressed as shown in Equation (6.1).

$$wG = \Sigma i \in N(ni \times wi)/\Sigma i \in Nni \qquad (6.1)$$

where n_i indicates the number of data points in dataset D_i, and w_i signifies the local model parameters of data owner i. By combining the relative sizes of the datasets, the aggregation guarantees a fair representation of the whole dataset while ensuring anonymity.

Fig. 6.1 demonstrates a typical design and training procedure of an FL system. The data owners operate as FL participants, cooperatively training an ML model required by an aggregate server. A crucial assumption underlying FL is the integrity of the data owners, meaning they employ their private data for training and transmit accurate local models to the FL server (Savazzi et al., 2020). This trust, although not always safe, is vital for the success and efficacy of the FL system.

Understanding the concept of FL and its operational dynamics, especially the aggregation formula, establishes the framework for understanding its applications in healthcare. In healthcare, FL offers major advantages by letting many healthcare professionals or health institutions collaborate and train a robust model without sacrificing patient privacy. This dispersed strategy guarantees that sensitive medical data remains on the local devices and is not transferred directly, complying with privacy requirements and safeguarding patient anonymity (Qian & Zhang, 2021). By employing FL in healthcare, enterprises may tap into similar knowledge hidden in heterogeneous datasets while protecting data sovereignty.

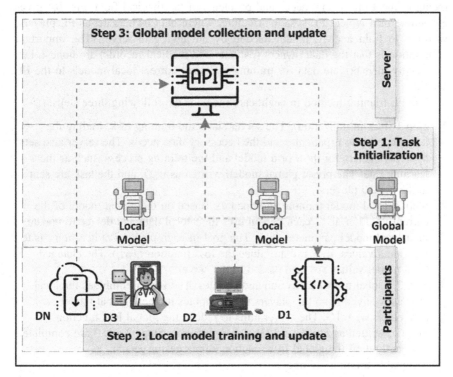

FIGURE 6.1

Typical design and training procedure of an FL system.

The collaborative model training approach allows healthcare practitioners to construct accurate and individualized models that can help in illness diagnosis, treatment planning, and predictive analytics. Furthermore, FL promotes continuous learning and model development by permitting frequent updates and adding fresh data from participating devices or universities.

As the area of FL continues to expand, addressing the particular difficulties and possibilities in healthcare settings becomes crucial. Ensuring the security and privacy of patient data, streamlining the aggregation process for diverse datasets, and establishing effective techniques for managing adversarial assaults are significant areas of attention. One such possible solution lies in the use of blockchain-based security measures for detecting malicious loopholes in the global model (Ramasamy, 2021; Awotunde, Imoize, et al., 2023).

6.3.2 System architectures of federated learning used in healthcare systems

In healthcare applications, the design and training method of an FL system is meant to meet the special needs and problems of the medical field. Fig. 6.1 gives an outline of a typical FL system structure, where data owners act as FL players, working to train an ML model needed by an aggregate server. The important assumption is that the data owners (patients and medical records) are honest, i.e., they apply their private data for training and send correct local models to the FL server.

The FL training method in healthcare includes the following three steps:

1. Step 1 (task initialization): The server starts the training task, finding the target application (patients) and the necessary data needs. The server also sets hyperparameters for the world model and the training process, such as the learning rate. The preset global model, written as $w_0 G$, and the task are sent to the picked players.
2. Step 2 (local model training and update): Based on the global model of the current iteration, $w_t G$, each person uses their local data and device to update their local model parameters, $w_t i$. The goal of each person i at iteration t is to pick ideal values, $w_t \times i$, that reduce the loss function $L(wt_i)$. The changed local model values are then passed to the server.
3. Step 3 (global model collection and update): The server combines the local models received from the players and computes the new global model parameters $w_t + 1G$. The server aims to reduce the global loss function $L(w_t G)$, written as $L(w_t G) = 1/N \times N_i = 1L(wt\ i)$. Steps 2 and 3 are completed frequently until the global loss function converges or gets an acceptable training accuracy.

It is important to emphasize that the FL training method may be used for many ML models that utilize the stochastic gradient descent (SGD) approach, including support vector machines (SVMs), neural networks, and linear regression (Lim et al., 2020). A typical training dataset in healthcare includes a set of data feature vectors, $x = \{x1, ..., xn\}$, and their related labels, $y = \{y1, ..., yn\}$.

6.3.3 Existing frameworks of federated learning in healthcare systems

Several open-source frameworks have been developed, providing tools and libraries to facilitate the integration of FL into healthcare systems. These frameworks allow researchers and developers to utilize the advantages of FL while preserving data privacy and security.

1. Federated AI technology enabler (FATE) is an open-source framework established by WeBank. FATE offers the federated and secure deployment of many ML models. It employs privacy-preserving methods, such as safe computing and encryption, to maintain the security of data throughout the FL process (WeBank, 2018). Healthcare systems may employ FATE to construct privacy-preserving FL models, enabling collaboration across numerous healthcare institutions while safeguarding patient data.

2. Another notable framework is LEAF, which contains benchmark datasets for FL, including federated extended MNIST (FEMNIST) and Sentiment140 (Caldas et al., 2018). These datasets duplicate the distribution of data held by members in FL, allowing researchers to assess the performance of newly created FL algorithms. LEAF allows meaningful comparisons across research and helps the building of powerful FL solutions specific to healthcare applications.

3. PySyft, a framework based on PyTorch, focuses on encrypted and privacy-preserving DL methods. By retaining the original Torch interface, PySyft assures that tensor operations may be done easily. Participants in the FL simulation are represented as virtual workers, and data is distributed using SyftTensor and PointerTensor (Ryffel, 2018). This technology offers secure and privacy-conscious FL applications in healthcare, concealing sensitive patient data while permitting collaborative model training.

4. TensorFlow federated (TFF), invented by Google, is a framework based on TensorFlow that delivers a high-level interface for decentralized ML and distributed computations (Zhu et al., 2019). TFF consists of two layers: FL and federated core (FC). The FL layer accelerates the introduction of FL into the existing TensorFlow models, while the FC layer allows users to experiment with unique FL methods. TFF may be leveraged in healthcare systems to train FL models on distributed healthcare data while safeguarding privacy and guaranteeing data ownership.

As shown in Fig. 6.2, the variety of frameworks makes different FL algorithm types and approaches available for use in a plethora of fields. These frameworks

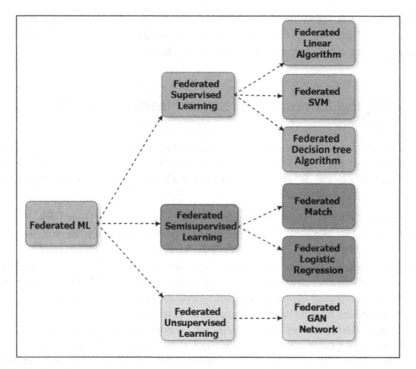

FIGURE 6.2

Classification of machine learning-based federated learning.

collectively provide valuable tools for researchers and developers in the healthcare domain. By embracing these technologies, healthcare organizations may explore the potential of FL, harnessing the collaborative power of distributed data while ensuring privacy norms and data security. The creation and application of FL frameworks in healthcare show significant promise for improving healthcare outcomes and encouraging secure data-driven relationships across healthcare organizations.

6.3.4 Underlying technologies in federated learning in healthcare systems

In healthcare systems FL relies on a combination of underlying technologies to ensure efficient and privacy-preserving model training and inference. While specific implementations may vary, the following technologies are commonly used in FL:

1. DL: FL leverages the power of DNNs for model training and inference. DNNs are adept at learning intricate patterns from data, making them well-suited for a wide range of FL applications in healthcare.

2. Mobile edge computing (MEC): FL takes advantage of the computing capabilities of mobile devices in edge networks. By utilizing MEC, FL algorithms can be executed directly on edge devices, minimizing the reliance on centralized cloud servers. This approach enhances efficiency and reduces latency in healthcare applications.
3. Communication Protocols: FL requires robust and efficient communication protocols to facilitate the exchange of model updates between devices and the FL server. Various protocols, including hypertext transfer protocol, remote procedure calls, and message queuing telemetry, can be employed to ensure secure and reliable communication in FL systems deployed in healthcare environments.
4. Privacy-preserving techniques: Privacy is a critical concern in healthcare systems, and FL incorporates privacy-preserving techniques to safeguard sensitive user data. DP, secure aggregation, and encryption methods are among the techniques utilized to protect privacy while enabling collaborative model training and inference in FL.

These underlying technologies have already found applications in healthcare systems in the past (Chen et al., 2021). The advantage of FL technology is that it has the ability to encompass all these technologies and many others, within a unified framework.

6.3.5 Challenges in implementing federated learning

The application of FL in healthcare systems as in other applications is characterized by various basic issues. These limitations affect the usefulness and economy of FL in digital healthcare systems:

1. High communication cost: Communication is a drawback in shared networks, worsened by privacy concerns surrounding the transfer of raw data. As shared networks could span across a number of devices, such as millions of smartphones, contact within the network might be substantially slower compared with local processing. To overcome this issue, communication-efficient methods are crucial, focusing on the frequent sharing of small signs or model changes rather than sending the whole dataset (Posner et al., 2021). Reducing the number of contact processes and the volume of exchanged messages present significant challenges in the healthcare applications of FL.
2. System variability: Federated networks exhibit variability in terms of storage, computing power, and communication skills across devices. Variability in hardware, network connections, and power levels, coupled with the constraint that only a subset of devices are active at any given time, adds to the complexity of FL in healthcare systems (Yu et al., 2021). FL methods must adjust for low involvement rates, accept hardware changes, and handle lost devices in the network to offer robustness and stability.

3. Statistical heterogeneity: Data made by devices in shared networks occasionally shows nonidentically distributed traits. Variations in language use, different data points across devices, and underlying structural imbalances between device groups add to statistical variation. This departure from widely accepted independent and identically distributed (IID) assumptions in distributed optimization presents difficulties in modeling, analysis, and evaluation. While the basic FL issue aims to train a single global model, newer methods, such as multitask learning and metalearning, suggest strategies to address statistical variation by simultaneously learning several local models or allowing personalized device-specific modeling (Sun, 2020).

4. Privacy concerns: Privacy is a major problem in FL uses, especially in healthcare systems. FL mitigates privacy risks by giving model changes instead of raw data. However, publishing model changes during the training process can still potentially expose sensitive information (Jere et al., 2021). Techniques such as safe multiparty computing and DP have been developed to improve privacy in FL. However, these methods generally come at the cost of lower model performance or system economy. Balancing the trade-offs between privacy protection and system efficiency is a key problem when using private shared learning systems in healthcare.

6.3.6 Privacy-preservation methods in federated learning

FL has emerged as a promising method to tackle the privacy dilemma in ML. As a distributed learning method that allows a network of client devices to collectively train a DL model while retaining the training data preserved on host devices themselves, it has provided substantial grounds for solving security problems in healthcare data, especially in cases where blockchain technology is incorporated into the systems (Imoize & Irabor, 2022; Abikoye, 2023). By conducting model training at the network edge, FL enables aggregate analytics without the requirement for centralizing the data. Achieving the optimal balance between data privacy and model usefulness is a vital challenge that depends on numerous elements such as data type, model architecture, and intended use. Additionally, safeguarding the security and privacy of the FL infrastructure is crucial for generating trust among participating clients (Jere et al., 2021). Fig. 6.3 is a depiction of FL architecture devoid of security protocols.

Addressing the outstanding problems regarding security, privacy, and effectiveness of FL systems will have huge societal impacts. Doing so will help the development of secure and privacy-preserving FL methodologies, assuring the responsible and effective use of distributed ML in different healthcare applications. Fig. 6.4 shows the first step involving the detection and diagnosis of a disease secured by the adoption of blockchain technology with FL technology. In this setup, the first step involves downloading an initial model from the blockchain-secured global model through a verified node. Next, each participating node aggregates the data from the pool of available information, and lastly, the

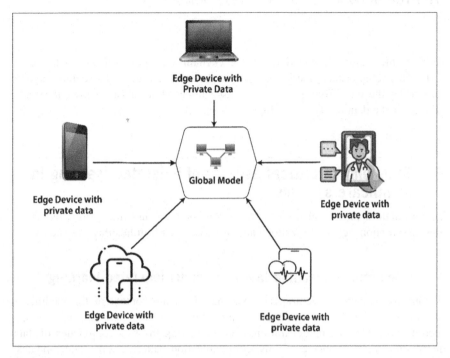

FIGURE 6.3

Nonsecured federated learning for disease detection.

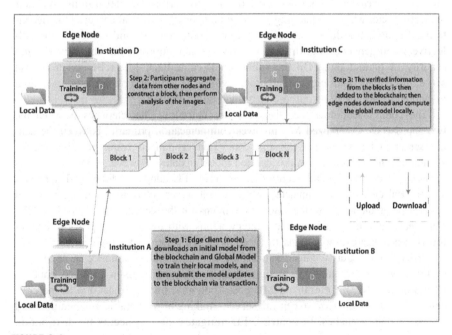

FIGURE 6.4

Secured (blockchain-based) federated learning for disease detection.

verified information is added to the blockchain via uploads from each secure node. This setup ensures the global and local models are trained to detect the disease using the data from each participating node while ensuring the safety and privacy of the data used for training.

6.4 Taxonomy of critical aspects of federated learning in healthcare systems

To formulate a functional taxonomy, we will now discuss the various technologies, assumptions, characterizations and use cases of FL in healthcare systems.

6.4.1 Centralized machine learning versus federated learning

A comparison between centralized ML and FL in the context of the healthcare system is crucial for informed decision-making, ensuring patient privacy, data security, and efficient model training. By examining the two approaches of data distribution, privacy considerations, communication patterns, and model updates, it becomes apparent that each method offers distinct advantages and limitations when applied to healthcare systems.

Efficient model training is a critical aspect of healthcare systems, as it directly impacts timely and accurate decision-making. In this regard, FL emerges as a promising approach. FL allows local devices to perform model training on their respective data while exchanging model updates with a central server. This decentralized approach reduces network bandwidth requirements and harnesses the collective intelligence of multiple devices without compromising the privacy and security of patient data. On the other hand, centralized ML relies on a central server to collect and process all data, which may introduce privacy risks and resource challenges in resource-constrained environments.

The communication aspect also exhibits contrasting characteristics between the two approaches. Centralized ML involves communication primarily between the central server and devices that transmit data. This centralized communication pattern streamlines the data collection process and model updates. However, it raises concerns regarding data privacy and security. In contrast, FL adopts a distributed approach where local devices communicate with a central server solely for exchanging model updates or gradients while the raw data remains on the devices (Asad et al., 2021). This decentralized communication strategy aligns with the stringent privacy requirements in healthcare, emphasizing the need to uphold patient data confidentiality.

Privacy and security considerations are of paramount importance in healthcare systems, where safeguarding patient data confidentiality is a primary concern (Rufai et al., 2023). Centralized ML poses potential privacy risks due to the concentration of all data on a central server, making it vulnerable to unauthorized access or data breaches. In contrast, FL mitigates these concerns by distributing

Table 6.2 Centralized machine learning versus federated learning.

References	Aspect	Centralized Machine Learning	Federated Learning
Drainakis (2023)	Approach	Central server	Collaborative devices/entities
Fauzi et al. (2022)	Data distribution	Centralized storage	Distributed across local devices
Asad et al. (2020)	Data privacy	Privacy concerns	Enhanced privacy protection
Asad et al. (2021)	Communication	Central server communication	Local device-to-central server communication
Mar'i, Supianto and Bachtiar (2023)	Model updates	Central server updates	Local device updates aggregated by central server
Savazzi et al. (2020)	Scalability	Dependent on central server	Improved scalability through distributed devices
Lim et al. (2020)	Network Bandwidth	Higher requirements	Lower requirements due to selective information sharing
Jere et al. (2021)	Failure Resilience	Vulnerable to central server	Local training provides better resilience to failures and attacks

data across devices, reducing the risk of a single point of failure. By keeping sensitive patient information on local devices, FL ensures enhanced privacy protection and compliance with regulatory requirements. Thus FL provides a more privacy-preserving framework for healthcare applications compared with centralized ML. These aspects and their relative differences are tabulated in Table 6.2.

6.4.2 Cloud computing versus edge computing

Cloud computing has drawn substantial attention due to its ability to address the challenges arising from exponential growth of data across various sectors. It allows internet-based computing where shared resources such as storage, computing powers, software, data, and applications may be viewed and abused on demand in a simple "pay-as-you-go" way. The cloud computing model gives users high-quality services at a lower cost, making it an attractive choice for students and businesses. Major firms, such as Amazon, Google, Salesforce, IBM, and Microsoft have developed cloud computing systems to cater to global demands (Asad et al., 2021).

In addition to its impact on numerous industries, cloud computing has also changed manufacturing by providing a cloud-based production model. This model virtualizes and includes all manufacturing resources and capabilities as services, which can be controlled, distributed, and consumed on demand via the cloud. By combining the Internet of Things (IoT), cyber-physical systems (CPS), big data analytics (BDA), and cloud computing tools, the cloud-based manufacturing model allows smart production and maintenance for numerous applications. Furthermore, cloud computing finds use in the Industrial Internet of Things (IIoT), AI-Enabled Internet of Medical Things (AioMT), and CPS areas. For instance, (Salem et al., 2022) introduced a cloud-based design for remote tracking and control of industrial tools in IIoT.

Despite its benefits, cloud computing in some cases may lead to delayed response times and big internet needs (Uppal, 2023). In circumstances where data received from machines and sensors is moved to the cloud for processing and the necessary result is given back to the devices, delay difficulties may occur. Additionally, the transfer of huge data needs significant bandwidth, which may be costly. Moreover, network failure might cause substantial problems, especially for software programs with severe delay limits.

In contrast to cloud computing, edge computing has grown as a response to these problems by putting processes closer to the data sources. Edge computing supports cloud computing by allowing data processing and analysis to be performed at the edge of the network, closer to IoT devices and sensors (Kumar, 2021). By exploiting local working capabilities, edge computing reduces delay, boosts real-time data analysis, and overcomes internet limits. In the context of collaborative learning in healthcare systems, edge computing allows for local data processing and model training, decreasing the need to move private healthcare data to the cloud while maintaining privacy and security (Salem et al., 2022).

When considering the applications of shared learning in healthcare systems, the choice between cloud computing and edge computing depends on different factors. Cloud computing excels in offering flexible infrastructure, high processing power, and unified control of resources and data. It allows fast data handling and powerful ML methods, which may result in strong models in healthcare applications. On the other hand, edge computing provides benefits such as reduced delay, greater privacy, and the ability to handle real-time data processing, making it well-suited for shared learning in healthcare settings.

To maximize the benefits of both approaches, a mixed method that mixes cloud computing with edge computing may be chosen. This method allows for a global and shared model training process, harnessing the assets of both cloud and edge resources. Edge devices may perform initial data preparation, feature extraction, and local model training, while the cloud design allows global model aggregation, changes, and sharing of knowledge across multiple edge devices. By finding a balance between data protection, computing speed, and joint learning, this mixed approach may successfully allow FL in healthcare systems.

6.4.3 Assumptions and characterization of federated learning

6.4.3.1 Assumptions

The following assumptions and their consequences provide insights into the challenges and opportunities inherent in leveraging the FL model within the healthcare system.

6.4.3.1.1 Availability of sufficient data

FL requires that a considerable amount of data is accessible across various health-care facilities. This assumption suggests that healthcare institutions contain extensive datasets, including a wide variety of patient groups, medical ailments, and treatments. However, in practice, the availability and accessibility of healthcare data may vary greatly among various organizations and areas. Disparities in data availability might impair the representativeness of the federated dataset and induce bias or restrictions in the resultant models.

6.4.3.1.2 Data privacy and security

One of the essential assumptions of FL in healthcare is the preservation of patient privacy. Healthcare systems deal with sensitive and secret patient information, making data privacy a vital problem. FL leverages privacy-preserving approaches, such as local data storage, encryption, and safe aggregation, to guarantee that patient data stays private throughout the training process. However, the installation of comprehensive privacy safeguards needs careful study and respect for legislative rules, such as the Health Insurance Portability and Accountability Act in the United States, to secure patient information successfully (Act, 2023).

6.4.3.1.3 Data heterogeneity and quality

Healthcare datasets generally demonstrate heterogeneity in terms of data kinds, formats, and quality. FL presupposes that the participating universities have standardized and well-curated data. However, healthcare systems may meet issues linked to data heterogeneity, including variances in data representation, missing values, or discrepancies. These concerns may impair the performance and generalizability of the federated approach. Therefore, initiatives to standardize data collection, harmonize variables, and solve data quality concerns are important to optimize the advantages of FL in healthcare.

6.4.3.1.4 Trust and cooperation

Successful deployment of FL in healthcare depends on trust and cooperation among participating institutions. Healthcare organizations need to create agreements and norms to enable a transparent and equal exchange of resources and expertise. This assumption underscores the necessity of developing relationships, fostering trust, and promoting a culture of cooperation across healthcare stakeholders. Establishing rigorous governance structures and mechanisms for data sharing, model training, and outcome assessment may boost the efficacy and acceptability of FL in healthcare systems.

6.4.3.1.5 Regulatory and ethical compliance

Healthcare is a highly regulated business, and FL must comply with appropriate rules and ethical principles. This assumption acknowledges the requirement for adherence to legislative frameworks controlling data exchange, patient consent, and research ethics. For example, gaining informed permission from patients for data consumption and maintaining compliance with data protection legislation are key issues. Healthcare institutions must traverse these legal and ethical settings, ensuring that FL projects conform with the current frameworks and principles to ensure patient rights and autonomy.

6.4.3.2 Characterization

The essential characteristics of FL and their relevance to healthcare applications are outlined below.

6.4.3.2.1 Decentralized data ownership

FL is characterized by the decentralized ownership of data within the healthcare system. In the healthcare domain, patient data resides in various institutions, such as hospitals, clinics, and research centers. This distribution of data ownership aligns with the federated approach, where each institution retains control over its data while participating in collaborative model training. By preserving data ownership, FL enables healthcare organizations to comply with privacy regulations and maintain patient trust, as sensitive data remains within the institutions' control.

6.4.3.2.2 Privacy-preserving collaboration

Privacy is a paramount concern in healthcare, and FL addresses this challenge effectively. Its mode of operation allows healthcare institutions to collaborate and train models without sharing raw patient data. Instead, only model updates, aggregated and anonymized, are exchanged during the training process. This privacy-preserving characteristic ensures that patient data remains protected and minimizes the risk of privacy breaches or unauthorized access, fostering trust among institutions and patients in healthcare systems.

6.4.3.2.3 Data diversity and representativeness

FL benefits from the diversity and representativeness of data across multiple healthcare institutions. In healthcare, different patient populations, geographic locations, and medical conditions contribute to the heterogeneity of data. By leveraging FL, healthcare systems can tap into this diverse data landscape and build models that are more robust, generalizable, and applicable to a wide range of patients and medical scenarios. This characteristic is particularly valuable in healthcare, where data heterogeneity can impact the accuracy and reliability of models trained on single-site datasets.

6.4.3.2.4 Resource efficiency and scalability

FL offers resource efficiency and scalability, making it well-suited for healthcare systems that often operate with limited resources. Training ML models on large centralized datasets can be computationally intensive and resource demanding. In contrast, FL allows institutions to collaboratively train models while leveraging their local computational resources. This characteristic reduces the burden on individual institutions and promotes scalability, enabling healthcare systems to efficiently utilize their resources for model development and deployment.

6.4.3.2.5 Transferability and knowledge sharing

FL facilitates the transferability of knowledge and insights gained from one institution to others within the healthcare ecosystem. In the context of healthcare, this characteristic enables the sharing of best practices, treatment guidelines, and predictive models across institutions. By fostering knowledge exchange, FL contributes to the collective advancement of healthcare research and clinical decision-making. It enables healthcare systems to harness the expertise and experience of different institutions, ultimately improving patient care and outcomes on a broader scale.

6.4.3.2.6 Collaboration with regulatory frameworks

Healthcare systems operate within a heavily regulated environment to ensure patient privacy, data protection, and ethical conduct. FL must align with these regulations and adhere to ethical guidelines, such as obtaining informed consent from patients for data utilization and ensuring compliance with data governance protocols.

6.4.4 Applications and trending use cases of federated learning in healthcare systems

FL offers significant potential to transform healthcare by leveraging distributed data sources and ensuring privacy, enabling advancements in patient care, medical research, and decision-making processes. Here, we discuss the applications and trending use cases of FL in healthcare systems.

6.4.4.1 Personalized medicine

FL can facilitate the development of personalized medicine by training models using patient data from multiple healthcare providers without sharing sensitive information. By leveraging FL, healthcare systems can collectively learn from diverse populations, leading to more accurate disease diagnostics, treatment recommendations, and patient outcome predictions. The collaborative nature of FL allows models to be continuously updated with real-world data, leading to improved accuracy and adaptability.

6.4.4.2 Disease surveillance and early detection

FL can enhance disease surveillance systems by enabling the aggregation and analysis of data from various healthcare institutions, including hospitals, clinics, and wearable devices. By combining data from multiple sources, FL can help identify patterns, detect outbreaks, and provide early warnings for infectious diseases. This collaborative approach improves the timeliness and accuracy of disease surveillance, allowing for proactive measures to prevent and control the spread of diseases.

6.4.4.3 Medical imaging analysis

FL holds great potential in medical imaging analysis, where privacy concerns often limit the sharing of sensitive patient images. With FL, models can be trained collaboratively using distributed image datasets from different healthcare institutions while ensuring patient privacy. This enables improved diagnostic accuracy, automated anomaly detection, and the development of predictive models for conditions such as cancer, cardiovascular diseases, and neurological disorders.

6.4.4.4 Drug discovery and development

FL can accelerate the drug discovery and development process by enabling pharmaceutical companies and research institutions to collaborate and train models on distributed datasets. By securely aggregating data from multiple sources, FL can facilitate the identification of novel drug targets, prediction of drug efficacy, and optimization of treatment regimens. This collaborative approach enhances the efficiency and effectiveness of the drug discovery process, potentially leading to the development of new therapies and improved patient outcomes.

6.4.4.5 Remote monitoring and telemedicine

FL can support remote monitoring and telemedicine applications by enabling the analysis of patient data collected from various devices, such as wearables, sensors, and mobile health apps. Unlike in AIoMT (Awotunde et al., 2023), by applying FL techniques, healthcare providers can obtain aggregated insights without compromising individual data privacy. This facilitates remote disease management, real-time monitoring of patient vitals, and personalized interventions, enhancing the delivery of healthcare services to remote or underserved populations.

6.4.4.6 Clinical decision support systems

FL can empower clinical decision support systems by training models on diverse patient data from multiple healthcare providers. By collectively learning from distributed datasets, FL can improve the accuracy of clinical decision-making, assist in diagnosis, recommend treatment options, and predict patient outcomes. This collaborative approach enhances the overall quality of care and reduces the potential biases that may arise from training on limited data sources.

6.4.5 Classification and clustering of federated learning in digital healthcare systems

FL in healthcare shows enormous potential for increasing innovation and medical advancement. However, a comprehensive categorization of the different techniques, architecture, and use cases of various FL models is fundamental to the continuity of FL systems in healthcare systems. The following categorizations show the various types of FL models adopted in medical systems. This chapter categorizes FL in healthcare systems into groups discussed below.

6.4.5.1 Approach or data type

Based on the approach, FL can be categorized into three major groups: horizontal FL, vertical FL, and federated transfer learning. Simply put, horizontal FL is widely related to datasets with different sources, which do not have similarities in the sample space but have similarities in the feature space. An example of this type of FL in medical systems is the scenario where datasets from different health facilities correspond to similar features (for example, hair diseases in patients), but are completely unrelated in the specific information for each patient.

Vertical FL, incontrast to horizontal FL, identifies with clients with data sources; such as clients who possess common sample space but exhibit different characteristics. An example of vertical FL is an FL system that shows the feeding patterns and shopping history of the same group of people. Federated transfer learning holds information about different characteristics of different sample spaces. It is for aggregating data from varying sources and fusing them to generate relatively personalized models for patients without compromising their security.

6.4.5.2 Federated learning architecture

The structure and architectural framework of the FL system used in healthcare is also a substantial dimension for segmenting the various types of FL systems and models used in the healthcare industry. Generally, FL architecture comprises parties, a manager, and the communication-computation framework (Moshawrab, 2023). While the parties represent the client and users of the FL model, the manager is the server and acts as an aggregator of the data collected from the parties. The communication-computation framework, however, is the combination of software and hardware responsible for handling the model training, data manipulation, and information sharing between the parties and the manager. Interestingly, FL models may have parties and managers (single or multiple). However, two communication-computation FL concepts are in use today, such as centralized and decentralized design. In centralized design, also known as client-server architecture, the direction of data movement is asymmetric as the manager aggregates data from the parties and sends the updated information to the model, resulting in a fine-tuned model. For decentralized design (also known as peer-to-peer architecture), information exchange is between the parties without the need for a manager (aggregator). This architecture allows for faster global parameter updates as the updates do not have to wait for approval from the central server for implementation (Zhang et al., 2021).

6.4.5.3 Application type

In healthcare systems, FL can also be categorized based on their application types: disease detection, treatment optimization, and clinical trial design. Disease detection FL systems, such as Google's automated retina disease assessment, develop models to detect diseases such as glaucoma, heart diseases, cancer, and other terminal diseases, which are difficult to diagnose at early stages. Treatment optimization FL models are designed to modify existing treatment plans to suit the condition and response of individuals for optimal recovery. At the Massachusetts General Hospital in 2021, Google AI researchers discussed the successful implementation of FL in optimizing the treatment for heart disease (Boukhatem et al., 2022). In the case of clinical trial design, FL models can be soft-margin L 1-regularized SVMs that can be used in conjunction with the cluster primal-dual splitting algorithm to both detect and plan for new ways to approach the treatment of existing diseases.

6.4.5.4 Data modality

The mode of data aggregation and processing for clinically applied FL models greatly affects their efficiency and is therefore an important categorization of FL models. The types of FL models based on modality are imagery, electronic health records (EHRs), wearable devices, text data, and audio data FL models. While imagery-based models are efficient at disease detection through medical images, EHR-based models are better at predicting diseases. Wearable data-based models monitor more clinical parameters, such as sleep, heart, or fall-related conditions, in patients. Medical audio data, used by algorithms such as FedAudio (Zhang & Feng, Alam, et al., 2022), is also a robust yet untapped FL data modality. The modality finds direct use in phonocardiography and auscultation for the diagnosis and monitoring of cardiovascular and respiratory issues, pain assessment, and other medical applications. Data sourced in text format may be sourced from patient medical history and self-evaluation that NLP FL algorithms can use for diagnosis, treatment suggestions, and disease predictions.

6.4.5.5 Privacy level

Three basic categories have been identified based on the privacy level of FL applications in healthcare systems. They include DP, complete zero knowledge, and raw data DP involves adding carefully calibrated noise to the gradients or model parameters of the FL during aggregation. The addition of this noise ensures that no single patient's data significantly influences the model, providing a strong privacy guarantee (Sei et al., 2023). Examples of DP applications in healthcare systems are FedAvg-DP and DP similarity weight aggregation (DP-SimAgg) methods (Khan et al., 2023). To reduce the loss of data and ensure the high utility of the dataset, an approach that adapts differential private stochastic gradient descent (DP-SGD) specifically for medical datasets has been developed. Rather than adding noise to the data, SGD adds some degree of randomness to the dataset, allowing a trade-off between privacy and utility. Although DP algorithms reduce the risk of data exposure, they do not prevent leakage of the data itself. SMC for FL, a complete zero-knowledge privacy level FL mechanism can be used to ensure that individual datasets are secured through advanced cryptographic techniques. Other examples of zero-knowledge privacy level FL in healthcare applications are homomorphic encryption FL and zero-knowledge proof FL. At the highest privacy level is the raw data protection FL algorithms. At the privacy level, the entire data protection is targeted at the patient's data itself while preprocessed information about the images, videos, X-ray scans, etc. is used to train the other models. One such example is the secure aggregation FL, where only the aggregated updates from each model are used for training. Another example of this level of privacy is federated transfer learning, where participants leverage pretrained models of local data to train the entire system. In this scenario, training parameters or gradients are transferred, and not the patient data.

6.4.5.6 Task

According to Liu et al. (2021), FL-related tasks in medical applications include but are not limited to patient similarity learning (Lee et al., 2018), patient representation learning and phenotyping (Kim et al., 2017; Liu et al., 2019), predictive or classification modeling (Huang et al., 2019), and biomedical named entity recognition (BioNER). In patient similarity learning, the FL algorithms are tasked with identifying what patients share in common, especially for uncharted diseases or areas where bottlenecks have been reached in the development of cures. These types of FL algorithms are also efficient for recommending treatments based on the experiences of similar patients in personalized medicine. For patient representation learning and phenotyping, the FL algorithms focus on learning about meaningful representation of patient data and how to categorize such data (phenotyping). By accurately representing patient data, more targeted interventions can be made for specific phenotypes (Liu et al., 2019), dividing FL into a two-stage federated approach for medical record classification. Similarly, predictive classification modeling FL algorithms also perform the categorization of patients based on their data. However, these algorithms classify patients based on their risk of exposure and disease susceptibility to plan for treatment responses or other clinical outcomes. While other tasks are patient-centric, FL, tasked with BioNER focuses on categorizing diseases and conditions mentioned in EHRs and other medical records. They are used to map diseases and medical conditions and examine existing medical knowledge bases.

6.4.5.7 Learning mode

Based on the learning mode, FL algorithms can be categorized into supervised, semisupervised, and unsupervised learning. The use of labeled data for FL training is known as supervised learning, and linear supervised FL algorithms base their predictions on the linear relationship between input features and target variables. SVM algorithms are also supervised FL algorithms used for tasks such as patient classification or predicting disease outcomes based on distributed data. Another type of supervised FL is decision tree algorithms that incorporate recursive partitioning of data based on features, creating a tree-like structure for decision-making. In-between supervised and unsupervised learning is semisupervised learning, where portions of the data used for training are labeled. FL techniques, such as federated match and logic regression, are common examples. The federated match is useful where labeled data is scarce. Logic regression is a statistical modeling semisupervised FL technique that finds use in scenarios where the relationships between variables are not strictly linear and logic rules can contribute to model interpretability. Lastly, GAN is an unsupervised FL method useful for data augmentation, privacy-preserving data generation, or exploring latent representations of medical data. In this type of learning, labeled data is not used in the training of the model, but the model is encouraged to generate synthetic data and categorization for the dataset without destroying the integrity of the underlying data distribution (Fig. 6.5).

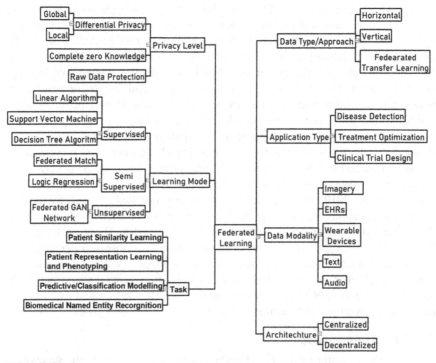

FIGURE 6.5

Proposed taxonomy for FL applications in digital healthcare systems. *FL*, Federated learning.

6.4.6 Future directions and challenges in federated learning within digital healthcare systems

The area of FL in healthcare shows enormous potential for increasing patient care, medical research, and decision-making processes. As this technology continues to improve, various future paths and difficulties must be explored to guarantee its successful adoption and maximize its potential within the healthcare system. This section highlights the projected future paths of FL in healthcare and the significant difficulties that need to be solved.

6.4.6.1 Enhanced privacy-preserving techniques

While FL allows privacy-preserving cooperation, more developments in privacy approaches are essential in handling the expanding world of data privacy legislation and dangers. Future initiatives should concentrate on establishing strong encryption techniques, secure aggregation algorithms, and improved DP measures to preserve patient data throughout the collaborative training process. These upgrades will boost confidence among healthcare organizations and patients, eventually encouraging the widespread use of FL in healthcare.

6.4.6.2 Federated transfer learning

The combination of transfer learning approaches with FL has the potential to improve healthcare applications. Future research should study approaches to transfer information and pretrained models between institutions while maintaining privacy concerns. By exploiting information collected from one institution to boost the performance of models at another, federated transfer learning may increase the efficiency, accuracy, and generalizability of healthcare models learned by FL.

6.4.6.3 Standardization of data representation and exchange

Data heterogeneity is a concern in FL since healthcare datasets can differ in terms of data kinds, formats, and quality. To solve this difficulty, future efforts should concentrate on building standardized data representation formats and sharing protocols particular to healthcare. Establishing common data standards and interoperability frameworks would enable smooth data sharing and aggregation, allowing more robust and trustworthy models across healthcare organizations.

6.4.6.4 Resource optimization and scalability

To assure the scalability and broad use of FL in healthcare, future approaches should research resource optimization strategies. Efficient usage of computing resources, network bandwidth, and energy consumption is crucial for large-scale federated training. Advances in distributed computing, edge computing, and adaptive resource allocation techniques will be crucial in maximizing resource use and addressing the computational needs of FL in healthcare systems.

6.4.6.5 Ethical and regulatory frameworks

As FL continues to be used in healthcare, it is necessary to build detailed ethical principles and regulatory frameworks particular to federated contexts. Future paths should concentrate on building open and responsible governance frameworks, addressing concerns such as data ownership, consent management, algorithmic fairness, and accountability in collaborative decision-making. Ensuring compliance with current healthcare rules and ethical norms will create confidence among stakeholders and reduce any ethical difficulties linked with FL.

6.4.6.6 Addressing communication and network constraints

FL depends on good communication and network infrastructure to support cooperation across scattered healthcare organizations. Future initiatives should examine strategies to solve communication obstacles originating from unpredictable network connections, latency concerns, and capacity restrictions. Innovative technologies, such as adaptive communication protocols, edge computing, and effective data compression techniques, will be important in overcoming these limits and allowing smooth federated training across geographically distant healthcare systems.

6.4.6.7 Validation and interpretability of federated models

Validating and interpreting models taught by FL is critical for getting insights, assuring transparency, and developing confidence in healthcare systems. Future studies should concentrate on establishing strong validation and interpretation tools that can account for the scattered character of FL. This covers ways to assess model performance, discover biases, explain model choices, and evaluate the influence of federated training on model robustness and fairness. Establishing assessment criteria and benchmarks relevant to federated models would allow their reliable application in healthcare.

6.4.7 Lessons learned

In this chapter, we have navigated the fundamentals of FL systems and identified their unique role in digital health systems. While discussing the principal technologies and architectures of FL algorithms, we also identified a plethora of FL algorithms that can be applied to healthcare systems. Thus, there is a need for a categorization of diverse designs, use cases, and adoption of FL systems across different healthcare scenarios. In its simplest form, we have categorized FL-based technologies applied to healthcare systems based on their approach or data type, architecture, application type, data modality, privacy level, task, and learning mode. From the taxonomy proposed, readers can easily select the type of FL method or algorithm suitable for various healthcare applications based on the categories and subcategories presented. Furthermore, the chapter has highlighted the challenges and future trends of FL in healthcare systems. This taxonomy is by no means exhaustive, as the field of FL continually evolves with the development of new techniques, algorithms, and modalities. Thus, readers are to use this taxonomy as a fundamental guide in identifying the type of FL currently in use, especially in healthcare applications. Observing the rapid growth rate in the development and implementation of various FL types, we can also infer that more FL types and categories will be required in the future.

6.5 Conclusion and future scope

In this chapter, we have presented a fundamental examination of FL in healthcare systems. We have covered the fundamental principles, advantages, and concerns related to FL in digital healthcare systems. Through our investigation, we have also uncovered the critical characteristics that contribute to the successful implementation of FL, including data privacy, data heterogeneity, trust and cooperation, and regulatory compliance. We also proposed a taxonomy that methodically evaluates and categorizes the different characteristics of FL in healthcare based on chosen criteria. This taxonomy serves as an important reference point for researchers and practitioners, encouraging a broader grasp of FL and its uses in healthcare contexts. Furthermore, we have underlined the multiple applications and emerging use cases of FL in healthcare. From customized medicine to disease surveillance, medical imaging analysis, pharmaceutical creation, remote monitoring, and clinical decision support systems, FL has the potential to alter patient care, medical research, and decision-making processes.

In conclusion, this chapter contributes to the field of FL in healthcare by presenting a detailed structure and taxonomy. We have shown the potential of FL to revolutionize healthcare systems, enhance patient care, and foster advancement in medical research. Moving forward, it is necessary to focus on establishing privacy-preserving tactics, exploring federated transfer learning systems, standardizing data format and sharing, boosting resource efficiency, and constructing ethical and legal frameworks.

References

Abdulrahman, S., Tout, H., Ould-Slimane, H., Mourad, A., Talhi, C., & Guizani, Mohsen (2021). A survey on federated learning: The journey from centralized to distributed onsite learning and beyond. *IEEE Internet of Things Journal, 8*(7), 5476–5497. Available from https://doi.org/10.1109/JIOT.2020.3030072.

Abikoye, O. C., Oladipupo, T., Imoize, A. L., Awotunde, J. B., Lee, C., & Li, C. (2023). Securing critical user information over the internet of medical things platforms using a hybrid cryptography scheme. *Future Internet, 15*(3), 99. Available from https://doi.org/10.3390/fi15030099.

Act, A. (2023). *Health insurance portability and accountability act. Public Law.*

Allam, Z., & Jones, D. S. (2020). On the coronavirus (COVID-19) outbreak and the smart city network: Universal data sharing standards coupled with artificial intelligence (AI) to benefit urban health monitoring and management. *Healthcare, 8,* 46.

Amin, S. U., & Hossain, M. S. (2020). Edge intelligence and Internet of Things in healthcare: A survey. *IEEE Access, 9,* 45—59.

Asad, M., Moustafa, A., & Ito, T. (2021). *Federated learning versus classical machine learning: A convergence comparison.* Available from http://arxiv.org/abs/2107.10976.

Asad, M., Moustafa, A., & Yu, C. (2020). A critical evaluation of privacy and security threats in federated learning. *Sensors, 20*(24), 7182. Available from https://doi.org/10.3390/s20247182.

Awotunde, J. B., Folorunso, S. O., Imoize, A. L., Odunuga, J. O., Lee, C., Li, C., & Do, D. (2023). An ensemble tree-based model for intrusion detection in industrial internet of things networks. *Applied Sciences, 13*(4), 2479. Available from https://doi.org/10.3390/app13042479.

Awotunde, J. B., Imoize, A. L., Jimoh, R. G., Adeniyi, E. A., Abdulraheem, M., Oladipo, I. D., & Falola, P. B. (2023). *AIoMT enabling real-time monitoring of healthcare systems security and privacy considerations. Handbook of security and privacy of AI-enabled healthcare systems and internet of medical thing* (pp. 97—133). Taylor and Francis. Available from https://doi.org/10.1201/9781003370321-5.

Boukhatem, C., Youssef, H. Y., & Nassif, A. B. (2022). Heart disease prediction using machine learning. 2022 *Advances in* science and engineering technology international conferences *(ASET),* pp. 1—6. https://doi.org/10.1109/ASET53988.2022.9734880

Caldas, S., Duddu, S. M. K., Wu, P., Li, T., Konečný, J., McMahan, H.B., Smith, V., Talwalkar, A. (2018) 'LEAF: A benchmark for federated settings'. Available from http://arxiv.org/abs/1812.01097.

Cao, L., Liu, X., Li, J., Qu, T., Chen, L., Cheng, Y., Hu, J., Sun, J., & Guo, J. (2021). A study of using a deep learning image reconstruction to improve the image quality of extremely low-dose contrast-enhanced abdominal CT for patients with hepatic lesions. *The British Journal of Radiology, 94*(1118), 20201086. Available from https://doi.org/10.1259/bjr.20201086.

Chai, D., Wang, L., Chen, K., & Yang, Q. (2021). Secure federated matrix factorization. *IEEE Intelligent Systems, 36*(5), 11—20. Available from https://doi.org/10.1109/MIS.2020.3014880.

Chen, M., Shlezinger, N., Poor, H. V., Eldar, Y. C., & Cui, S. (2021). Communication-efficient federated learning. *Proceedings of the National Academy of Sciences, 118*(17). Available from https://doi.org/10.1073/pnas.2024789118.

Chen, Y., Qin, X., Wang, J., Yu, C., & Gao, W. (2020). FedHealth: A federated transfer learning framework for wearable healthcare. *IEEE Intelligent Systems, 35*(4), 83—93. Available from https://doi.org/10.1109/MIS.2020.2988604.

Cheng, Y., Liu, Y., Chen, T., & Yang, Q. (2020). Federated learning for privacy-preserving AI. *Communications of the ACM, 63*(12), 33—36. Available from https://doi.org/10.1145/3387107.

Darzidehkalani, E., Ghasemi-rad, M., & van Ooijen, P. M. A. (2022). Federated Learning in Medical Imaging: Part I: Toward Multicentral Health Care Ecosystems. *Journal of the American College of Radiology, 19*(8), 969—974. Available from https://doi.org/10.1016/j.jacr.2022.030.015.

Dimitrov, D. V. (2016). Medical internet of things and big data in healthcare. *Healthcare Informatics Research, 22*(3), 156—163.

Drainakis, G., Pantazopoulos, P., Katsaros, K. V., Sourlas, V., Amditis, A., & Kaklamani, D. I. (2023). From centralized to Federated Learning: Exploring performance and end-to-end resource consumption. *Computer Networks, 225*, 109657. Available from https://doi.org/10.1016/j.comnet.2023.109657.

Fauzi, M. A., Yang, B., & Blobel, B. (2022). Comparative analysis between individual, centralized, and federated learning for smartwatch based stress detection. *Journal of Personalized Medicine, 12*(10), 1584. Available from https://doi.org/10.3390/jpm12101584.

Guo, L., Lei, Y., Xing, S., Yan, T., & Li, N. (2019). Deep convolutional transfer learning network: A new method for intelligent fault diagnosis of machines with unlabeled data. *IEEE Transactions on Industrial Electronics, 66*(9), 7316−7325. Available from https://doi.org/10.1109/TIE.2018.2877090.

Hadzima-Nyarko, M., Nyarko, E. K., Ademović, N., Miličević, I., & Kalman Šipoš, T. (2019). Modelling the influence of waste rubber on compressive strength of concrete by artificial neural networks. *Materials, 12*(4), 561. Available from https://doi.org/10.3390/ma12040561.

Hagos, D. H., Jha, D., Håkegård, J.E., Bagci, U., Rawat, D.B., & Vlassov, V. (2024). *Federated learning for medical applications: A taxonomy, current trends, challenges, and future research directions*, preprint arXiv:2208.03392.

Hao, M., et al. (2020). Privacy-aware and resource-saving collaborative learning for healthcare in cloud computing. in ICC 2020 - 2020 IEEE international conference on communications (ICC). IEEE, pp. 1−6. Available from https://doi.org/10.1109/ICC40277.2020.9148979.

Huang, L., Shea, A., Qian, H., Masurkar, A., Deng, H., & Liu, D. (2019). Patient clustering improves efficiency of federated machine learning to predict mortality and hospital stay time using distributed electronic medical records. *Journal of Biomedical Informatics, 99*, 103291. Available from https://doi.org/10.1016/j.jbi.2019.103291.

Imoize, A. L., Gbadega, P. A., Obakhena, H. I., Irabor, D. O., Kavitha, K. V. N., & Chakraborty, C. (2022). Artificial Intelligence-enabled Internet of Medical Things for COVID-19 pandemic data management. *Explainable Artificial Intelligence in Medical Decision Support Systems*, 357−380.

Imoize, A. L., Hemanth, J., Do, D. T., & Sur, S. N. (2022). Explainable artificial intelligence in medical decision support systems. In A. L. Imoize, & J. Hemanth (Eds.), *Institution of Engineering and Technology*. IET. Available from https://doi.org/10.1049/PBHE050E.

Imoize, A. L., Irabor, D. O., Gbadega, P. A., & Chakraborty, C. (2022). Blockchain technology for secure COVID-19 pandemic data handling. *Smart health technologies for the COVID-19 pandemic: Internet of medical things perspectives*, 141−179. Available from https://doi.org/10.1049/PBHE042E_ch6, Institution of Engineering and Technology.

Jere, M. S., Farnan, T., & Koushanfar, F. (2020). 'A taxonomy of attacks on federated learning'. *IEEE Security & Privacy, 19*(2), 20−28.

Jere, M. S., Farnan, T., & Koushanfar, F. (2021). A taxonomy of attacks on federated learning. *IEEE Security & Privacy, 19*(2), 20−28. Available from https://doi.org/10.1109/MSEC.2020.3039941.

Kang, J., Xiong, Z., Niyato, D., Xie, S., & Zhang, J. (2019). Incentive mechanism for reliable federated learning: A joint optimization approach to combining reputation and contract theory. *IEEE Internet of Things Journal, 6*(6), 10700−10714. Available from https://doi.org/10.1109/JIOT.2019.2940820.

Khan, M.I., Alhoniemi, E., Kontio, E., Khan, S.A., & Jafaritadi, M. (2023). *Differential privacy for adaptive weight aggregation in federated tumor segmentation*. http://arxiv.org/abs/2308.00856

Kim, Y., Sun, J., Yu, H., & Jiang, X. (2017). Federated tensor factorization for computational phenotyping. *Proceedings of the 23rd ACM SIGKDD* international conference on knowledge discovery and data mining, pp. 887−895. https://doi.org/10.1145/3097983.3098118

Kumar, R. L., Wang, Y., Poongodi, T., & Imoize, A. L. (2021). Internet of Things, artificial intelligence and blockchain technology. In R. L. Kumar (Ed.), Cham: Springer International Publishing. Available from https://doi.org/10.1007/978-3-030−74150-1.

Lee, J., Sun, J., Wang, F., Wang, S., Jun, C.-H., & Jiang, X. (2018). Privacy-preserving patient similarity learning in a federated environment: Development and analysis. *JMIR Medical Informatics*, 6(2), e20. Available from https://doi.org/10.2196/medinform0.7744.

Li, D., Han, D., Weng, T. H, Zheng, Z., Li, H., Castiglione, A., & Li, K. C. (2021). Blockchain for federated learning toward secure distributed machine learning systems: A systemic survey. *Soft Computing*, 26(9), 4423−4440. Available from https://doi.org/10.1007/s00500-021-06496-5.

Li, T., Sahu, A. K., Talwalkar, A., & Smith, V. (2020). Federated learning: Challenges, methods, and future directions. *IEEE Signal Processing Magazine*, 37(3), 50−60. Available from https://doi.org/10.1109/MSP.2020.2975749.

Li, Z., Sharma, V., & P. Mohanty, S. (2020). Preserving data privacy via federated learning: Challenges and solutions. *IEEE Consumer Electronics Magazine*, 9(3), 8−16. Available from https://doi.org/10.1109/MCE.2019.2959108.

Lim, W. Y. B., Luong, N. C., Hoang, D. T., Jiao, Y., Liang, Y. C., Yang, Q., Niyato, D., & Miao, C. (2020). Federated learning in mobile edge networks: A comprehensive survey. *IEEE Communications Surveys & Tutorials*, 22(3), 2031−2063. Available from https://doi.org/10.1109/COMST.2020.2986024.

Liu, D., Dligach, D., & Miller, T. (2019). *Two-stage federated phenotyping and patient representation learning*. http://arxiv.org/abs/1908.05596

Liu, M., Ho, S., Wang, M., Gao, L., Jin, Y., Zhang, H. (2021). *Federated* learning meets natural language processing: *A survey*. Available from https://arxiv.org/abs/2107.12603.

Mar'i, F., Supianto, A. A., & Bachtiar, F. A. (2023). Comparison of federated and centralized learning for image classification. *PIKSEL: Penelitian Ilmu Komputer Sistem Embedded and Logic*, 11(2), 393−400. Available from https://doi.org/10.33558/piksel.v11i2.7367.

Massaoudi, M., Abu-Rub, H., Refaat, S. S., Chiihi, I., & Oueslati, F. S. (2021). Deep learning in smart grid technology: A review of recent advancements and future prospects. *IEEE Access*, 9, 54558−54578. Available from https://doi.org/10.1109/ACCESS.2021.3071269.

Moshawrab, M., Adda, M., Bouzouane, A., Ibrahim, H., & Raad, A. (2023). Reviewing federated machine learning and its use in diseases prediction. *Sensors*, 23(4), 2112. Available from https://doi.org/10.3390/s23042112.

Nguyen, D. C., Ding, M., Pham, Q. V., Pathirana, P. N., Le, L. B., Seneviratne, A., Li, J., Niyato, D., & Poor, H. V. (2021). Federated learning meets blockchain in edge computing: Opportunities and challenges. *IEEE Internet of Things Journal*, 8(16), 12806−12825. Available from https://doi.org/10.1109/JIOT.2021.3072611.

Oyebiyi, O. G., Abayomi-Alli, A., Arogundade, O. T., Qazi, A., Imoize, A. L., & Awotunde, J. B. (2023). A systematic literature review on human ear biometrics: Approaches, algorithms, and trend in the last decade. *Information*, *14*(3), 192. Available from https://doi.org/10.3390/info14030192.

Pfitzner, B., Steckhan, N., & Arnrich, B. (2021). Federated learning in a medical context: A systematic literature review. *ACM Transactions on Internet Technology*, *21*(2), 1−31. Available from https://doi.org/10.1145/3412357.

Posner, J., Tseng, L., Aloqaily, M., & Jararweh, Y. (2021). Federated learning in vehicular networks: Opportunities and solutions. *IEEE Network*, *35*(2), 152−159. Available from https://doi.org/10.1109/MNET.011.2000430.

Qadri, Y. A., Nauman, A., Zikria, Y. B., Vasilakos, A. V., & Kim, S. W. (2020). The future of healthcare internet of things: a survey of emerging technologies. *IEEE Communications Surveys & Tutorials*, *22*(2), 1121−1167.

Qian, F., & Zhang, A. (2021). The value of federated learning during and post-COVID-19. *International Journal for Quality in Health Care*, *33*(1). Available from https://doi.org/10.1093/intqhc/mzab010.

Rahman, A., Hossain, M. S., Muhammad, G., Kundu, D., Debnath, T., Rahman, M., Khan, M. S. I., Tiwari, P., & Band, S. S. (2023). Federated learning-based AI approaches in smart healthcare: Concepts, taxonomies, challenges and open issues. *Cluster Computing*, *26*(4), 2271−2311. Available from https://doi.org/10.1007/s10586-022-03658-4.

Ramasamy, L. K., KP, F. K., Imoize, A. L., Ogbebor, J. O., Kadry, S., & Rho, S. (2021). Blockchain-based wireless sensor networks for malicious node detection: A survey. *IEEE Access*, *9*, 128765−128785. Available from https://doi.org/10.1109/ACCESS.2021.3111923.

Ramesh, N., & Tasdizen, T. (2019). Cell segmentation using a similarity interface with a multi-task convolutional neural network. *IEEE Journal of Biomedical and Health Informatics*, *23*(4), 1457−1468. Available from https://doi.org/10.1109/JBHI.2018.2885544.

Rieke, N., Hancox, J., Li, W., Milletari, F., Roth, H. R., Albarqouni, S., Bakas, S., Galtier, M. N., Landman, B. A., Maier-Hein, K., & Ourselin, S. (2020). 'The future of digital health with federated learning'. *NPJ Digital Medicine*, *3*(1), 119.

Roscher, R., Bohn, B., Duarte, M. F., & Garcke, J. (2020). Explainable machine learning for scientific insights and discoveries. *IEEE Access*, *8*, 42200−42216. Available from https://doi.org/10.1109/ACCESS.2020.2976199.

Rufai, A. U., Fasina, E. P., Uwadia, C. O., Rufai, A. T., & Imoize, A. L. (2023). *Cyberattacks against Artificial Intelligence_Enabled Internet of Medical Things. Handbook of security and privacy of AI-enabled healthcare systems and internet of medical things* (pp. 191−216). Taylor and Francis.

Ryffel, T., Trask, A., Dahl, M., Wagner, B., Mancuso, J., Rueckert, D., Passerat-Palmbach, J. (2018). *A generic framework for privacy preserving deep learning*. Available from http://arxiv.org/abs/1811.04017.

Salem, R. M. M., Saraya, M. S., & Ali-Eldin, A. M. T. (2022). An industrial cloud-based IoT system for real-time monitoring and controlling of wastewater. *IEEE Access*, *10*, 6528−6540. Available from https://doi.org/10.1109/ACCESS.2022.3141977.

Savazzi, S., Nicoli, M., & Rampa, V. (2020). Federated learning with cooperating devices: A consensus approach for massive IoT networks. *IEEE Internet of Things Journal*, *7*(5), 4641−4654. Available from https://doi.org/10.1109/JIOT.2020.2964162.

Sei, Y., Ohsuga, A., Onesimu, J. A., & Imoize, A. L. (2023). *Local Differential privacy for artificial intelligence of medical things. Handbook of security and privacy of AI-enabled healthcare systems and internet of medical things* (pp. 241−270). Taylor and Francis. Available from https://doi.org/10.1201/9781003370321-10.

Shaheen, M., Farooq, M. S., Umer, T., & Kim, B. S. (2022). Applications of federated learning; taxonomy, challenges, and research trends. *Electronics, 11*(4), 670. Available from https://doi.org/10.3390/electronics11040670.

Shi, X., He, J., Lin, M., Liu, C., Yan, B., Song, H., Wang, C., Xiao, F., Huang, P., Wang, L., & Li, Z. (2021). Comparative effectiveness of team-based care with a clinical decision support system versus team-based care alone on cardiovascular risk reduction among patients with diabetes: Rationale and design of the D4C trial. *American Heart Journal, 238*, 45−58. Available from https://doi.org/10.1016/j.ahj.2021.040.009.

Steinberg, E., Jung, K., Fries, J. A., Corbin, C. K., Pfohl, S. R., & Shah, N. H. (2021). Language models are an effective representation learning technique for electronic health record data. *Journal of Biomedical Informatics, 113*, 103637. Available from https://doi.org/10.1016/j.jbi.2020.103637.

Sun, X., Bommert, A., Pfisterer, F., Rähenfürher, J., Lang, M. and Bischl, B. (2020). 'High dimensional restrictive federated model selection with multi-objective Bayesian optimization over shifted distributions', in *Proceedings of SAI Intelligent Systems Conference*, pp. 629−647. Available from https://doi.org/10.1007/978-3-030-29516-5_48.

Tanuwidjaja, H. C., Choi, R., Baek, S., & Kim, K. (2020). Privacy-preserving deep learning on machine learning as a service—A comprehensive survey. *IEEE Access, 8*, 167425−167447.

Uppal, M., Gupta, D., Goyal, N., Imoize, A. L., Kumar, A., Ojo, S., Pani, S. K., Kim, Y., & Choi, J. (2023). A real-time data monitoring framework for predictive maintenance based on the Internet of Things. In M. Ahmadieh Khanesar (Ed.), *Complexity* (2023, pp. 1−14). Hindawi. Available from https://doi.org/10.1155/2023/9991029.

Voigt, P., & dem Bussche, A. (2017). *'The EU general data protection regulation (GDPR)', A Practical Guide* (1st Ed., p. 10), Cham: Springer International Publishing.

Wang, C., Yang, G., Papanastasiou, G., Zang, H., Rodrigues, J. J., & De Albuquerque, V. H. C. (2021). Industrial cyber-physical systems-based cloud IoT edge for federated heterogeneous distillation. *IEEE Transactions on Industrial Informatics, 17*(8), 5511−5521. Available from https://doi.org/10.1109/TII.2020.3007407.

WeBank. (2018) *Fate: An industrial grade federated learning framework.*

Yu, S., Chen, X., Zhou, Z., Gong, X., & Wu, D. (2021). When deep reinforcement learning meets federated learning: Intelligent multitimescale resource management for multiaccess edge computing in 5G ultradense network. *IEEE Internet of Things Journal, 8*(4), 2238−2251. Available from https://doi.org/10.1109/JIOT.2020.3026589.

Zhang, C., Xie, Y., Bai, H., Yu, B., Li, W., & Gao, Y. (2021). A survey on federated learning. *Knowledge-Based Systems, 216*, 106775. Available from https://doi.org/10.1016/j.knosys.2021.106775.

Zhang, P., Wang, C., Jiang, C., & Han, Z. (2021). Deep reinforcement learning assisted federated learning algorithm for data management of IIoT. *IEEE Transactions on Industrial Informatics, 17*(12), 8475−8484. Available from https://doi.org/10.1109/TII.2021.3064351.

Zhang, T., Feng, T., Alam, S., Lee, S., Zhang, M., Narayanan, S., & Avestimehr, S. (2022). *FedAudio: A* federated learning benchmark for audio tasks.

Zhang, W., & Li, X. (2022). Federated transfer learning for intelligent fault diagnostics using deep adversarial networks with data privacy. *IEEE/ASME Transactions on Mechatronics*, *27*(1), 430−439. Available from https://doi.org/10.1109/TMECH.2021.3065522.

Zhu, X., Wang, J., Hong, Z., Xia, T. & Xiao, J. (2019). Federated learning of unsegmented Chinese Text recognition model. In 2019 IEEE 31st International Conference on Tools with Artificial Intelligence (ICTAI). IEEE, pp. 1341−1345. Available from https://doi.org/10.1109/ICTAI.2019.00186.

IoHT-FL model to support remote therapies for children with psychomotor deficit

7

Jaime Muñoz-Arteaga[1], María Libertad Aguilar Carlos[1] and José Rafael Rojano-Cáceres[2]

[1]Information System Department, Autonomous University of Aguascaleintes,
Aguascalientes, Mexico
[2]Faculty of Statistics and Informatics, University of Veracruz, Xalapa, Mexico

7.1 Introduction

A family unit that includes one or more members with impairments is a unique system that requires regulations that go beyond the medical or psychological realms. It also requires orientation that allows for the ongoing treatment of the persons with disabilities within the family unit, particularly in the early stages of life when a psychomotor impairment is present. In this sense, the follow-up of the comprehensive management of children with disabilities should be supported by the appropriate medical and educational assessments for their development, as well as additional support for their families in the form of technological resources and the necessary knowledge to use them for analytical and predictive purposes. The 2030 Agenda for Sustainable Development emphasizes that expanding information and communication technologies and global interconnection offer great possibilities to accelerate human progress, overcome the digital divide, and develop knowledge societies (World Health Assembly WHO, 2023).

Remote psychomotor therapy, also known as online therapy or teletherapy, combines elements of psychotherapy and movement therapy delivered remotely through video conferencing or other forms of online communication. Psychomotor therapy involves physical movement and sensory awareness to explore and address emotional and psychological issues. Remote psychomotor therapy aims to provide individuals with the same level of care and support as traditional in-person therapy while also offering the convenience and accessibility of receiving treatment from the comfort of home. In remote psychomotor therapy, a trained therapist guides the patient through various movement exercises and encourages them to express their feelings and emotions through physical movements. This can include activities such as stretching, breathing exercises, and body awareness exercises, as well as more dynamic movements such as dance or yoga. Remote psychomotor therapy has become increasingly popular in recent years (Venturo-Conerly et al., 2022), especially in light of the COVID-19 pandemic (Watson et al., 2023), which has made in-person therapy more difficult or impossible for some people (Appleton et al., 2021; Barnett et al., 2021). However, it is essential to note that remote therapy may not be suitable for everyone, and individuals should discuss their specific needs and concerns with a mental health professional to determine if it is a good fit for them.

There is no doubt that coordinating the provision of services to the patient comprehensively poses a significant challenge. Nevertheless, COVID-19 has brought about a transformation, forcing organizations to establish, for the first time, a system for collecting and integrating electronic medical records across the healthcare ecosystem. This has facilitated the remote connection of medical specialists with patients, providing medical professionals with benefits such as infrastructure management, growth solutions, and customization of services. This enables individuals to get more information about their bodies, helping to maintain or identify health problems (Barbuti et al., 2020).

Given the above context, the current approach proposes an alternative solution. Accordingly, the chapter is divided into the following sections: Section 7.2 provides the background, containing the definitions of related technologies to be used. Section 7.3 discusses related works that propose a healthier system of care using federated learning (FL). Section 7.4 talks about the problem, including works based on the technologies that make up this proposal. Section 7.5 recommends an architectural model that describes the general components of FL to support distance therapies for children with psychomotor deficits. Section 7.6 presents a case study applying various Internet of Health Things (IoHT) technologies. Section 7.7 provides a broader view of how this approach can be used beyond the specific study. The last section presents the conclusions of the research work.

7.2 Background

7.2.1 Cloud computing

The distribution of computational resources called cloud computing, which provides infrastructure, services, platforms, and applications as required in the networks, is increasingly being used. The benefits of using cloud computing are that there is no need for the user to know its infrastructure. It becomes an abstraction, "a cloud" where applications and services can quickly grow to become more efficient, reliable, and transparent (Maksimović & Vujović, 2017). Cloud computing is an action that is in charge of executing a specific workload in a cloud. The design and use of clouds employ technological elements such as software and hardware systems.

7.2.2 **Federated learning**

FL is a method of machine learning (ML) that trains an ML algorithm with local data samples distributed over multiple edge devices or servers without any exchange of data (Zhang et al., 2021). It allows multiple nodes to collaboratively create a learning model without the necessity of exchanging their data samples. A typical architecture in an IoT system consists of three layers of hierarchy, where the middle layer is data storage in distributed learning, and in FL, this layer refers to a training model (Wen et al., 2023). FL introduces an innovative training method for constructing personalized models without compromising user privacy. This is achieved through the exchange of encrypted processed parameters, ensuring that attackers cannot access the source data (Konečný et al., 2016).

According to Phd and Jeno (2022), there are two types of FL, horizontal and vertical. In horizontal FL the local datasets have the same features but with different samples. While in the vertical FL, the datasets have different features for the same samples.

For example, in a horizontal approach, we can define that each mobile device has its own local data corresponding to each patient and their therapies. On the local devices, an algorithm is trained to acquire parameters specific to the patient and the therapy. At regular intervals the parameters from the training model are transmitted to the server, where an aggregation process, such as averaging, is used to generate a global model. Most importantly, this is done without compromising the data from patients. Once a new global model is established, it is distributed between the devices to leverage specific tasks from collective information.

On the other hand, in a vertical FL, we could imagine a therapy process where data is captured from an electromyography sensor during the therapy sessions. Initially, the data is categorized according to attributes or relevant features, wherein each local process is responsible for the selected dimension. In this way, each device runs a training model that is specialized in learning its dimension. Afterward, each device sends the parameters from the model so that the global model can be updated by aggregating this knowledge. Finally, the result of the aggregation is a new global model that can be distributed between the devices to conduct specific therapies.

7.2.3 **Children with disabilities**

Children with disabilities face unique challenges in their daily lives. They may encounter physical, cognitive, or sensory limitations affecting their mobility, communication, or learning abilities. It is essential to provide them with equal opportunities, promote their independence, and foster a supportive and inclusive environment that values their strengths and unique perspectives. Thus, with the right resources, inclusive education, and specialized assistance, children with disabilities can thrive and reach their full potential.

7.2.4 **COVID-19 contingency**

The world has been experiencing a health contingency since March 2019, originating from a disease known as COVID-19, caused by the SARS-CoV-2 virus (Nguyen et al., 2021). As of 2023, the pandemic has officially come to an end. The consequences worldwide have been catastrophic in the health, economy, industry, and education sectors (WHO, 2020) (Dasaradharami Reddy & Gadekallu, 2023). One of the most significant global repercussions of the pandemic has been the widespread closure of schools, resulting in the loss of learning, increased rates of school dropout, and heightened inequity, including economic crisis in households, contributing to a lower educational attainment level (Shaheen et al., 2022a). The sudden confinement measures suggested a change for all stakeholders in the educational sector—schools, teachers, and students—involving both children and adolescents. The shift involved the adoption of information and communication technologies as tools to adapt to the new ways of working, impacting both teachers and their students. In inclusive education, children suffer from stagnation due to this abrupt change, particularly students with learning deficits or special educational needs, as they require specialized attention from teachers. For students facing challenges such as language disorder, autism or attention deficit, and hyperactivity, discontinuing classes and the corresponding support they receive to improve their symptoms represents a setback in their education.

7.2.5 **Psychomotor deficit**

The monitoring plan for psychomotor development is considered a critical mission for ensuring early childhood health (Shaheen et al., 2022a). It helps in the timely detection of a possible diagnostic of psychomotor deficit observed from the prenatal phase through postnatal and infantile time, facilitating the correction or attenuation of some alterations. Psychomotor deficit is a clinical manifestation of pathologies of the central nervous system distorted by genetics or the environment, affecting children's psychomotor development, especially in the first months of life (24−26 months). This results in changes that affect progress in motor skills, language, manipulation, and social areas (Manickam et al., 2022). Some features of psychomotor deficit present in children are muscle weakness, abnormal muscle tone, decreased range of motion of the joints, and decreased balance and coordination (Shaheen et al., 2022b). In such instances, brain plasticity plays a role in reducing a part of the alterations through rehabilitation procedures. Brain plasticity involves the reorganization of the neural capacity to compensate for or restore lost functions. Therefore with a correct diagnosis, a set of rehabilitation therapy experiences can modify neural circuits because of brain plasticity. In most cases this often leads to changes in efficacy and the number of synaptic connections (Zhang et al., 2021).

7.2.6 **Internet of Health Things**

The IoHT has the potential to revolutionize healthcare by enabling new ways of monitoring, diagnosing, and treating a variety of conditions (Firouzi et al., 2020). Some examples of IoT devices that can assist children with psychomotor deficits are discussed below. It is important to note that the effectiveness of these devices can vary depending on the specific needs of each child and the severity of their condition. It is always best to consult a healthcare professional to determine the best approach for each individual.

7.2.6.1 *Smart glasses*

These are wearable devices that can help children with visual impairments see more clearly. Reality virtual headsets are possible to take into account. They can also be used to track eye movements and detect abnormalities that may be related to psychomotor deficits.

7.2.6.2 *Smart sensors*

These are small, wireless devices that can be attached to clothing or other objects to track movement and activity. They can help children with psychomotor deficits improve their balance and coordination by providing feedback on their movements.

7.2.6.3 *Virtual reality therapy*

Virtual reality can provide therapy for children with psychomotor deficits by simulating real-life scenarios and environments. This can help children to improve their motor skills and coordination.

7.2.6.4 *Wearable trackers*

These devices can be worn on the wrist or ankle to track activity levels, heart rate, and other metrics. They can monitor physical activity and help children with psychomotor deficits stay active and healthy.

7.2.6.5 *Smart toys*

These are interactive toys that can help children with psychomotor deficits improve their cognitive and motor skills. They can also be used to provide feedback on progress and encourage children to engage in physical activity.

7.2.6.6 *Smart home devices*

Smart home devices, such as voice-activated assistants and smart lighting, can be used to make it easier for children with psychomotor deficits to navigate their environment. They can also be used to provide reminders and prompts to help children stay on track with their therapy and other activities.

7.3 Related works

There are existing works (Matsangidou et al., 2020; Shaheen et al., 2022b) related to the application of FL in different domains such as industry, education, finances, and health. However, there are not many studies in the domain of remote psychomotor therapies. Table 7.1 shows some related works.

Matsangidou et al. (2020) developed a technology supporting remote therapies using virtual reality; however, the study did not make use of FL and IoHT. Dasaradharami Reddy & Gadekallu (2023) proposed utilizing IoHT for remote health monitoring, particularly for COVID-19 detection. The work by Maksimović & Vujović (2017) also incorporated biomedical-assisted methods, but neither of these studies considered the potential advantages of FL. It is crucial to highlight that the current approach is more comprehensive, addressing the future requirements for the development of remote psychomotor therapy for children. In addition, several papers (Kholod et al., 2021; Shaheen et al., 2022b) discuss the taxonomy, challenges, and research trends related to FL applications in healthcare. However, there is a notable absence of applications in the specific domain of remote therapies in psychomotor development.

7.4 Problem outline

Traditional health services have difficulties ensuring the quality of services demanded by people with population growth (Tunç et al., n.d.) and lack of medical staff, resulting in late diagnoses and inadequate treatment processes for patients. Several papers in the literature on e-learning (Mortis-Lozoya et al., 2017; Dubey et al., 2023) and software engineering (Yang et al., 2019; Martin, 2017) have presented successful experiences using technologies based on a cloud approach to assist remote therapies. Unfortunately, in underdeveloped countries, there is a lack of infrastructure and computational resources for this type of therapy, whereas most interactive health applications designed to support distance modality do not provide effective use.

On the other hand, a significant problem is encountered when discussing therapies for children with disabilities, specifically focusing on the resistance exhibited by parents, caregivers, and participants toward distance psychological, motor, and early education therapies. This resistance toward remote therapies is due to several factors, including limited technological literacy, privacy concerns, skepticism about the efficacy of remote interventions, and perceived lack of personal connection compared with in-person sessions.

In the case of data-driven approaches, people are reluctant to adopt technological devices that capture and store data in the cloud, primarily due to concerns about data security, potential misuse of sensitive information, and lack of trust in data handling processes. However, it is necessary to propose new ways of

Table 7.1 Collection of works related to therapies, Internet of Health Things (IoHT), and federated learning.

Criteria	Matsangidou et al. (2020)	Dasaradharami Reddy and Gadekallu (2023)	Shaheen et al. (2022b)	Maksimović and Vujović (2017)	Kholod et al. (2021)	Current approach
Area	Eating disorders	Covid-19 detection	Multiple areas	Biomedical systems	Multiple areas	Motor and cognitive disabilities
Patients	Young people	For all	For all	For all	For all	Children
Therapies	Remote psychotherapy	Remote health monitoring.	Remote health monitoring.	Remote health monitoring.	N/A	Remote psychomotor therapy
IoHT		IoHT	IoT	IoHT	IoT	IoHT
Technology	Multiuser VR	Smart healthcare applications	Smart healthcare applications	Smart healthcare applications	Smart applications	Smart healthcare applications
Federated learning			X		X	X

performing psychomotor therapies that enhance both the therapist's effectiveness and the quality of patient care.

This highlights the potential for enhancing the effectiveness of therapies for children with disabilities, emphasizing the implications for therapeutic interventions and the potential benefits of adopting distance service delivery. This is relevant not only in health emergencies such as COVID-19 but also in broader contexts.

7.5 Federated learning architectural model with Internet of Health Things

The current work proposes FL as an architectural model for the IoHT to support remote therapies for children with psychomotor deficits (See Fig. 7.1). It is important to note that the digital transformation of healthcare can be disruptive; however, technologies such as the Internet of Things, virtual assistance, remote

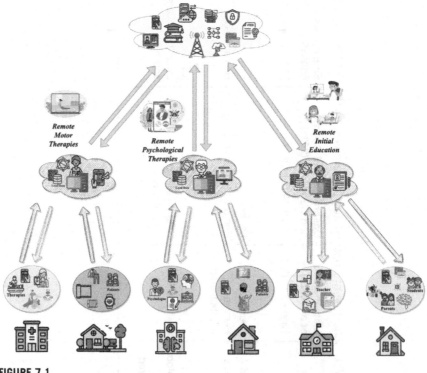

FIGURE 7.1

Federated learning architectural model for the Internet of Health Things to support remote therapies for children with psychomotor deficits.

monitoring, artificial intelligence, big data analytics, blockchain, smart wearables, and platforms are tools that enable the exchange and storage of data. They are tools that allow remote data capture and data and information exchange within the health ecosystem, leading to continuity, improving medical diagnosis, data-based treatment decisions, digital therapies, clinical trials, self-management, and people-centered care, as well as expanding evidence-based knowledge, skills, and competencies of professionals to deliver health services (World Health Assembly, WHO, 2023).

According to Fig. 7.1, remote motor and psychological therapies, as well as inclusive education are possible because of FL. FL makes online services, cloud-based content, and the use of algorithms for personalizing remote therapy online services feasible. Some benefits of this architectural model are as follows:

- Facilitate communication among stakeholders in therapy services, which include parents, therapists, managers, doctors, specialists, and even children involved in the process.
- Assist parents or caregivers in managing tailored and personalized care, incorporating therapies for their sons and daughters.
- Serve as a guide for the rehabilitation process or diagnosis of psychomotor deficit levels.
- Implement remote therapies with the support of virtual meeting platforms and other digital tools.
- Develop schedules that integrate both remote and in-person services, including therapies, medical consultations, evaluations, training courses, and other relevant activities.
- Ensure privacy and safety for local data throughout the aforementioned processes, while models are trained and distributed between participants to achieve specific goals.

One of the common behavioral therapy techniques applied to the treatment of various psychological problems, especially those related to fear and anxiety, is exposure therapy. Through repeated exposures to a feared stimulus, new learning is achieved in relation to the threat, which in turn leads to better management of anxiety and fear. During remote psychological therapy sessions, individuals typically meet with a licensed mental health professional, such as a psychologist or therapist, who uses various therapeutic techniques and interventions to help them address their mental health concerns. By connecting people, resources, data, and solutions, the healthcare of the future will be able to streamline operations and reduce risk, while increasing operational efficiency and profitability. The field can harness the potential of remote therapies and data-driven approaches to provide accessible and personalized interventions for children with disabilities. This kind of remote therapy may involve discussing past experiences, exploring thoughts and emotions, setting goals, and developing coping strategies. The specific approach used may vary depending on the individual's needs and the therapist's training and expertise.

7.5.1 **Remote motor therapy with Internet of Health Things**

Home-based healthcare provides a viable and cost-effective method of delivery for resource- and labor-intensive therapies, such as rehabilitation therapies. Home healthcare provides a simple extension of the clinical sessions to enable patients to be active participants in their own treatment; this is particularly attractive for low-risk but resource-intensive rehabilitation therapies. The rapid adoption of telemedicine during the COVID-19 global pandemic has also seen a surge in consumer demand for technologies to support treatment from home. However, despite its benefits, home healthcare has seen slow adoption into clinical practice due to provider concerns over patient compliance and optimal performance of therapies at home. To truly integrate clinics with patients in patient-centered care, there is an opportunity to leverage emerging technologies such as the IoT (Matsangidou et al., 2020) to address these barriers to adoption. IoT offers a seamless platform where digital devices can make sense of information without human intervention.

Remote motor therapy for children with disabilities is a therapeutic approach to provide motor intervention and enhance the physical development of children. It offers some advantages due to its accessibility and convenience. In a remote session, a therapist specializing in occupational therapy or physical therapy works with the child through video conferencing or other technological communication tools in real time. The therapeutic session is tailored to the child's specific needs and focuses on developing motor skills, strengthening muscles, improving balance and coordination, and promoting functional independence. The therapist guides the child through a series of exercises and activities designed to address his/her motor challenges and promote the development of specific skills, providing verbal instructions, visual demonstrations, and using additional resources, such as pictures or videos, to facilitate the child's understanding and active participation, with parents or caregiver's support if necessary. To maintain a child's interest and motivation, it may include interactive online games, rhythmic movement activities, ball exercises, balance and coordination activities, and other playful activities tailored to the child's needs and abilities. During remote motor therapy, the therapist monitors the child's progress, provides ongoing feedback, and adjusts exercises and activities as needed. Parents or caregivers could offer recommendations for home practice and provide additional resources to ensure continuity of treatment outside of sessions.

7.5.2 **Remote psychological therapy with Internet of Health Things**

During remote psychological therapy sessions, individuals typically meet with a licensed mental health professional, such as a psychologist or therapist, who uses various therapeutic techniques and interventions to help them address their mental health concerns. These sessions might encompass conversations about past experiences, delving into thoughts and emotions, establishing goals, and formulating coping strategies. The particular approach employed can differ based on the individual's requirements and the therapist's training and expertise.

One of the common behavioral therapy techniques applied to the treatment of various psychological problems, especially those related to fear and anxiety, is exposure therapy. Through repeated exposures to a feared stimulus, new learning is achieved in relation to the threat, which in turn leads to better management of anxiety and fear.

7.5.3 Remote initial education with Internet of Health Things

In a remote pedagogical class designed for children with disabilities at the initial stage of education, technology, and distance communication are employed to deliver customized teaching and educational support tailored to each child's needs. The approach is rooted in John Dewey's theory that learning is not only about acquiring theoretical knowledge but also about developing practical real-world object skills and promoting personal growth. This approach can be justified in the context of planning remote initial education therapies using devices with sensors and information-gathering algorithms to enrich the knowledge of education professionals in several ways, including the integration of theoretical and practical knowledge that allows to obtain objective and quantitative information about children's performance in different areas, such as motor, cognitive, and social skills. This information complements theoretical knowledge and helps professionals better understand each child's individual needs and strengths, enabling them to plan more effective and personalized interventions. The focus on integral development involves thoroughly and accurately assessing progress and designing intervention strategies that holistically address the diverse needs of children with disabilities. The feedback is based on the objectives set for the collected data in remote therapy.

These remote classes are specially designed to address the challenges that children with disabilities may face in their learning process and ensure their participation and inclusion in the educational environment. Also, through online platforms, video conferences, or other communication tools in real time, specialized educators work directly with the child and their parents or caregivers. The goal is to provide an individualized education that meets the child's abilities and needs and fosters their academic development and learning skills.

Some of the common activities may include educational lessons by teachers with adapted resources to foment the child's participation; tasks and activities adapted to the needs and abilities of the child with exercises to promote the physical, cognitive, social, and emotional development, for example, sensorial stimuli; motor exercises, such as crawling, rolling, climbing, jumping, throwing, or in reading, writing, basic mathematics; and imitation and symbolic games; visual and auditory discrimination; language and communication; and of course, the socialization. In addition to the classes, the teacher may work closely with the child's parents or caregivers, providing additional guidance and resources to support learning at home. This may include specific strategies, work patterns, and recommendations for adapting the study environment to the child's needs. It is important to emphasize that remote pedagogical classes for children with disabilities must be flexible and adapt to the individual needs of each child. Constant communication between the educator, the child, and their parents or caregivers is essential to ensure an effective and successful teaching-learning process.

7.6 Real-world scenario

Implementation of the proposed architectural model allows the children's educational and rehabilitative development process to be modified in accordance with the constant changes. Feedback sessions are essential, and a group of professionals, parents, and children is taken into consideration to create a learning ecosystem that is continuously evolving and improving. This real scenario has been presented in APAC (2023) a Mexican institution for cerebral palsy care which was attending 30 children with autism spectrum disorder level III at the age of 4. Though they are identified as the same chronological age as 4-year-olds, they have different neurological ages in manual, verbal, and tactile abilities that condition them in mobility and social development. They are patients with psychomotor deficits. They lived in the desert villages located in the northern region of the state of Zacatecas, far from the capital city of Mexico. Hence, the architectural approach was applied in conjunction with experts and institutions such as the APAC Association, an institution dedicated to treating children and youth with disabilities with services of physiotherapy, hydrotherapy, occupational therapy, language therapy, and psychological and pedagogical sessions. In this scenario, the FL model was applied remotely to three main services: psychological, motor (physiotherapy), and initial education (pedagogical sessions).

7.6.1 Remote psychological therapy

During remote sessions of educational psychology therapy aimed at improving cognitive intelligence, some children were diagnosed with cerebral palsy. Some materials and resources with incorporated sensors were used. These materials are instruments or toys of different colors and shapes, serving the purpose of tracking instructions, recognizing shapes and colors, and aiding in memory learning, among other objectives. FL was used to gather information through the rehabilitation equipment, which enhanced the quality of the remote therapy service provided.

The remote therapy session begins with a video conferencing platform. The therapist ensures that the technology is properly set up and that both parties can see and hear each other. The session is personalized to address the specific needs and goals of the child, considering their unique abilities and challenges associated with cerebral palsy (Fig. 7.2).

The therapist introduces the interactive sensor-based toys to the child and the adults who accompany them. These toys are equipped with sensors that capture data regarding the child's interaction, such as their responses to instructions, recognition of shapes and colors, and memory learning. The therapist explains the purpose of the toys and guides the child and parents on how to use them effectively, presenting various tasks and activities designed to improve cognitive

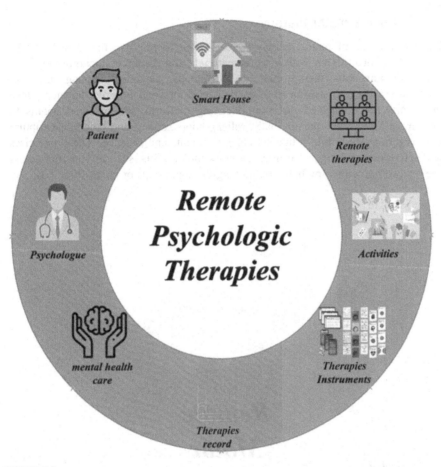

FIGURE 7.2

Set of FL services for remote psychological therapies. *FL*, Federated learning.

abilities. The data collected by the sensors are used to personalize the profile, which generates a trained model. The model parameters are transmitted securely through FL (Ludwig & Baracaldo, 2022), allowing other therapists to receive an improved model as a result of the individual parameters from different patients for analysis without compromising children's privacy. The learning model can be updated to monitor the progress of the children, identify areas of improvement, and customize therapy planning accordingly. The families of the children may receive an update on their progress, along with recommendations for continued practice and learning at home.

The session is concluded with a warm and encouraging note, emphasizing the child's strengths and potential for growth. The therapist schedules the next session and ensures that the necessary materials and resources are readily available for continued remote therapy.

7.6.2 **Remote motor therapy**

Remote sessions of motor therapy were planned to improve the gross and fine motor skills of children with various materials and resources incorporating sensors, including mats, therapy balls, rollers, hoops, and cones, among others, to enhance crawling, transition to a standing position, and improve balance. Once both parts are connected on a conferencing platform, the therapist begins to encourage using mats, therapy balls, rollers, hoops, cones, and others that capture data regarding the child's interaction, movement, and balance. These exercises and activities are designed to improve gross motor skills, such as crawling across the mat, rolling a therapy ball, and moving through hoops or cones (Fig. 7.3).

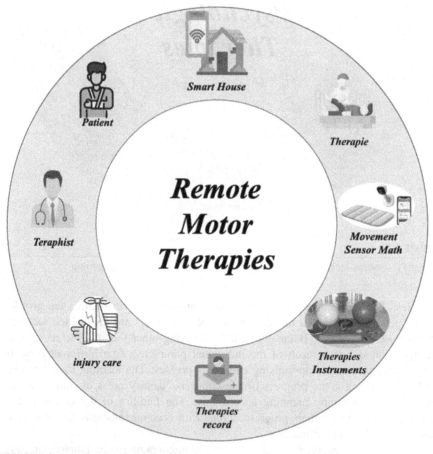

FIGURE 7.3

Set of FL services for remote motor therapies. *FL,* Federated learning.

Additionally, they may incorporate fine motor activities, such as picking up objects or manipulating small items. Progress is summarized during the session, and schedules and recommendations are provided for continued practice and exercises at home. The use of materials with sensor movements promotes their physical development and enhances balance, coordination, and overall motor abilities, providing them with valuable support and guidance in their motor skills journey.

7.6.3 Remote initial education

Didactic materials with sensors also serve the purpose of enhancing reading, writing, and logical and mathematical skills. Therefore remote initial education therapy sessions begin with interactive books, writing tablets, number cards, or other educational

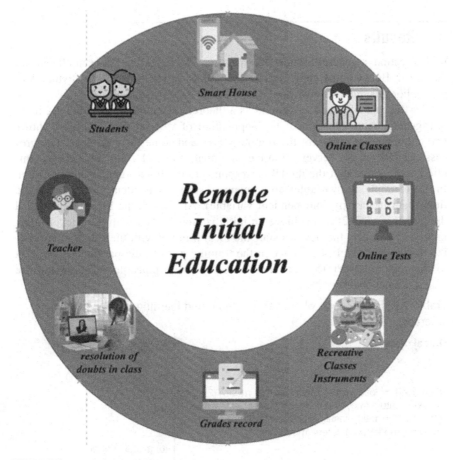

FIGURE 7.4

Set of FL services for remote initial education. *FL,* Federated learning.

tools equipped with sensors that capture data regarding the child's interaction and progress in various activities, such as involvement in reading passages and answering questions, practicing letter formation or word recognition, or working on simple addition or subtraction problems. The remote early education therapy sessions offer a personalized and interactive approach for improving the pedagogical skills of children with cerebral palsy. It promotes their cognitive development and enhances their reading, writing, and basic mathematical abilities, providing them with valuable support and guidance in their educational journey (Fig. 7.4).

Provide families with resources, including seeds for planting, pots for gardening, paper and pencils for creative activities, and open-ended play materials. This initiative aims to support children in families, particularly those in disadvantaged areas, who may not have many materials at home to play with.

7.7 **Results**

In this section, a general pseudocode specifies the interaction between clients and the server. Below is the implementation of the said pseudocode in the Python language. For this, the flower frame (Beutel et al., 2020) has been used.

For the case of pseudocode, and as a theoretical example, in Table 7.2 there is a general representation of the application of vertical FL. The table shows two columns that represent the remote server and some local devices. The process starts when the server initializes a global model (1). For the sake of simplicity, we define that the model corresponds to the linear regression algorithm. In this example, the variables are movement speed (dependent variable) and the time spent in therapy (independent variable). Afterward, the local devices load the global model (2) and achieve an initialization of data with their local information (3). Hence, the devices start the local training with their data to specialize the global model (4), finally in the training process the parameters are sent to the remote server (5). The remote server is responsible for updating the

Table 7.2 Pseudocode where vertical federated learning is applied in motor therapy.

Local device	Remote server
	(1) global_model = init_linear_regression()
(2) model1 = load_global_model (global_model) (3) data1 = load_feature (movement_speed-time_spent_session1) (4) model1 = train_ linear_regression (model1, data1) (5) send_model_parameters (model1)	
	(6) global_model = aggregate_local_models (model1, model2, ..., modeln)
(7) model1 = load_global_model(global_model)	

Table 7.3 Python code using the flower framework.

Local device	Remote server
| ```python
import flwr as fl
dataset = pd.read_csv('../data.csv')
X = dataset.drop(['Patient', 'session'], axis=1)
X_train, X_test = train_test_split(X, test_size=0.2,
random_state=42)
num_clusters = 6
kmeans = KMeans(n_clusters=num_clusters,
random_state=42, init="k-means++", n_init="auto")

class KmeansClient(fl.client.NumPyClient):
    def get_parameters(self, config):
        return kmeans.get_params()
    def fit(self, parameters, config):
        X_train['cluster'] = kmeans.fit_predict
            (X_train)
        centroides = kmeans.cluster_centers_
        return [centroides], len(X_train), {}
    def evaluate(self, parameters, config):
        X_test['cluster'] = kmeans.predict(X_test)
        inertia = kmeans.inertia_
        dbi = davies_bouldin_score(X_test, X_test
            ['cluster'])
        return inertia, len(X_test), {"dbi": dbi}

fl.client.start_numpy_client(server_address="ip:port",
client=KmeansClient())
``` | ```python
import flwr as fl
class ACS(fl.server.strategy.FedAvg):
    def aggregate_evaluate(
        self,
        server_round: int,
        results: List[Tuple[ClientProxy,
            EvaluateRes]],
        failures: List[Union[Tuple[ClientProxy,
            FitRes], BaseException]],
    ) -> Tuple[Optional[float], Dict[str, Scalar]]:
        if not results:
            return None, {}
        aggregated_loss, aggregated_metrics =
            super().aggregate_evaluate
            (server_round, results, failures)
        accuracies = [r.metrics["dbi"] *
            r.num_examples for _, r in results]
        examples = [r.num_examples for _, r in
            results]
        aggregated_accuracy = sum(accuracies) /
sum(examples)
        return aggregated_loss, {"dbi":
            aggregated_accuracy}
strategy = ACS(
    fraction_fit=0.1,
    min_fit_clients=2,
    min_available_clients=2,
)
``` |
| | ```python
fl.server.start_server(
config=fl.server.ServerConfig(num_rounds=30),
    strategy=strategy
)
``` |

global model through an aggregation process. Finally, the devices get an update from the global model (7).

Thus each specialist has registered their data, and as a consequence, we have used FL to share the parameters between different client components. The implementation in Python code is shown in Table 7.3.

The data used during the three sessions for the 30 children was defined by the therapy specialist. These helped to identify the dataset for FL with the following key parameters:

- Patient: Representing the child attending the sessions
- Session: Number of the sessions
- Movement-speed: Time spent and volunteer-movements
- Color-correct answer: Latest percentage and memory-correct answers
- Textures: Aspects such as similarity and reaction.
```

This dataset was applied with the K-means clustering algorithm where the centroids were defined from the original data. The following subsections show the results obtained through a 3D representation.

7.7.1 Remote psychological therapies

The structure of the variables is as follows:

- Patient (*integer*): ID of each patient.
- Session (*integer*): Number of sessions that a patient attends

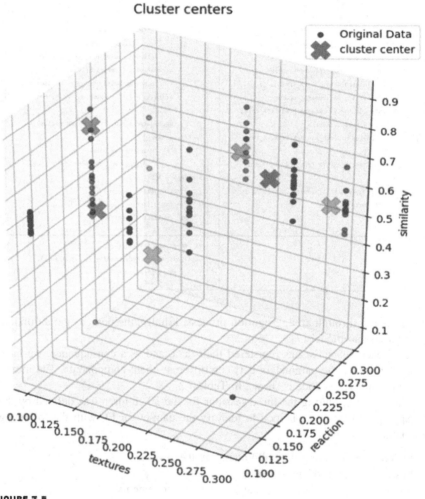

FIGURE 7.5

3D representation related to the status of remote psychology therapies

- Textures (string): Includes one of three types of textures (smooth, wrinkled, and sandy).
- Reaction (*string*): Identifies the reaction to textures; has three different values (positive, neutral, and negative)
- Similarity (*float*): Percentage of similarity of the written letters.

Through psychological therapies, it has been detected that there is a group of students who identify the similarity of objects through their texture, while another subgroup identifies through similarity with these objects (see Fig. 7.5).

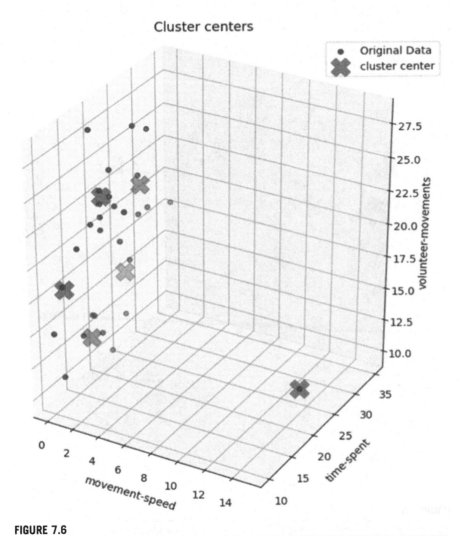

FIGURE 7.6

3D representation related to the status of remote motor therapies

7.7.2 **Remote motor therapies**

The structure of the variables is as follows:

- Patient (*integer*): ID of each patient.
- Session (*integer*): Number of sessions that a patient attends

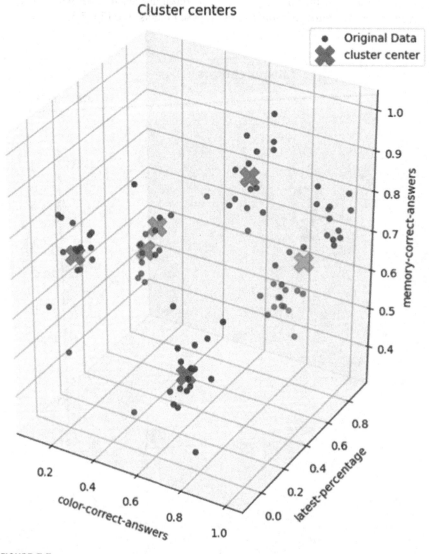

FIGURE 7.7

3D representation related to the status of remote initial education therapies.

- Movement speed (*float*): Speed of movements.
- Time spent (*float*): Time spent developing the movements.
- Voluntary movements (*integer*): Number of voluntary movements.

According to the motor therapist (see Fig. 7.6), there is a group of children with guided motor movements and another group with free-will motor movements.

7.7.3 Remote initial education therapies

The structure of the variables is as follows:

- Patient (*integer*): ID of each patient.
- Session (*integer*): Number of sessions that a patient attends.
- Correct color answers (*float*): Correct answers related to color recognition.
- Last percentage (*float*): The number of correct answers related to quantity recognition
- Correct-memory answers (*float*): The number of correct answers related to memory.

The didactic strategies applied allowed us to observe that one group of children was facilitated to memorize and another group was facilitated to memorize the proportion of the quantities (see Fig. 7.7).

7.8 Discussion

The proposed FL model for the IoHT, aimed at supporting children with psychomotor deficits, proves to be extremely helpful in terms of rehabilitation and inclusion. The model ensures comprehensive attention for patients, irrespective of their place of living, by offering remote services. However, the feasibility of the approach warrants consideration. The need for equipment transfer to different locations, coupled with the necessity for periodic maintenance and updates, raises practical challenges. Additionally, addressing the technological unfamiliarity of users demands a more personalized remote support system. Even with access to all the basic resources and materials at home, optimal therapy execution requires continuous training, updated equipment, data monitoring, documentation, and feedback. To tackle these challenges, the FL architectural model is designed to understand the environment, reinforcing relationships to achieve inclusion, cooperation, and equality.

The resistance faced in implementing remote psychological or motor therapies and early education poses a challenge for effective therapeutic interventions. The adoption of data monitoring and performance tracking of children with disabilities requires access to census data for data processing. Capitalization on this data supports the development of a global model and contributes to the creation of algorithms that provide enhanced information, help in predictive programs of integral attention to children, and adequate training for parents in technological education, while at the same time ensuring the protection of sensitive data. The IoHT-powered e-health systems, using small, lightweight wireless approaches connected through the IoHT, have the potential to enable remote monitoring and secure capturing of a substantial amount of patient health data and fitness information (Kholod et al., 2021). Thus medical devices, such as wearables and home health monitoring devices (e.g., wirelessly connected thermometers, glucometers, heart rate or blood pressure monitors), can be connected to the IoHT technology, enabling remote, timely, and comfortable monitoring of patients from a rehabilitation center environment.

By leveraging sensor-based toys and the data obtained through FL, the remote educational psychology therapy sessions offer a tailored and interactive approach to enhance the cognitive intelligence of children with cerebral palsy, promoting their overall development and providing them with valuable support and guidance in their educational journey. Those who intend to pass down digital cultural content must ensure that it still has meaning and value for future users (Barbuti et al., 2020). Efforts should be made to improve technological literacy among these stakeholders, provide transparent information regarding data privacy and security measures, and emphasize the potential benefits of remote therapies and FL architecture to enhance therapeutic outcomes.

7.9 Conclusions

Technologies such as the FL, along with the use of the IoT, virtual assistance, remote monitoring, artificial intelligence, big data analytics, blockchain, smart wearables, platforms, tools to exchange and store data, and tools that allow remote data capture and information exchange within the health ecosystem, lead to continuity of care and can improve health outcomes by improving medical diagnoses, data-based treatment decisions, digital therapies, clinical trials, self-care and people-centered care, in addition to expanding the evidence-based knowledge, skills, and competencies of professionals to deliver health services (Digital Health Strategy [World Health Assembly WHO, 2023]). This research encourages the adaptation of parents' needs in many ways. It offers communication viability regardless of the distance or the health circumstances of the world, facilitating access to information that previously took longer to obtain, which is undoubtedly useful for the treatment of patients who need rehabilitation at all ages and as training for parents or guardians.

It is important to note that remote therapies are not intended to replace face-to-face therapy completely but may be a complementary or temporary option in situations where physical assistance is not possible or convenient. The effectiveness of remote therapies can vary depending on the needs and abilities of the child, as well as the quality of the connection and the technological tools used.

References

Appleton, R., Williams, J., Vera San Juan, N., Needle, J. J., Schlief, M., Jordan, H., Sheridan Rains, L., Goulding, L., Badhan, M., Roxburgh, E., Barnett, P., Spyridonidis, S., Tomaskova, M., Mo, J., Harju-Seppänen, J., Haime, Z., Casetta, C., Papamichail, A., Lloyd-Evans, B., Simpson, A., Sevdalis, N., Gaughran, F., & Johnson, S. (2021). Implementation, adoption, and perceptions of telemental health during the COVID-19 pandemic: systematic review. *Journal of Medical Internet Research*, *23*(12), e31746.

Asociación Pro Paralítico Cerebral A.C. (APAC), https://apac.mx/ (accessed on May 27, 2023).

Barbuti, N., de Felice, G., Zanni, A. di, Russo, P., Barbuti, N., de Felice, G., Zanni, A. di, Russo, P., & Valentini, A. (2020). Creating digital culture by digitizing cultural heritage: The Crowd dreaming living lab method. *Umanistica Digitale*, *4*(9), 19−34. Available from https://doi.org/10.6092/issn.2532-8816/9956.

Barnett, P., Goulding, L., Casetta, C., Jordan, H., Sheridan-Rains, L., Steare, T., Williams, J., Wood, L., Gaughran, F., & Johnson, S. (2021). Implementation of telemental health services before COVID-19: Rapid umbrella review of systematic reviews. *Journal of Medical Internet Research*, *23*(7), e26492.

Beutel, D. J., Topal, T., Mathur, A., Qiu, X., Fernandez-Marques, J., Gao, Y., Sani, L., Kwing, H. L., Parcollet, T., Gusmão, P. P. B. de, & Lane, N. D. (2020). Flower: A friendly federated learning research framework. *arXiv preprint arXiv*, *2007*, 14390.

Dasaradharami Reddy, K., & Gadekallu, T. R. (2023). A comprehensive survey on federated learning techniques for healthcare informatics. *Computational Intelligence and Neuroscience*, *2023*, 19. Available from https://doi.org/10.1155/2023/8393990, Article ID 8393990.

Dubey, P., Pradhan, R. L., & Sahu, K. K. (2023). Underlying factors of student engagement to E-learning. *Journal of Research in Innovative Teaching & Learning*, *16*(1), 17−36.

Firouzi, F., Chakrabarty, K., & Nassif, S. (Eds.), (2020). *Intelligent internet of things: from device to fog and cloud*. Springer International Publishing. Available from https://doi.org/10.1007/978-3-030-30367-9.

Kholod, I., Yanaki, E., Fomichev, D., Shalugin, E., Novikova, E., Filippov, E., & Nordlund, M. (2021). Open-source federated learning frameworks for IoT: A comparative review and analysis. *Sensors (Switzerland)*, *21*(1), 1−22. Available from https://doi.org/10.3390/s21010167.

Konečný, J., McMahan, H. B., Yu, F. X., Richtárik, P., Suresh, A. T., & Bacon, D. (2016). Federated learning: Strategies for improving communication efficiency. *arXiv preprint arXiv*, *1610*, 05492.

Ludwig, H., & Baracaldo, N. (2022). *Federated Learning: A Comprehensive Overview of Methods and Applications*. Springer.

Maksimović, M., & Vujović, V. (2017). *Internet of things based e-health systems: Ideas, expectations and concerns* (pp. 241−280). Available from https://doi.org/10.1007/978-3-319-58280-1_10.

Manickam, P., Mariappan, S. A., Murugesan, S. M., Hansda, S., Kaushik, A., Shinde, R., & Thipperudraswamy., S. P. (2022). Artificial intelligence (AI) and internet of medical things (IoMT) assisted biomedical systems for intelligent healthcare. *Biosensors*, *12*(8), 562. Available from https://doi.org/10.3390/bios12080.

Martin, R. (2017). *Clean architecture: A Craftsman's guide to software structure and design*. Pearson, ISBN 9780134494272.

Matsangidou, M., Otkhmezuri, B., Ang, C. S., Avraamides, M., Riva, G., Gaggioli, A., Iosif, D., & Karekla, M. (2020). "Now I can see me" designing a multi-user virtual reality remote psychotherapy for body weight and shape concerns. *Human-Computer Interaction*, 1532–7051. Available from https://doi.org/10.1080/07370024.2020.1788945, ISSN 0737-0024. E-ISSN.

Mortis-Lozoya S., Muñoz-Arteaga J., & Zapata-González A. (2017). Reducción de brecha digital e inclusión educativa: Experiencias en Norte, Centro y Sur de México, Porrua, ISBN: 978-607-9239-96-1

Nguyen, D. C., Ding, M., Pathirana, P. N., Seneviratne, A., Li, J., & Vincent Poor, H. (2021). Federated learning for internet of things: A comprehensive survey. *in IEEE Communications Surveys & Tutorials*, 23(3), 1622–1658. Available from https://doi.org/10.1109/COMST.2021.3075439.

Phd, K. N., & Jeno, G. (2022). *Federated learning with Python: Design and implement a federated learning system and develop applications using existing frameworks*. Packt Publishing.

Shaheen, M., Farooq, M. S., Umer, T., & Kim, B.-S. (2022a). Applications of federated learning; taxonomy, challenges, and research trends. *Electronics*, 11(4), 670. Available from https://doi.org/10.3390/electronics11040670.

Shaheen, M., Farooq, M. S., Umer, T., & Kim, B. S. (2022b). Applications of federated learning; taxonomy, challenges, and research trends. *Electronics (Switzerland)*, 11(4). Available from https://doi.org/10.3390/electronics11040670.

Tunç, M.A., Gures, E., & Shayea, I. (n.d.). *A survey on IoT smart healthcare: Emerging technologies, applications, challenges, and future trends.*

Venturo-Conerly, K. E., Fitzpatrick, O. M., Horn, R. L., Ugueto, A. M., & Weisz, J. R. (2022). Effectiveness of youth psychotherapy delivered remotely: A meta-analysis. *American Psychologist*, 77(1), 71.

Watson, J. D., Pierce, B. S., Tyler, C. M., Donovan, E. K., Merced, K., Mallon, M., & Perrin, P. B. (2023). Barriers and facilitators to psychologists' telepsychology uptake during the beginning of the COVID-19 pandemic. *International Journal of Environmental Research and Public Health*, 20(8), 5467.

Wen, J., Zhang, Z., Lan, Y., et al. (2023). A survey on federated learning: challenges and applications. *International Journal of Machine Learning and Cybernetics*, 14, 513–535. Available from https://doi.org/10.1007/s13042-022-01647-y.

World Health Assembly (WHO), https://www.who.int/, (accessed on May 27, 2023).

Yang Z., Cui, Y., Li, B., Liu, Y., & Xu, Y. (2019). Software-defined wide area network (SD-WAN): Architecture, advances and opportunities, 2019 28th International Conference on Computer Communication and Networks, Spain, pp. 1–9. Available from https://doi.org/10.1109/ICCCN.2019.8847124.

Zhang, C., Xie, Y., Bai, H., Yu, B., Li, W., & Gao, Y. (2021). A survey on federated learning. *Knowledge-Based Systems*, 216. Available from https://doi.org/10.1016/j.knosys.2021.106775.

Blockchain-based federated learning in internet of health things

B. Akoramurthy[1], B. Surendiran[1], K. Dhivya[2], Subrata Chowdhury[3], Ramya Govindaraj[4], Abolfazl Mehbodniya[5] and Julian L. Webber[5]

[1]*Department of CSE, National Institute of Technology, Puducherry, India*
[2]*Department of CSE, Pondicherry University, Puducherry, India*
[3]*Department Of CSE, Sreenivasa Institute of Technology and Management Studies, Chittoor, Andhra Pradesh, India*
[4]*Department of IT, Vellore Institute of Technology, Vellore, Tamil Nadu, India*
[5]*Department of Electronics and Communication Engineering, Kuwait College of Science and Technology (KCST), Doha, Kuwait*

8.1 Introduction

The increasing accessibility to health information through the Internet of Health Things (IoHT) is a positive development in the healthcare sector (Alamri et al., 2020; Yu et al., 2021). The latest advances in deep learning (DL) applications have made it possible to autonomously analyze IoHT data with great accuracy, allowing for the autonomous monitoring of the well-being of individuals. By using cutting-edge DL algorithms for screening and assessment, hospitals can reduce waiting time for patients. Because the IoHT collects sensitive information about people's health and the environment, strict security measures are required (Tai et al., 2021). Regulations on centralized data collecting have been enacted by several countries throughout the world. As a result, typical DL applications are running into legal roadblocks (Khalil et al., 2022) because they rely on a centralized, sophisticated cloud platform that gathers a massive quantity of IoHT data to train a model appropriately. Cloud and centralized machine learning (ML) solutions (Zhu et al., 2022) may not be conducive to the training of a large volume of IoHT data due to legal constraints. Recent developments in edge learning make it possible to train confidential health data on-site in hospitals, mitigating concerns about the safety and confidentiality of information in the IoHT (Lăzăroiu et al., 2022).

Federated learning (FL) is an instructional model that aims to solve the issue of data privacy and security by educating algorithms in a group setting without any data sharing. While FL was initially developed for applications in other domains, such as smartphone and edge device instances, it is now experiencing success in the medical field. FL enables collaborative insight-gathering, without transferring patient data outside of their respective institutions' firewalls, such as in the form of a consensus model. Instead, as shown in Fig. 8.1, ML is performed discreetly at every entity of the organization, and only model properties (e.g., variables, biases) are shared. Algorithms prepared with FL have been proven to outperform those developed on data sets housed in one place and to outperform those trained on data sets maintained at one location.

As a result, FL holds the potential to significantly enhance the accessibility of personalized healthcare by producing unbiased models that accurately portray each patient's unique biological functioning and can address rare medical conditions

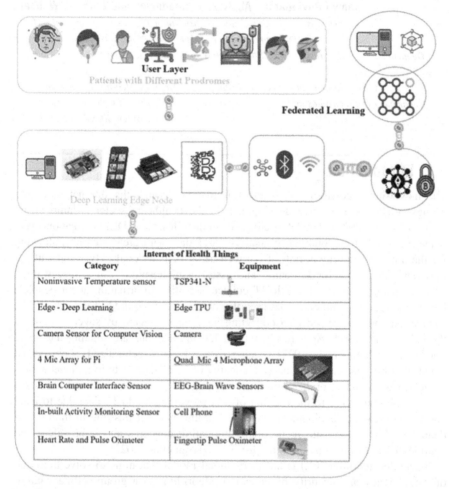

FIGURE 8.1

Applications based on DL use Internet of Medical Things benefit from high levels of safety, confidentiality, and lineage. *DL*, Deep learning.

without compromising patient privacy (Vizitiu et al., 2021). However, FL still needs careful technological study to make sure the algorithm is running smoothly without jeopardizing the confidentiality or safety of patients. Nonetheless, it has the ability to get around the problems with methods that rely on a centralized database.

The chapter introduces several original contributions:

- Implementation of blockchain and off-chain techniques to ensure provenance data against alteration and unauthorized access, allowing IoHT stakeholders and federated nodes to instill trust in the origins of training data and shared models. Blockchain facilitates decentralized consensus and tamper-proof gradient mining, substituting the central gradient aggregator.
- Development of a data lineage model and management system for the DL process.
- Creation of a DL model that classifies the IOHT data.
- Design and implementation of a secure transfer learning model for training COVID-19 datasets.

8.2 Background

Several approaches (Samarati, 2001; Shields & Levine, 2000) may be adopted to achieve confidentiality and information lineage in the IoHT for DL applications. Scholars have attempted to develop methods to guarantee confidentiality during group training. In an FL setting, some researchers (Serjantov & Danezis, 2002) created a private data leaking mechanism. Zhu et al. (2004) defined a numerically trustworthy metric by introducing secrecy and obscurity to evaluate the lineage of IoHT data. If we compare a single training node to a federation of nodes, with some nodes having higher quality and more datasets, we can see that the resulting training data is more accurate. Each federated node's privacy can be protected with some edge-level training. Even though FL can protect sensitive information, additional methods, such as differential privacy (DP) or homomorphic encryption (HE), may be used if the dataset, in whole or in part, must be shared with another party during training and anonymization or confidentiality must be maintained (LeFevre et al., 2006). Weighted FLs were proposed in a recent work (Zhu et al., 2020) that employed multiparty computation (MPC) protocol multiplication and addition across encrypted space. In local gradient updates on a batch of gradients (Heikkilä et al., 2020), additive HE can be used to hide model parameters. In addition, it is possible to encrypt a model that was learned on a central node or edge/federated node (Kamal & Tariq, 2018). De-identification can be performed on trained models used by multiple entities across a network of dispersed nodes. Given that the FL model permits multiple nodes to participate in training, it is important to identify malicious nodes that take part in the process and compromise training, model aggregation, or inferencing so that suitable, safe defense mechanisms can be used. The trustworthiness of the centralized aggregation server is an incorrect assumption in vanilla FL. A decentralized gradient aggregation approach has been presented as a solution by Guo et al. (2005). Blockchain's cryptographic prowess is formidable, and its decentralized method to ensure safe transactions between untrusted parties is highly attractive (Kesidis et al., 2006; Norros & Reittu, 2008). Both blockchain and off-chain solutions have shown promising results in providing data lineage for the IoHT.

Training datasets are often stored off-chain using a system such as the BitTorrent file system (BTFS), and blockchain has gained traction to oversee the trust and lineage of trustworthy federated nodes, their datasets, the credibility of each node, the reliability of the models each node produces, the immutability of the overall model, etc. (Chen et al., 2009). A data lineage management strategy for Ethereum-based blockchains is presented by Zhu et al. (2020). The study discussed in LeFevre et al. (2006) suggests a trusted execution environment (TEE), in which a hardware-based enclave attests the confidentiality, integrity, privacy, security, and authentication of both the data and the parties involved in the secure computation, among other things. Nodes with software/hardware implementations such as Intel's Software Guard Extensions (SGX) and Arm TrustZone may be found within the FL environment. In Kamal and Tariq (2018), the authors describe a lightweight security and provenance protocol that uses the wireless connection strength of IoMT nodes as a metric.

Ruan et al. (2021) explored the decision lineage of a DL application as a process to verify the accuracy of AI algorithms, adding another layer to the concept of lineage. Siddiqui et al. (2021) examined the use of blockchain and ML for securing and protecting IoHT data. The most recent study by Farhad et al. (2021) focuses on the topic of applying a DL technique (Rbah et al., 2022) to healthcare (Christopherson, 2007; Lea et al., 2001; Qian & Scott, 2007) and IoHT (Rahman & Jahankhani, 2021) data at the edge. In response to a heightened security concern surrounding health data privacy and strict security standards, scientists are increasingly turning to DL models that permit secure training, model aggregation, and distribution of the models as well as the deduced findings. With FL, many private nodes with sensitive IoHT data may collectively share a single encrypted DL model. Each node can then retrain the model using local data to create an individualized model. Subsequently, all associated local models can be aggregated either centrally or decentrally, depending on factors such as the reputation of the locally trained model, the duration of the training, the quality and accuracy of the trained model, etc.

In their pursuit to develop a collaborative radiometric diagnostic tool for COVID-19, (Alam & Rahmani, 2021) introduced a decentralized FL approach. Others have also explored the efficacy of a decentralized FL by incorporating techniques such as federated transfer learning (Xu et al., 2022). For instance, Rehman et al. (2022) offered a model in which each federated node could maximize its own benefit, keep local data private and secure, and reduce communication costs, all while the system as a whole guaranteed the security of the global model. In another aspect, Lakhan et al. (2023) proposed a deep reinforcement learning model that could learn such requirements and provide an optimal solution to the problem of the wide variety of federated nodes, their security and privacy needs, and the limitations they impose on performance and resources. Depending on whether the FL model is trained using independent and identically distributed or nonindependent and identically distributed health information, researchers have developed different classifications for the privacy and security of the IoHT (Kamal & Tariq, 2018).

Having the DL model and the training dataset within the private network of a user is necessary to enable the advances of the IoHT in a safe and privacy-protected way (Ahmed et al., 2023). Yet, getting a good model is usually beyond grasp for the data owner because of the complexity of the DL application life cycle, such as the high-quality training and testing of an IoHT model using a substantial training and validation dataset. In this context, "data owner" refers to whoever physically has the IoHT. Despite the need for protecting sensitive information, the model used to train this data must either be made accessible to the private data on the edge from an externally trained model or the private data must be submitted to the owner of the model. Privacy and/or safety concerns arise in both cases. Encrypting the model means the owners of private data can use the trained model without disclosing their data to the model's creator. However, if the private data is encrypted before being shared with the model owner, then the model owner cannot access either the private training dataset or the interpreting result while working on the secret dataset. To apply a well-known DL algorithm to private datasets, Microsoft scientists developed CrypTFlow (Ali et al., 2023), a tool that converts a standard TensorFlow model to the MPC cryptographic protocol. Several investigators have put forward a hybrid approach in which secure MPC and DP may be implemented to improve accuracy and data quality (Christopherson, 2007). However, it is worth noting that secure MPC is still susceptible to inferencing exploits.

The latest advances in DL for encrypted computing have boosted the profile of cryptographic protocols, such as complete or partial HE and secure MPC. As a result, companies, for instance, Google and Facebook, have developed encrypted versions of their DL frameworks, such as TensorFlow and Keras, and opaque versions of PyTorch. These models enable both the owner of private data and the owner of an external model to encrypt their data and model, respectively, and to carry out secure training and model inferencing in a distributed setting without having to rely on a trusted third party. To identify malicious attackers, scientists propose using secret tokens generated by a differentially private generative adversarial network (GAN). Whereas to deal with privacy leaks, scientists propose using differentially private stochastic gradient descent. Scholars have developed an encryption approach for effective interaction and additive HE and DP for data and model security and privacy (Han et al., 2023) since FL requires dispersed federated nodes to interact in a safe and security-oriented fashion. Invasive edge nodes can be blacklisted, while harmless training results are discarded before the FL aggregation process begins (Mahmood et al., 2020). Another study (Khalil et al., 2022) advocated a two-stage FL procedure, with the first stage being a vote for a trustworthy committee member from the federation and the second stage involving the real FL itself, with the committee members acting as guardians and privacy watchdogs. In light of the recent COVID-19 pandemic (Azeem et al., 2021; Mahmood et al., 2020; Rahman & Jahankhani, 2021), the widespread use of the IoHT has become a real possibility. The present FL approaches that prioritize security and privacy have bright futures. Few studies, however, have demonstrated the viability of the innovations in controlling COVID-19.

8.2.1 Safe in-depth models of learning

Yazdinejad et al. (2021) observe the state of research on the privacy and safety of DL applications running on 5G edge networks. Cui et al. (2019) presented a protocol on top of FL within 5G networks to secure the sensitive data accessible to network functions or network slices, as well as to secure the confidentiality of local DL model updates. Dehez Clementi et al. (2019) used FL in IoT-based traffic data, where the private data from a single train line was trained using an SVM RBF kernel function. The global training module was managed by a trusted blockchain smart contract. Verifying the validity and integrity of FL models stored locally on a device is a growing use case for blockchain (Ruan et al., 2019). Cui et al. (2019) presented a byzantine-resilient distributed learning approach that is well-suited to 5G networks since it allows a large number of federated clients to collaborate on model construction. Ruan et al. (2020) employ a permissioned blockchain and a locally directed acyclic network to protect the DL model parameters, and it is implemented in the vehicular IoT arena. Data utilized in the FL network's deep reinforced learning process must be trusted, and the blockchain provides that assurance. To filter out IoT nodes or mobile federated nodes with malicious or low-quality data, researchers have suggested a consortium blockchain for discovering the reputation of each federated node (Kannan et al., 2020). Ruan et al. (2021) secured a DL (Aceto et al., 2019; Kaissis et al., 2021; Karanov et al., 2018; Zhang et al., 2019) model that may be used without a trusted third party. Before being placed on the blockchain for immutability and provenance, the IoHT data is encrypted using the Paillier homomorphic cryptosystem. Garcia et al. (2022) examine data poisoning and inferencing threats against FL algorithms. The study also provides recommendations for building a robust FL model. Aceto et al. (2018), David et al. (2016) and Wang et al. (2018) mainly concentrated on classification and automated learning concepts for different applications based on DL.

8.3 Proposed framework

The proposed framework is about a historical acquisition and control system for DL that makes use of blockchain and off-chain techniques to document the origins of all of the necessary data, models, and transactions. The big picture of the suggested framework is depicted in Fig. 8.1. As can be seen in Fig. 8.1, we have chosen a collection of IoHT sensors that can aid us in handling COVID-19 prodromes and diagnosis or pandemic preparedness. The edge nodes interact with the IoHT devices. The graphics processing unit (GPU)-equipped edge nodes can store and process their personal information locally for inference and training. In addition to k-anonymity and homomorphic operations, the edge nodes can function as a personalized blockchain user or federated workforce. Prodromes associated with COVID-19 can be inferred by the nodes closer to the edges of the network. The k-anonymity is executed by the interaction module of the edge nodes, and then the encoded architecture as well as training data is safely transferred to the blockchain dApp (decentralized application) for additional analysis. To facilitate worldwide assessment, blockchain clients process block formation and disseminate the blended ledger node's smart contract. The k-anonymity IoHT raw data is stored in the BTFS repository, and then, for lineage and shared model training, a hash of the simulated data or model region from the BTFS is stored on the blockchain. We investigated Intel SGX to supply a TEE with MPC to offer an additional level of protection at the distributed aggregating nodes. For the purpose of hosting the system nodes and smart contracts, we may simply utilize a standard cloud service, such as Google Cloud.

The ghetto of the TEE, where model pooling takes place, receives security updates from the federated nodes containing their trained system. The characteristics of the system are secured against theft in this way using cryptography. Users who have been verified as legitimate can safely inquire about the DL technique, datasets used, as well as training procedures. The course of study is educated through supervised learning to identify hostile attacker networks by keeping track of trust ratings. When it comes to FL, we built in each of the federated probabilistic gradient-descending and averaging methods. Federated averaging places additionally work on edge nodes but improves training precision. We presumptively deployed tensor processing units (TPUs) (e.g., cloud-based tensor processing units) to our edge nodes (e.g., the Raspberry PI) assuming they have sufficient memory.

Specifically, we focused on a few COVID-19 programs that made use of DL to categorize IoHT data. Upon entering the DL environment, the sensory data of the IoHT is transmitted via a chain of modules that includes a lineage module, a k-anonymity module, and a secret module. We employed chock-full HE so that the aggregating or decentralized hostile federated nodes would not be able to learn about our gradient.

While the activation function initially needs to be transformed to polynomials, the addition and multiplication procedures for convolution, including maximal pooling, and different other kinds, are homomorphic. We implemented an encrypted transfer of learning model, which allows a blockchain-based edge node to securely download a globally trained model for COVID-19 assessment over safe FL. We then use that model to create a unique, locally tailored model.

Lastly, we put safety and lineage techniques through their paces on 22 deep learning applications linked to COVID-19, using a wide range of information sets, scenarios, edge-federated node types, and efficacy measures. We accounted for both nonidentical and unequal data distributions as a design parameter. We additionally looked at the effort required to apply these safeguards to edge devices, such as handsets and Raspberry Pi. We took into account measures such as federated training length, secure communication additional costs, energy consumption of edge devices in transit, and edge memory requirements. Our goal was to reduce the communication cost between IoHT devices and TPU-based devices at the edge, even while dealing with thousands of learning parameters. A prototype architecture for a federated DL application that supports immunization certificate origin, k-anonymity production, preservation, and exchange with authorized partners is presented below. There is a possibility for safe distribution to several IoHT nodes along the edge.

8.4 Internet of Health Things secure and source-aware context design

The COVID-19 (Samarati, 2001) pandemic has severely impacted our everyday activities. Although there is currently a vaccine, it is imperative that all possible measures must be exercised to prevent infection. This includes maintaining social isolation everywhere, isolating identified patients from the general population, and continually updating the condition of individuals with signs and observations regarding COVID-19 infection. The inability to determine someone's COVID-19 status has long been a security risk for businesses, including stores, terminals, colleges and universities, and eateries. A popular way of spotting people with COVID-19, using a heat imaging camera, falls short when dealing with asymptomatic cases. In addition, evaluations show that it is not possible to share the patient's condition among regulators and suppliers of services because of the lack of security surrounding the exchange of medical information. In the event of a COVID-19 outbreak, disclosing safely and confidentially kept health records is essential in demonstrating one's health status. Technologies such as blockchain, the IoT, and social media have recently advanced to the point where such a medical system is feasible. Smart contracts trained by DL applications can be deployed on blockchains and off-chains. In our FL design, we deployed the extended-optimal identical generalization hierarchy (OIGH) algorithm for k-anonymity operations to secure data privacy.

In this research, we put forward an all-encompassing answer to the issues mentioned above through the creation of an improved system (as depicted in Fig. 8.2) that can be used by the COVID-19 subjects, healthcare providers, and participants, including government agencies, companies, and other institutions, to ensure the protection of their employees and customers. In this instance, the

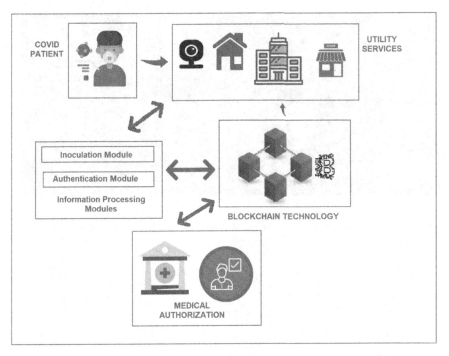

FIGURE 8.2

System design.

people in charge of keeping tabs on everyone give out permission to view their health records while also allowing the people who actually have those records—the citizens—to view and share them with others.

Using the healthcare authority dApp, the authorized healthcare agency creates a QR code-based profile of each patient who visits. The dApp lets users upload authorization data, including photos of their faces and information about their signs of COVID-19, for our DL algorithm to analyze. However, encoded information, including biometric traits, is held separately from the blockchain. The person's health condition can be confirmed and updated at any time by the relevant health officials.

The general population dApp employs the user's QR code and facial scan data for medical assessments in open settings, where it may be accessed by participants, sometimes known as providers of services, to verify the well-being of their clients. As a demonstration of the feasibility of the system's basic ideas, we're presenting the setup we've built.

Three apps and two interconnected data-analysis components form the basis of the demo model developed. The system's overall structure is depicted in Fig. 8.3. Each user can make a citizen dApp profile, which is complete with information

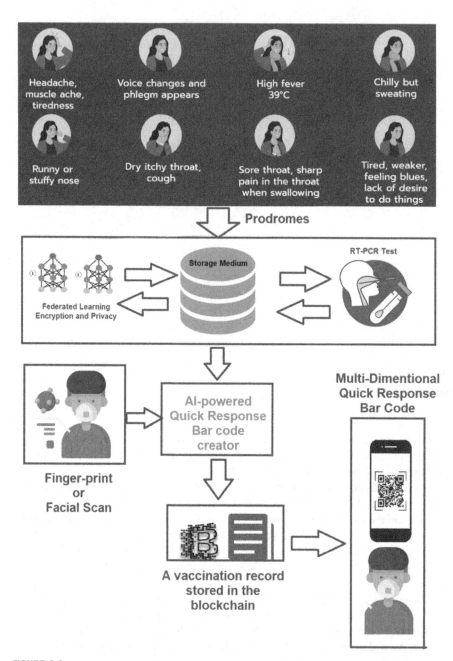

FIGURE 8.3

Vaccination plugin for quick response code generation. Client devices are not required to transmit data to global servers, which signifies the noninvolvement of client or end-user devices. In contrast, the unprocessed data residing on edge devices are utilized for training the model within a localized environment, hence enhancing the confidentiality of the data.

about biometric identities and prodromes uploaded to the chain and beyond block-chain. Medical immunization records and a citizen's COVID status can be uploaded to the blockchain by approved healthcare providers. Fig. 8.3 depicts the steps required to create a medical immunization certificate and then display the most up-to-date medical information using QR codes in different colors.

The third app is designed for public service providers such as medical facilities, shopping centers, educational institutions, eateries, etc. It allows them to verify a customer's vaccination status against COVID by either (1) scanning a QR code on a health immunization certificate or (2) verifying the customer's identity using DL-based fingerprint attributes. Fig. 8.2 depicts the process for submitting an application to corporate stakeholders. Two distinct components power these software suites. The immunization component, for example, can identify COVID from pro-drome data using DL techniques. The other is a biometric component that employs

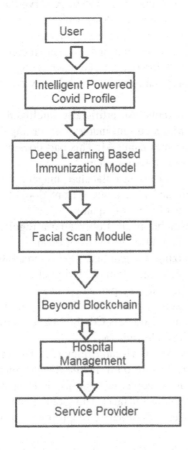

FIGURE 8.4

User interaction flow module.

convolutional neural networks (Wang et al., 2020) to identify people. We built the DLT infrastructure with Ether and Hyperledger Fabric and used BTFS for beyond-chain storage. The relationship between them is depicted in Fig. 8.4.

8.5 Implementation

8.5.1 Edge applications for safe and anonymous deep learning with Internet of Health Things

We put the subsequent programs in our safe cloud server as part of the demonstration of the concept.

Access control based on fingerprints or facial scans without physical contact: To stop the spread of the COVID-19 virus, this program went through extensive development to permit a contactless entryway. A surveillance module was interfaced with an entrance key sensor. A Raspberry Pi is the central component of both. It is the Raspberry Pi that hosts the DL software.

Text recognition from handwriting and medical record processing: This section makes use of Tesseract for optical character recognition. With this attachment, the user of an ordinary camera or a mobile device can read signs, such as handwritten scripts.

AI-powered clinical chatbots: An artificially intelligent chatbot is utilized to back up and chase along with a current inquiry concerning COVID-19.

Data gathering of prodromes using AI: This part does natural language processing on a variety of prodrome questionnaires.

Distinguishing gender subjects in unlabeled imagery: Utilizing a typical open surveillance camera positioned at a home's front door, this software employed DL to determine the gender of the participants.

Different types of recyclable trash: Leveraging a regular video feed, this program can distinguish between different forms of trash.

Application for recognizing medical text: We've created a dedicated DL module that can identify data fields in a PDF/RTF digital copy or printed medical file and recommend the next steps for the patient's care.

Identifying several objects at once: YoloV8 was deployed on an edge device equipped with a TPU, such as Google's Coral TPU, to address the issue of multiple object detection with a single camera.

Vision aid powered by AI for visually impaired: The graphical user interface of the app uses the camera feature on a mobile device running YoLoV8 to identify and flag things, including people, face masks, and objects that are moving.

Identification of mask worn on face: This component was taught to identify mask-wearing and mask-free individuals through observation or using cell phone footage.

Sensing of damp and dry surfaces: This add-on for a public-facing webcam can identify moist and dry floors and trigger warning alarms.

Recognizing foods and fruits: Here, we have developed a computer vision (CV)-based program that can sort fruits based on their types and freshness.

Sensors that detect a user's fall: Using YoloV8 and AI, this component can identify a tumble. When a fall is identified, a warning is issued.

Analysis of expression: This module can identify a variety of facial expressions, including those indicating happiness, grief, rage, surprise, pain, and more (Mitra, 2022).

Safely tracking down a contact: In this section, we implement DL and the blockchain to track down lost contact.

Fatigue/weariness monitoring: This component can monitor for signs of fatigue, such as yawning and tiredness, by analyzing facial characteristics.

Recognizing pills and medications on site: This component can be used for pill registration and recognition.

Tracking movement from a distance: This software facilitates the detection of remote motion of targets of interest.

Thermographic visualization of fever: This component utilizes a thermal imaging camera to identify the faces of individuals and calculate fever levels.

Warning for breach of social distance: Using CV, the software can detect when people are violating their social distancing and send alerts.

Using quick response codes with dynamic color coding to monitor the progress of COVID-19: Fig. 8.2 depicts the DL application's architecture.

Blockchain-based smart contract verification of vaccination records: One example of a DL application's architecture is shown in Fig. 8.3.

8.5.2 Training

All of the applications mentioned above relied on free and publicly available datasets. We used custom datasets produced specifically for this study for certain niche applications. We built a federated network with three NVIDIA Jetson Nanos and four 4 GHz Raspberry Pi. Each edge node hosted multiple IoHT to provide various healthcare services connected to COVID-19. Both internal GPUs (e.g., the Jetson Nano) and external GPUs (e.g., Google Coral TPU) were used in the edge nodes. We used handsets and Note 10C 5G with middling GPUs and a laptop with NVIDIA GeForce RTX 3050Ti Graphics, with a 16-GB GPU as an edge client. All the nodes on the periphery were run as clients for the blockchain (Nguyen et al., 2023) and DL. Decentralized, k-anonymity, and multicomputing function and model sharing were made possible by the federated nature of these edge nodes and their communication with a full chain node located at the edge. We used Keras and a TensorFlow Encrypted backend to create the training model. Both the DP and non-DP versions of the model were trained. To prevent the model from remembering or leaking sensitive training information, k-anonymity was used during development. We evaluated a distributed secret communication cryptographic approach for aggregating model gradients from dispersed clients without the need for a centralized authority to verify the data.

During the training of PyTorch models, we extensively tested the AnonPy library for K-Anonymity algorithm through its paces. We utilized Autograd APIs to efficiently calculate per-sample gradients in bulk. Our setup involved implementing a concurrent CUDA 11 NVIDIA driver-based system on an NVIDIA GeForce RTX 3050Ti with support for AES-128-bit encryption keys using pyCANON. The framework allowed for the generation of confidential forecasts based on encrypted raw training datasets, ensuring end-to-end encryption throughout the process. Techniques such as Open FL enabled confidential forecasting.

By using Open FL, you can be rest assured that your model's variables and weights are also secure. Distributing the encoded model across several federated nodes, performing encoded calculations, and categorizing the outcomes based on secret information are all possible with the Open FL sequential class. We utilized a substrate adaptation of Intel SGX for the TEE enclave.

8.5.3 Environment setup

In a decentralized network, as shown in Fig. 8.5, nodes at the edge might be dynamic or static in nature (Yang & Guo, 2021). According to the task at hand, different edge nodes will have different GPUs, amounts of processing power, and kinds of IoHT sensors.

This code sample demonstrates how to set up a federated network with three Raspberry Pi 4 operating on Fedora and linked to a 5G router. The screenshots we captured in some of the chosen deep-learning apps are shown in Fig. 8.6. Fig. 8.6A. It depicts the frame rate at which an Intel Movidius GPU-equipped Raspberry Pi can recognize objects at the network's periphery.

Fig. 8.6B depicts the user interface for a Raspberry Pi with a connected Pi camera to perform tablet recognition in real-time video. Fever detection using a Fluke Ti25 Thermal Imager at 9 Hz and its accompanying Raspberry Pi host of IoHT nodes is depicted in Fig. 8.6C. Fig. 8.6D depicts a Raspberry Pi functioning as a federated node, aiding in a distant espionage activity in which individual behavior is being recognized. In Fig. 8.6E, we see a Raspberry Pi coupled with a Google Coral TPU wellness tracking application that records a Pi Camera stream to monitor things such as tiredness, muscle pain, groaning, etc. Lastly, in Fig. 8.6F, we see an NVIDIA edge node monitoring for and reacting to a human fall. All of the applications make use of the knowledge acquisition (KA), model configuration, and FL configurations described in Section 8.3.

8.5.4 Blockchain lineage

Figs 8.7 and 8.8 depict some of the dApps we created for patients, healthcare providers, and government agencies. Fig. 8.7 depicts the procedure by which a user's

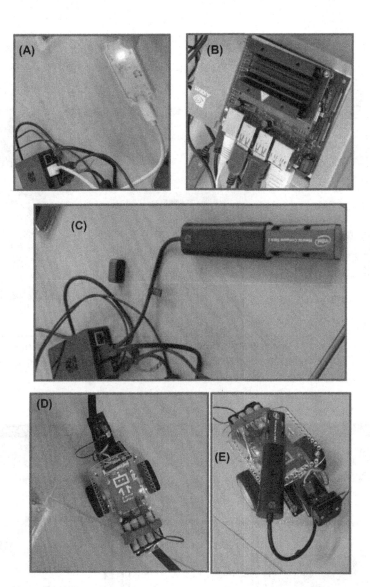

FIGURE 8.5

Configuration of edge nodes in various FL scenarios: (A) 3 NVIDIA Jetson Nanos with Raspberry Pi. (B) Jetson Nano and external GPU (C) NCS2 and Raspberry Pi 4. (D) Raspberry Pi with GoPiGo mechanical body, IoMT, and CV gear. (E) Extension of D with GPU. *FL*, Federated learning; *GPU*, graphics processing unit; *IoMT*, Internet of Medical Things.

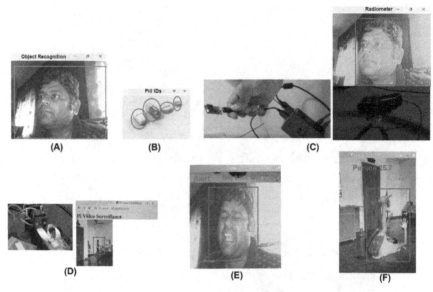

FIGURE 8.6

Edge nodes perform FL applications. (A) Person recognition. (B) Pill recognition. (C) Fever detection. (D) Human activity surveillance. (E) Person state monitoring. (F) Recognition, monitoring, and warning of a person's drop. *FL*, Federated learning.

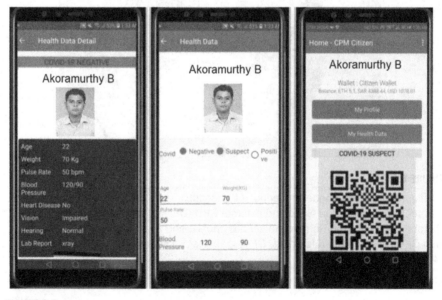

FIGURE 8.7

Blockchain lineage: Healthcare professional updates the COVID status, which appears on the client's application.

FIGURE 8.8

Healthcare professional checking the person's COVID status.

COVID status is verified by healthcare professionals and changed from negative to suspicious, with the change reflected in the user's app in real time. In Fig. 8.8, we see the same user uploading his fingerprint details and medical immunization certificate to the service provider's platform. For IoHT data origin, smart contracts built on Ethereum were employed. A lineage database layer, a general lineage layer, and a specific lineage layer make up the smart contract platform. As an illustration, we developed 22 DL-based applications to demonstrate our notion. Though each of these applications had unique attribution demands, they had some similar characteristics.

8.6 Testimonies and discussions

We review the testimonials and their discussion in the subsequent section. Fig. 8.9 displays the accuracy (both for testing and training) and the loss during the training period for the proposed system depicted in Figs. 8.7 and 8.8. It is also observed that the training loss remains within acceptable bounds considering the addition of KA noise, encryption of the local gradients, and an origin guarantee. In terms of accuracy, the proposed system achieved above 92% accuracy during training and above 88% during testing.

The performance indicators for the 6 DL work shown in Fig. 8.6 are displayed in Table 8.1. As can be observed, the average accuracy of all the tested apps is above 90%. It is due to the fact that all applications developed the convolution process, and forward and backward propagation of the layers, and estimation of the activation procedures are all carried out in the cryptography realm.

Furthermore, the KA necessitates the addition of an appropriate noise limit and searches for confidentiality constraints on funds, followed by denoises, all of which compromise the precision and loss statistics. Fig. 8.10 depicts the outcome of the safety versus precision observations and balance test. As the confidentiality cost rises, so does the test precision, up to a point. Yet another anonymity

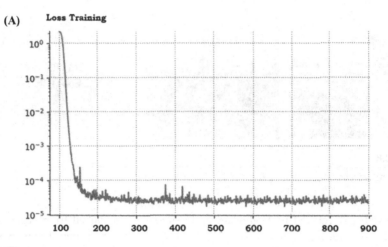

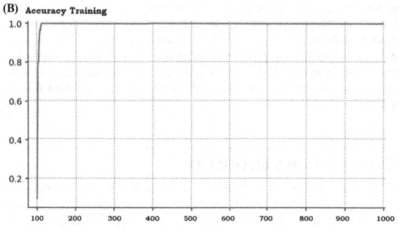

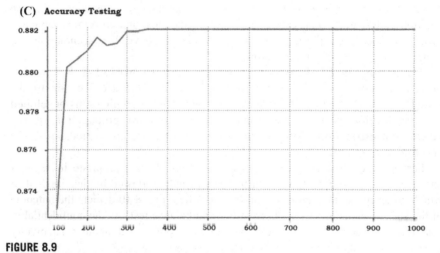

FIGURE 8.9

(A) Loss in training. (B) Accuracy in training. (C) Accuracy in testing data.

Table 8.1 Parameters for automatic scoring.

Applications developed	Loss (%)	Accuracy/precision (%)
A	0.08	92.34
B	0.09	90.15
c	0.02	90.02
d	0.03	94.87
e	0.03	92.77
f	0.01	90.11

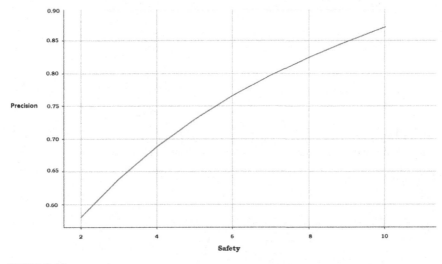

FIGURE 8.10

Safety versus precision observations and balance test.

measure discovered was the number of cycles in FL. As shown in the graph mentioned above, the KA guarantee diminishes as the number of cycles in the FL interaction cycle rises. This is due to the fact that each cycle introduces some kind of noise. We additionally attempted to quantify the power consumed by the IoHT as a result of the extra cost of cryptography and KA (Fig. 8.11). We attempted to construct an arbitrary arrangement of sensing devices in the federated nodes, as shown in Fig. 8.12.

Fig. 8.12 depicts an IoHT that captures essential data in an invasive manner, such as with electrocardiogram and electroencephalography sensors, as well as in a way that is not intrusive, such as with a thermal camera. The power utilization in each case indicates that developed applications needing security, origin, and anonymity use an average to a substantial degree of power. Fig. 8.12 depicts the

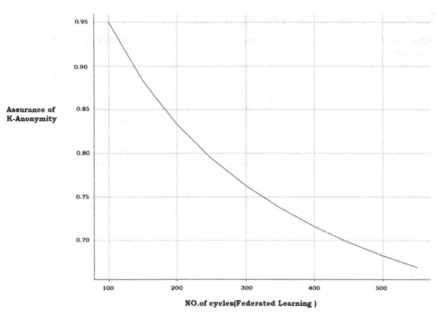

FIGURE 8.11

Assurance of k-anonymity versus number of cycles (federated learning).

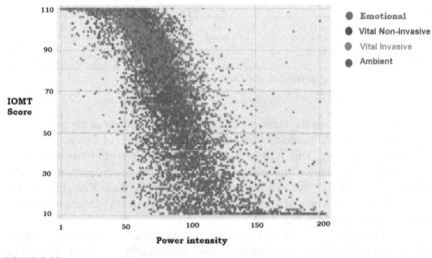

FIGURE 8.12

IoMT Score versus power intensity. *IoMT*, Internet of Medical Things.

expected versus actual real values for each application listed in Table 8.1. It is seen that the distribution as a whole has a broad continuous relationship to fewer anomalies. Also, Fig. 8.12 depicts the behavior and grade of Internet of Medical Things (IoMT) sensors in terms of power utilization. The result falls as the need for power rises. This demonstrates the limitations of employing edge nodes. We additionally attempted to determine the impact of variations in consumption, asset type, bandwidth at hand, and various other parameters on trustworthy FL systems. Fig. 8.12 depicts our findings over three fields: the intangible distinction among FL edge nodes, the geographical position of nodes, and the kind of IoMT utilized by nodes. We found no major implications for each of these variables. In a nutshell, the addition of security and confidentiality to DLFL applications has a minimal impact on these three parameters. For instance, a lack of sufficient interaction capacity for certain edge nodes had an impact on the whole interactive teaching session.

8.7 Conclusion and future prospects

We tackled the issue of introducing a lightweight security and privacy mechanism that may be employed inside the FL ecosystem in this work. We specifically targeted IoHT-powered edge devices that required a privacy guarantee for privately held health data as federated nodes during collaborative training. Each federated node used k-anonymity to prevent the leakage of raw data. Furthermore, the quantity of noise introduced for privacy protection was carefully selected to establish a compromise between the privacy budget and accuracy deterioration. During the FL process, HE on the edge node permitted additive and multiplicative matrix operations on DL processes. To eliminate bias and privacy leaks from a central aggregating body, the FL process was governed through a blockchain-based decentralized consensus method. Rather, blockchain nodes maintained the entities engaged in distributed learning, the reputation and quality of each client input, and the storage of intermediate and global models in a decentralized repository, namely BTFS. The results of the test reveal the collected accuracy and loss metrics, which we intend to enhance as future prospects.

References

Aceto, G., Ciuonzo, D., Montieri, A., & Pescapè, A. (2018). Mobile encrypted traffic classification using deep learning. *Network Traffic Measurement And Analysis Conference, 2018*.

Aceto, G., Ciuonzo, D., Montieri, A., & Pescapé, A. (2019). Mobile encrypted traffic classification using deep learning: Experimental evaluation, lessons learned, and challenges. *IEEE Transactions on Network and Service Management, 16*(2), 445−458.

Ahmed, J., Nguyen, T. N., Ali, B., Javed, M. A., & Mirza, J. (2023). On the physical layer security of federated learning based IoMT networks. *IEEE Journal of Biomedical and Health Informatics*, 27(2), 691−697. Available from https://doi.org/10.1109/JBHI.2022.3173947.

Alam, M. U., & Rahmani, R. (2021). Federated semi-supervised multi-task learning to detect COVID-19 and lungs segmentation marking using chest radiography images and Raspberry Pi devices: An Internet of Medical Things application. *Sensors (Basel, Switzerland)*, 21(15), 5025. Available from https://doi.org/10.3390/s21155025.

Alamri, A., Gumaei, A., Al-Rakhami, M., Hassan, M. M., Alhussein, M., & Fortino, G. (2020). An effective bio-signal-based driver behavior monitoring system using a generalized deep learning approach. *IEEE Access*, 8, 135037−135049, 9146560. Available from https://doi.org/10.1109/ACCESS.2020.3011003.

Ali, M., Naeem, F., Tariq, M., & Kaddoum, G. (2023). Federated learning for privacy preservation in smart healthcare systems: A comprehensive survey. *IEEE Journal of Biomedical and Health Informatics*, 27(2), 778−789.

Azeem, M., Ullah, A., Ashraf, H., Jhanjhi, N., Humayun, M., Aljahdali, S., & Tabbakh, T. A. (2021). FoG-oriented secure and lightweight data aggregation in IoMT. *IEEE Access*, 9, 111072−111082. Available from https://doi.org/10.1109/ACCESS.2021.3101668.

Chen, Y.-Y., Jan, J.-K., Chi, Y.-Y., & Tsai, M.-L. (2009). A feasible DRM mechanism for BT-like P2P system. *International Symposium on Information Engineering and, 2009*.

Christopherson, K. M. (2007). The positive and negative implications of anonymity in Internet social interactions: "On the Internet, Nobody Knows You're a Dog." *Computers in Human Behavior*, 23(6), 3038−3056. Available from https://doi.org/10.1016/j.chb.2006.09.001.

Cui, H., Chen, Z., Xi, Y., Chen, H., & Hao, J. (2019). IoT data management and lineage traceability: A blockchain-based solution. In: *2019 IEEE/CIC International Conference on Communications*.

David, O. E., Netanyahu, N. S., & Wolf, L. (2016). Deepchess: End-to-end deep neural network for automatic learning in chess. In: Artificial Neural Networks and Machine Learning−ICANN 2016: 25th International Conference on Artificial Neural Networks, Barcelona, Spain, September 6−9, 2016, Proceedings, Part II 25 (pp. 88−96). Springer International Publishing.

Dehez Clementi, M., Larrieu, N., Lochin, E., Kaafar, M. A., & Asghar, H. (2019). When air traffic management meets blockchain technology: A blockchain-based concept for securing the sharing of flight data. In *2019 IEEE/AIAA 38th digital avionics systems conference*.

Farhad, A., Woolley, S. I., & Andras, P. (2021). A preliminary scoping study of federated learning for the Internet of Medical Things. *Studies in Health Technology and Informatics*, 281, 504−505, PMID: 34042622. Available from https://doi.org/10.3233/SHTI210216.

Guo, L., Chen, S., Xiao, Z., Tan, E., Ding, X., & Zhang, X. (2005). *Measurements, analysis, and modeling of BitTorrent-like systems*.

Garcia, R. D., Sankar Ramachandran, G., Jurdak, R., & Ueyama, J. (2022). A blockchain-based data governance with privacy and provenance: a case study for e-prescription. *2022 IEEE International Conference on Blockchain and Cryptocurrency (ICBC)*, Shanghai, China, 1−5. Available from https://doi.org/10.1109/ICBC54727.2022.9805545.

Han, S., Ding, H., Zhao, S., Ren, S., Wang, Z., Lin, J., & Zhou, S. (2023). Practical and Robust federated learning with highly scalable regression training. *IEEE Transactions on Neural Networks and Learning Systems*. Available from https://doi.org/10.1109/TNNLS.2023.3271859.

Heikkilä, M. A., Koskela, A., Shimizu, K., Kaski, S., & Honkela, A. (2020). Differentially private cross-silo federated learning. arXiv preprint arXiv:2007.05553.

Kaissis, G., Ziller, A., Passerat-Palmbach, J., Ryffel, T., Usynin, D., Trask, A., Lima, I., Mancuso, J., Jungmann, F., Steinborn, M.-M., Saleh, A., Makowski, M. R., Rueckert, D., & Braren, R. (2021). End-to-end privacy preserving deep learning on multi-institutional medical imaging. *Nature Machine Intelligence*, *3*, 473−484. Available from https://doi.org/10.1038/s42256-021-00337-8.

Kamal, M., & Tariq, M. (2018). Light-weight security and data provenance for multi-hop Internet of Things. *IEEE Access*, *vol. 6*, 34439−34448. Available from https://doi.org/10.1109/ACCESS.2018.2850821.

Kannan, K., Singh, A., Verma, M., Jayachandran, P., & Mehta, S. (2020). Blockchain-based platform for trusted collaborations on data and AI models. *IEEE International Conference on Blockchain, 2020.*

Karanov, B., Chagnon, M., Thouin, F., Eriksson, T. A., Bülow, H., Lavery, D., Bayvel, P., & Schmalen, L. (2018). End-to-end deep learning of optical fiber communications. *Journal of Lightwave Technology*, *36*(20), 4843−4855.

Kesidis, G., Jin, Y., Mortazavi, B., & Konstopoulos, T. (2006). QRP03-1: An epidemiological model for file-sharing with BitTorrent-like incentives: The case of a fixed peer population. *IEEE Globecom 2006, San Francisco, CA, USA*, 1−2. Available from https://doi.org/10.1109/GLOCOM.2006.430.

Khalil, A. A., Ibrahim, F. E., Abbass, M. Y., Haggag, N., Mahrous, Y., Sedik, A., Elsherbeeny, Z., Khalaf, A. A. M., Rihan, M., El-Shafai, W., El-Banby, G. M., Soltan, E., Soliman, N. F., Algarni, A. D., Al-Hanafy, W., El-Fishawy, A. S., El-Rabaie, E.-S. M., Al-Nuaimy, W., Dessouky, M. I., & El-Samie. (2022). Efficient anomaly detection from medical signals and images with convolutional neural networks for Internet of Medical Things (IoMT) systems. *International Journal for Numerical Methods in Biomedical*, *38*(1), e3530, Epub 2021 Dec 18. PMID: 34506081. Available from https://doi.org/10.1002/cnm.3530.

Lakhan, A., Mohammed, M. A., Nedoma, J., Martinek, R., Tiwari, P., Vidyarthi, A., Alkhayyat, A., & Wang, W. (2023). Federated-learning based privacy preservation and fraud-enabled blockchain IoMT system for healthcare. *IEEE Journal of Biomedical and Health Informatics*, *27*(2), 664−672. Epub 2023 Feb 3. PMID: 35394919. Available from https://doi.org/10.1109/JBHI.2022.3165945.

Lăzăroiu, G., Andronie, M., Iatagan, M., Geamănu, M., Stefanescu, R., & Dijmărescu, I. (2022). Deep learning-assisted smart process planning, robotic wireless sensor networks, and geospatial big data management algorithms in the Internet of Manufacturing Things. *ISPRS International Journal of Geo-Information*, *11*(5), 277. Available from https://doi.org/10.3390/ijgi11050277.

Lea, M., Spears, R., & de Groot, D. (2001). Knowing me, knowing you: Anonymity effects on social identity processes within groups. *Personality and Social Psychology Bulletin*, *27*(5), 526−537. Available from https://doi.org/10.1177/0146167201275002.

LeFevre, K., DeWitt, D. J., & Ramakrishnan, R. (2006). Mondrian multidimensional K-anonymity. *22nd International Conference on Data Engineering (ICDE'06), Atlanta, GA, USA*, 25−25. Available from https://doi.org/10.1109/ICDE.2006.101.

Mahmood, K., Akram, W., Shafiq, A., Altaf, I., Lodhi, M., & Islam, S. K. (2020). An enhanced and provably secure multi-factor authentication scheme for internet-of-multimedia-things environments. *Computers and Electrical Engineering, 88*, 106888. Available from https://doi.org/10.1016/j.compeleceng.2020.106888.

Mitra, S. (2022). OConsent - Open Consent Protocol for Privacy and Consent Management with Blockchain. *ArXiv*, abs/2201.01326.

Nguyen, L. T., Nguyen, L. D., Hoang, T., Bandara, D., Wang, Q., Lu, Q., Xu, X., Zhu, L., Popovski, P., & Chen, S. (2023). Blockchain-empowered trustworthy data sharing: Fundamentals, applications, and challenges. *arXiv preprint arXiv:2303.06546*.

Norros, I., & Reittu, H. (2008). *Urn models and peer-to-peer file sharing.*

Qian, H., & Scott, C. R. (2007). Anonymity and self-disclosure on weblogs. *Journal of Computer-Mediated Communication, 12*(4), 1428−1451. Available from https://doi.org/10.1111/j.1083-6101.2007.00380.x.

Rahman, M., & Jahankhani, H. (2021). Security vulnerabilities in existing security mechanisms for IoMT and potential solutions for mitigating cyber-attacks. In H. Jahankhani, S. Kendzierskyj, & B. Akhgar (Eds.), *Information security technologies for controlling pandemics. Advanced sciences and technologies for security applications.* Cham: Springer. Available from https://doi.org/10.1007/978-3-030-72120-6_12.

Rbah, Y., Mahfoudi, M., Balboul, Y., Fattah, M., Mazer, S., Elbekkali, M., & Bernoussi, B. (2022). Machine learning and deep learning methods for intrusion detection systems in IoMT: A survey. *2nd International Conference on Innovative Research in, 2022*.

Rehman, A., Abbas, S., Khan, M. A., Ghazal, T. M., Adnan, K. M., & Mosavi, A. (2022). A secure Healthcare 5.0 system based on blockchain technology entangled with federated learning technique. *Computers in Biology and Medicine, 150*, 106019.

Ruan, P., Chen, G., Dinh, A., Lin, Q., Ooi, B. C., & Zhang, M. (2019). Fine-grained, secure and efficient data provenance for blockchain". *Proceedings of the VLDB Endowment, 12*, 975−988.

Ruan, P., Dinh, T. T. A., Lin, Q., Zhang, M., Chen, G., & Ooi, B. C. (2020). Revealing every story of data in blockchain systems. *ACM SIGMOD Record, 49*(1), 70−77. Available from https://doi.org/10.1145/3422648.3422665.

Ruan, P., Dinh, T. T. A., Lin, Q., Zhang, M., Chen, G., & Ooi, B. C. (2021). LineageChain: A fine-grained, secure and efficient data provenance system for blockchains. *The VLDB Journal, 30*, 3−24. Available from https://doi.org/10.1007/s00778-020-00646-1.

Samarati, P. (2001). Protecting respondents identities in microdata release. *IEEE Transactions on Knowledge and Data Engineering, 13*, 1010−1027. Available from https://doi.org/10.1109/69.971193.

Serjantov, A., & Danezis, G. (2002). *Towards an information theoretic metric for anonymity.*

Shields, C., & Levine, B. N. (2000). *A protocol for anonymous communication over the internet.*

Siddiqui, S., Khan, A. A., Dev, K., Dey, I. (2021). Integrating federated learning with IoMT for managing obesity in smart city. In: Proceedings of the 1st Workshop on Artificial Intelligence and Blockchain Technologies for Smart Cities with 6G (6G−ABS '21). Association for Computing Machinery:New York, NY, USA. pp. 7−1. Available from https://doi.org/10.1145/3477084.3484950.

Tai, Y., Gao, B., Li, Q., Yu, Z., Zhu, C., & Chang, V. (2021). Trustworthy and intelligent COVID-19 diagnostic IoMT through XR and deep-learning-based clinic data access. *IEEE Internet of Things Journal, 8*(21), 15965−15976. Available from https://doi.org/10.1109/JIOT.2021.3055804.

Vizitiu, A., Nita, C.-I., Toev, R. M., Suditu, T., Suciu, C., & Itu, L. M. (2021). Framework for privacy-preserving wearable health data analysis: proof-of-concept study for atrial fibrillation detection. *Applied Sciences, 11*(19), 9049. Available from https://doi.org/10.3390/app11199049.

Wang, P., Ye, F., Chen, X., & Qian, Y. (2018). Datanet: Deep learning based encrypted network traffic classification in SDN home gateway. *IEEE Access, 6*, 55380—55391. Available from https://doi.org/10.1109/ACCESS.2018.2872430.

Wang, X., Chen, S., & Su, J. (2020). *App-Net: A hybrid neural network for encrypted mobile traffic classification. IEEE Infocom 2020.* IEEE Conference on Computer.

Xu, Z., Guo, Y., Chakraborty, C., Hua, Q., Chen, S., & Yu, K. (2022). A simple federated learning-based scheme for security enhancement over Internet of Medical Things. *IEEE Journal of Biomedical and Health Informatics, 27*(2), 652—663. Available from https://doi.org/10.1109/JBHI.2022.3187471.

Yang, J., & Guo, Y. (2021). AEFETA: Encrypted traffic classification framework based on self-learning of feature, In *2021 6th international conference on intelligent computing.*

Yazdinejad, A., Parizi, R. M., Dehghantanha, A., & Karimipour, H. (2021). Federated learning for drone authentication. *Ad Hoc Networks, 120*. Available from https://doi.org/10.1016/j.adhoc.2021.102574.

Yu, Z., Amin, S. U., Alhussein, M., & Lv, Z. (2021). Research on disease prediction based on improved DeepFM and IoMT. *IEEE Access, 9*, 39043—39054. Available from https://doi.org/10.1109/ACCESS.2021.3062687.

Zhang, J., Li, F., Wu, H., & Ye, F. (2019). Autonomous model update scheme for deep learning based network traffic classifiers. *IEEE Global communications conference (GLOBECOM), 2019.*

Zhu, B., Wan, Z., Kankanhalli, M.S., Bao, F., & Deng, R.H. (2004). Anonymous secure routing in mobile ad-hoc networks. In *29th annual IEEE international conference on local computer.*

Zhu, H., Li, Z., Cheah, M., & Goh R. S. M. (2020). Privacy-preserving weighted federated learning within oracle-aided MPC framework. vol. 2, vol. 1, no. 1, pp. 1—10, arXiv:2003.07630. [Online]. Available: https://arxiv.org/abs/2003.07630.

Zhu, T., Kuang, L., Daniels, J., Herrero, P., Li, K., & Georgiou, P. (2022). IoMT-enabled real-time blood glucose prediction with deep learning and edge computing. *IEEE Internet of Things Journal, 10*(5), 3706—3719.

Further reading

Alam, I., & Kumar, M. (2023). A novel authentication protocol to ensure confidentiality among the Internet of Medical Things in Covid-19 and future pandemic scenario. *Internet of Things (Amsterdam, Netherlands), 22*, 100797. Available from https://doi.org/10.1016/j.iot.2023.100797.

Ali, Z., Naz, S., Zaffar, H., Choi, J., & Kim, Y. (2023). An IoMT-based melanoma lesion segmentation using conditional generative adversarial networks. *Sensors (Basel, Switzerland), 23*(7), 3548. Available from https://doi.org/10.3390/s23073548.

Anamika, A., & Tyagi, K. (2014). Secure approach for location aided routing in mobile ad hoc network. *International Journal of Computer Applications, 101*(8), 24—27. Available from https://doi.org/10.5120/17708-8711.

Bittner, K., Cock, M. D., & Dowsley, R. (2020). Private speech classification with secure multiparty computation. *arXiv preprint arXiv:2007.00253.*

Cano, M.-D., & Cañavate-Sanchez, A. (2020). Preserving data privacy in the Internet of Medical Things using dual signature ECDSA. *Security and Communication Networks*, *2020*(5), 1−9.

Chaganti, R., Mourade, A., Ravi, V., Vemprala, N., Dua, A., & Bhushan, B. (2022). A particle swarm optimization and deep learning approach for intrusion detection system in Internet of Medical Things. *Sustainability*, *14*(19), 12828. Available from https://doi.org/10.3390/su141912828.

Chaudhry, S. A., Irshad, A., Nebhen, J., Bashir, A. K., Moustafa, N., Al-Otaibi, Y. D., & Zikria, Y. B. (2021). An anonymous device to device access control based on secure certificate for internet of medical things systems. *Sustainable Cities and Society*, *75*, 103322, ISSN: 2210-6707.

Chauhan, J., Rajasegaran, J., Seneviratne, S., Misra, A., Seneviratne, A., & Lee, Y. (2018). Performance characterization of deep learning models for breathing-based authentication on resource-constrained devices. *Proceedings of the ACM on Interactive, Mobile, Wearable and Ubiquitous Technologies*, *2*(4), 24, 158. Available from https://doi.org/10.1145/3287036.

Egala, B. S., Pradhan, A. K., Badarla, V., & Mohanty, S. P. (2021). Fortified-chain: A blockchain-based framework for security and privacy-assured Internet of Medical Things with effective access control,". *IEEE Internet of Things the Journal*, *8*(14), 11717−11731. Available from https://doi.org/10.1109/JIOT.2021.3058946.

Garcia, R. D., Ramachandran, G., Jurdak, R., & Ueyama, J. (2022). A blockchain-based data governance with privacy and provenance: A case study for E-prescription. *IEEE International Conference on Blockchain and*, *2022*.

Kang, Y., Yang, G., Eom, H., Han, S., Baek, S., Noh, S., Shin, Y., & Park, C. (2023). GAN-based patient information hiding for an ECG authentication system. *Biomedical Engineering Letters*, *13*, 197−207. Available from https://doi.org/10.1007/s13534-023-00266-y.

Kesdogan, D., Egner, J., Büschkes, R. (1998). Stop-and-Go-MIXes providing probabilistic anonymity in an open system.

Khowaja, S. A., Lee, I. H., Dev, K., Jarwar, M. A., & Qureshi, N. M. F. (2022). Get your foes fooled: Proximal gradient split learning for defense against model inversion attacks on IoMT data. ARXIV-CS.CR.

Kim, K., Ryu, J., Lee, Y., & Won, D. (2023). An improved lightweight user authentication scheme for the Internet of Medical Things. *Sensors (Basel, Switzerland)*, *23*(3), 1122. Available from https://doi.org/10.3390/s23031122.

Mani, S., Sankaran, A., Tamilselvam, S., & Sethi, A. (2019). Coverage testing of deep learning models using dataset characterization. *arXiv preprint arXiv:1911.07309*.

Masud, M., Gaba, G. S., Alqahtani, S., Muhammad, G., Gupta, B. B., Kumar, P., & Ghoneim, A. (2021). A lightweight and robust secure key establishment protocol for Internet of Medical Things in COVID-19 patients care. *IEEE Internet of Things Journal*, *8*(21), 15694−15703. Available from https://doi.org/10.1109/JIOT.2020.3047662.

Praseetha, V. M., Bayezeed, S., & Vadivel, S. (2020). Secure fingerprint authentication using deep learning and minutiae verification. *Journal of Intelligent Systems*, *29*(1), 1379−1387. Available from https://doi.org/10.1515/jisys-2018-0289.

Rose, J., & Bourlai, T. (2019). Deep learning based estimation of facial attributes on challenging mobile phone face datasets. In 2019 IEEE/ACM international conference on advances in Social Network Analysis and Mining (ASONAM).

Sarasadat, A. (2020). Deep learning for smartphone security and privacy,

Sedik, A., Tawalbeh, L., Hammad, M., Abd El-Latif, A. A., El Banby, G. M., Khalaf, A. A. M., Abd El-Samie, F. E., & Iliyasu, A. M. (2021). Deep learning modalities for biometric alteration detection in 5G networks-based secure smart cities. *IEEE Access*, *9*, 94780−94788. Available from https://doi.org/10.1109/ACCESS.2021.3088341.

Singh, P., Singh Gaba, G., Kaur, A., Hedabou, M., & Gurtov, A. (2023). Dew-cloud-based hierarchical federated learning for intrusion detection in IoMT. *IEEE Journal of Biomedical and Health Informatics*, *27*(2), 722−731. Available from https://doi.org/10.1109/JBHI.2022.3186250.

Tanwani, A. K., Anand, R., Gonzalez, J. E., & Goldberg, K. (2020). RILaaS: Robot inference and learning as a service. *IEEE Robotics and Automation Letters*, *5*(3), 4423−4430. Available from https://doi.org/10.1109/LRA.2020.2998414.

Zhao, Z., Hsu, C., Harn, L., Yang, Q., & Ke, L. (2021). Lightweight privacy-preserving data sharing scheme for Internet of Medical Things. *Wireless Communications and Mobile Computing*, 1−13.

Integration of federated learning paradigms into electronic health record systems

Hope Ikoghene Obakhena[1,2], Agbotiname Lucky Imoize[3,4] and Francis Ifeanyi Anyasi[1]

[1]*Department of Electrical and Electronics Engineering, Faculty of Engineering and Technology, Ambrose Alli University, Ekpoma, Nigeria*
[2]*Department of Electrical and Electronics Engineering, Faculty of Engineering, University of Benin, Benin, Nigeria*
[3]*Department of Electrical and Electronics Engineering, Faculty of Engineering, University of Lagos, Lagos, Nigeria*
[4]*Department of Electrical Engineering and Information Technology, Institute of Digital Communication, Ruhr University, Bochum, Germany*

9.1 Introduction

Recent years have witnessed a tremendous explosion in the availability of large volumes of data. Particularly in healthcare, there has been an upward trajectory of efforts targeted toward data analytics, data harmonization, and ubiquitous access for both the provider and patient while maintaining the integrity, confidentiality, and nonrepudiation of healthcare data (Batko & Ślęzak, 2022). Prior to the advent of modular information technology (IT) systems and biomedical machines, paper-based records were widely adopted in healthcare institutions (Shahnaz et al., 2019). However, given the sensitive nature of healthcare data, the increased risk of data security compromises, poor penmanship, and several other mitigating factors of paper-based prescriptions and reports, there was an urgent need to develop a disruptive paradigm for handling and safeguarding sensitive data in any healthcare delivery setting (Coorevits, 2013). The concept of electronic health record (EHR) systems holds great promise for revolutionizing the healthcare sector by enabling computerized records, timely access to patient data, collaboration among hospitals, improved understanding of public health patterns and trends, heightened healthcare intelligence, user experience, operational performance, and related costs (Cowie, 2017).

Fundamentally, a longitudinal electronic version of a patient's health information generated over time during their visit to the hospital or any care provider is referred to as an EHR. The data that is maintained by the provider may fall into

Federated Learning for Digital Healthcare Systems. DOI: https://doi.org/10.1016/B978-0-443-13897-3.00017-5

one of the categories: demographic information, medical data, historical data, psychological data, genetic data, habitual data, radiology data, laboratory data, progress notes, vital signs, medications, and immunization (Kruse et al., 2018). Despite the promising potential presented by EHR systems, it is plagued by increased security and privacy concerns by the public, and their concerns are entirely justifiable due to rising actions of user information leakage, data breaches, cyberattacks, hefty payments to ransomware attackers, and disruptive services. EHR systems are also confronted with the challenges of medical malpractice, scalability, and access control (Pilares et al., 2022).

Although artificial intelligence (AI) techniques are of specific interest and a substantial figure-of-merit to maximize processes in the medical domain, implementing AI models with a number of stringent privacy-preserving models presents a performance overhead in terms of communication efficiency and confidentiality of personal data due to the centralized nature of cloud-based architectures (Amalraj & Lourdusamy, 2022). Due to the large, diverse, and unstructured medical dataset and difficulty obtaining data for certain tasks, open data sharing with the cloud or data centers may face stringent ethical, logistical, and regulatory restrictions, necessitating datasets to reside on one server. Thus, the trepidations of the public remain because most centralized AI functions are performed in a black-box manner, and health data are not centrally located but decentralized (Rieke, 2020). Lately, there has been a growing interest in the implementation of a disruptive learning paradigm that enables collaborative model training in a decentralized manner (Dasaradharami Reddy & Gadekallu, 2023).

Federated learning (FL) has emerged as a key candidate architecture to build ML (machine learning) models with fragmented, sensitive data across multiple clients and is gaining significant traction in the medical domain (Xu et al., 2021). In this regard, the ML process occurs using pseudo-data, and only the aggregated global model is broadcast to all participants. More precisely, FL operation involves the following techniques: system initialization and client selection; local model updates; model aggregation; and download. First, a client selection technique is determined for a healthcare analytic task, then each client at its end trains a local model, after which each client sends back their local model to the server for aggregation toward a fully trained global model without exchanging the data itself (Pfitzner et al., 2021). Given these characteristics, FL can potentially enable system heterogeneity, communication efficiency, precision medicine at large scale, and mitigate privacy leakage risks (Nguyen, 2022). Fig. 9.1 presents a typical illustration of the FL architecture.

A number of research studies have been conducted in this direction. Yang et al. (2019) presented a comprehensive discussion on the foundation, architecture, and techniques of FL. Gosselin et al. (2022) provided an insightful exposition of the privacy-preserving technique applicable to FL systems. Lim et al. (2020) have shown that considerable improvement in mobile edge networks is achievable with FL. Boobalan (2022) and Khan et al. (2021) accounted for the integration of FL in Internet of Things (IoT) networks. Rieke (2020) characterized

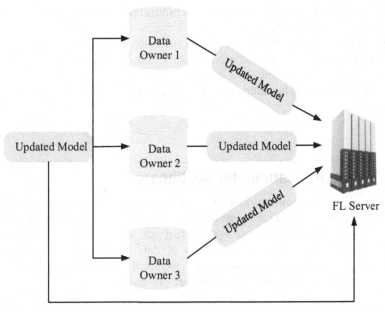

FIGURE 9.1

General overview of the federated learning architecture.

the integration of FL in digital healthcare. A detailed account of the requirements, technical concerns, and open research challenges of FL in digital health is reported. Moreover, a discussion on the benefits and drawbacks of EHR systems is presented in Menachemi and Collum (2011). Dang et al. (2022) proposed to solve the challenges of regulation, infrastructure, privacy, and data standardization in healthcare organizations using FL. As a step further, Antunes et al. (2022) presented a comprehensive survey of current FL efforts for EHR systems.

Nonetheless, this is an ongoing research direction, and the survey papers on FL in the context of EHR systems for smart healthcare are quite few. Moreover, FL-based EHRs still require further technical considerations to achieve optimal integrity and confidentiality of data. This chapter presents a comprehensive discussion on the distinctive integration of FL paradigms into EHR systems. Particularly, this chapter attempts to enhance the perspective and preliminary results on the synergistic interconnection of FL and EHR systems.

The noteworthy contributions of the chapter are outlined as follows:

1. Introduction to the context and details on the motivation, technical requirements, and types of FL for healthcare.
2. Discussion on the foundational background of EHRs, with special emphasis on their benefits and associated challenges.
3. Concise overview of existing proposals to solve EHR concerns.

4. Brief summary of selected learning algorithms that claim to incorporate heterogeneity in their design.
5. Initiation of an empirical discussion on statistical concerns, communication efficiency, data security, and privacy issues, followed by insightful propositions to overcome these concerns.
6. Identification and elucidation of open research issues intended to guide future research.

9.2 Federated learning for healthcare

FL is a disruptive framework with a recent surge in popularity as it brings together multiple healthcare institutions via collaborative learning without exchanging training data. These special features of FL make it appealing for healthcare applications as the limitations of data governance and privacy are addressed. In this context, the motivation, technical considerations, and types of FL architecture for maximized healthcare are presented.

9.2.1 Motivation

The key motivations for using FL in healthcare institutions are as follows:

1. By maintaining local data stores, FL holds great promise in enhancing data privacy compared with AI models that aggregate data captured from different sources on a central server.
2. FL is a potential candidate for enhanced system heterogeneity, where datasets from different healthcare organizations with heterogeneity in terms of communication and computational resources may be easily accessed.
3. FL offers great potential for reducing communication latency and network resources (transmit power and spectrum) since offloading health data to the cloud is eliminated.
4. FL is a more suitable model to handle massively nonidentically independent distribution (IID) among various stakeholder groups in the healthcare industry, including but not limited to researchers, healthcare specialists, pharmacists, and emergency facilities.
5. FL significantly improves the scalability of healthcare-based solutions, as data training can occur at multiple medical sites without centralizing datasets.

9.2.2 Requirements

The following sections discuss the technical requirements and performance metrics to be considered in deploying FL for healthcare (Nguyen, 2022).

9.2.2.1 Massive connectivity of multiple data sources

It is pertinent to ensure the availability of datasets from different silos (health data clients) to improve the quality of health data training while satisfying the low latency requirements of FL. Thus, smart healthcare environments are required to generate their own local updates, after which the aggregated global model is broadcast to all clients.

9.2.2.2 Reliable training at local sites

Every client at its end locally trains a model, which is aggregated to the server via wireless links. The convergence speed and the overall performance of the aggregated, fully trained global model remain major bottlenecks due to the shortage of resources and the unreliability of wireless links. It is therefore critical to maximize the reliability of limited wireless resources to realize the full potential of FL operations in healthcare.

9.2.2.3 Ultra-high privacy preservation

The risk of data security and privacy compromise may prove very costly. Although FL presents a promising alternative in enabling participants to collaborate and produce superior models without revealing updated information, it is necessary to ensure ultra-high privacy preservation between local data stores and the global server before FL operations are implemented in healthcare.

9.2.2.4 Trusted server

Unlike centralized AI functions located in the cloud, FL operations primarily involve an aggregation server where local models trained at the client level are uploaded to the server for aggregation. User information leakage and privacy breaches may occur during the data training and model aggregation processes. Consequently, a trusted server is paramount to ensuring safe and reliable client-server communication in healthcare.

9.2.3 Types of federated learning for healthcare

FL frameworks utilized in healthcare can be categorized into the following (Jawadur Rahman, 2021).

9.2.3.1 Horizontal federated learning

When each device possesses a dataset with identical feature spaces but distinct samples, horizontal federated learning (HFL), as presented in Fig. 9.2, is utilized. In this case, datasets are shared horizontally among healthcare clients, and each local FL participant may train their local model using their dataset. After that, the local updates are aggregated to generate a global update.

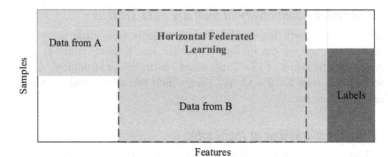

FIGURE 9.2

Horizontal federated learning architecture (large overlap of features of the two datasets).

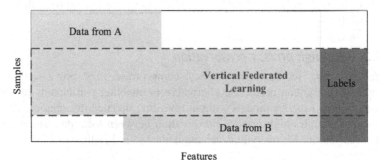

FIGURE 9.3

Vertical federated learning architecture (large overlap of sample identifiers of the two datasets).

9.2.3.2 Vertical federated learning

When each device possesses a dataset with similar sample instances but distinct feature spaces, vertical federated learning (VFL), as illustrated in Fig. 9.3, is utilized. In this context, a central server or another neutral entity may or may not be involved. For instance, if two individual parties in a healthcare organization have datasets with identical sample space but different data features, model training may be achieved using VFL to cooperatively build a shared AI model.

9.2.3.3 Hybrid/federated transfer learning

Federated transfer learning (FTL) is a promising paradigm to address the challenge of providing an extension to VFL (hybridizing scattered data during data federation). It is dimensioned to handle datasets with different feature and sample spaces. Thus the transfer learning technique provides training solutions in difficult situations where datasets differ not only in features but also in sample values. Fig. 9.4 illustrates the architecture of FTL.

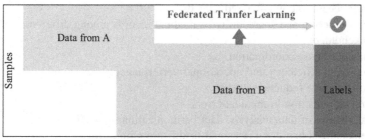

FIGURE 9.4

Federated transfer learning architecture.

9.3 Foundational background of electronic health records

Before the emergence of biomedical machines and modular IT systems, paper-based healthcare records were widely adopted in healthcare institutions (Shahnaz et al., 2019). However, with the unprecedented efforts by healthcare organizations toward data security and harmonization, there is an urgent need to develop disruptive technologies for handling healthcare data analytics (Garg & Chari, 2020). EHR is an innovative technology seeking to address the problems of security and privacy concerns, increased costs and time, ordering, and poor penmanship of physicians presented by paper-based healthcare records. It involves the synergistic interconnection of paper-based and electronic medical records (EMRs). Simply put, a longitudinal electronic version of a patient's health information generated over time during their visit to the hospital or any care provider is referred to as an EHR (Cowie, 2017). The data maintained by the provider may fall into one of the following categories: demographic information, medical history, historical records, psychological assessments, genetic information, lifestyle habits, radiology images, laboratory test results, progress notes, vital signs, medications, and immunization records. EHRs provide the much-needed functionality to the healthcare sector in terms of confidentiality, integrity, availability, and shareability medical data across various stakeholders, such as hospitals, pharmacists, specialists, researchers, government entities, insurance providers, laboratories, and emergency facilities (Kruse et al., 2018).

9.3.1 Benefits of electronic health records

The benefits of EHRs include (Menachemi & Collum, 2011):

1. Elimination of poor penmanship predominant in paper-based healthcare records.
2. Improved patient healthcare due to the availability of large volumes of data.

3. Easy access to computerized records.
4. Easy ordering and shareability of medical records among different healthcare stakeholders.
5. Enhanced care coordination.
6. Improved efficiency and operational performance.
7. Considerable reduction in cost.
8. Drastic reduction in medical errors.
9. Facilitation of data analysis, data harmonization, and ML.
10. Personalized health services and healthcare monitoring.
11. Ability to better conduct research, leading to maximized population health.

9.3.2 Issues associated with electronic health records

Despite the expanding body of research on the potential benefits presented by various EHR functionalities, it is characterized by increased security and privacy concerns, medical malpractices, scalability issues, disruptive services, changes in workflow, related costs, and several unintended consequences. Specifically, the topmost concerns of the healthcare sector as it relates to EHR systems are security and privacy issues; data storage; data integrity; access control; authentication; data ownership; confidentiality; interoperability; anonymity; accessibility; auditability; and availability (Pilares et al., 2022). Fig. 9.5 is a pictorial representation of the top-most concerns of EHR systems. These challenges are further compounded by increased susceptibility to cyberattacks, the availability of large and diverse datasets, and massive interconnectivity between these systems.

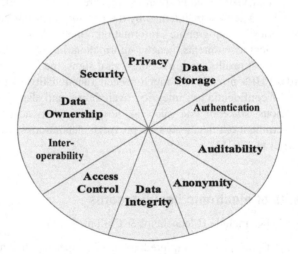

FIGURE 9.5

Topmost concerns associated with EHR systems. *EHR*, Electronic health record.

9.3.3 Review of existing solutions to solve electronic health record concerns

Several proposals aimed at solving EHR concerns have been presented in the literature. These solutions, range from storage mechanisms (interplanetary file system (IPFS), off-chain edge node, cloud storage, blockchain-based data storage), blockchain-based models (Consortium, Ethereum, Permissioned, Hyperledger fabric), Cryptography (decentralizing attribute-based signature (DABS), Asymmetric key, Hash table, symmetric searchable encryption), to General Technologies [ML, deep learning (DL), Routing overlay, mobile applications, Django], have attempted to address multiple challenges that exist in EHR (Poongodi & Imoize, 2021; Abikoye et al., 2023). A comprehensive summary of selected literature that investigates EHR optimization with a focus on the technologies used, key contributions, and limitations is presented in Table 9.1.

Table 9.1 Existing solutions to solve electronic health records concerns.

Reference	Technologies used	Main contributions	Limitations
Zhao et al. (2017)	BSN and SKE	Proposed an effective key management scheme to protect privacy data on health blockchain. The study is also aimed at maximizing private physiological data in the block	1. Risk of privacy compromises in the public ledger 2. Exposure of health records in the event where corresponding symmetric key is lost
Guo et al. (2018)	Attribute-based signature scheme with multiple authorities.	Attempts to guarantee the validity of patient's data encapsulated in blockchain, overcome forgeability of the verifier, and resist collision attacks	1. Model does not support general nonmonotone predicates 2. Computation cost is relatively high.
Lee and Yang (2018)	Public-key encryption	Presented a nail analysis management technique using microscopy and blockchain technology to ensure the confidentiality of sensitive data. Accurate biometric authentication is realized using deep neural network, support vector machine, and random forest tree	1. Privacy concerns in terms of nail data leakage in the public blockchain ledger. 2. Resource-limited IoT devices remain a prime challenge

(Continued)

Table 9.1 Existing solutions to solve electronic health records concerns. *Continued*

Reference	Technologies used	Main contributions	Limitations
Al Omar et al. (2017)	Public-key encryption (Elliptic curve cryptography)	Proposed a privacy preservation technique to protect sensitive health data. Cryptographic function of blockchain technology is exploited to ensure pseudonymity	1. Bottlenecks may appear in terms of complex key management, risk of data leakage, and related cost.
Zheng et al. (2018)	Symmetric key encryption (Advanced encryption standard)	Considers the distinctive interplay between blockchain, cloud storage, and ML techniques to enable data accountability and shareability of personal health data derived from wearable devices. Users hold the right to own, control, and share their personal health data in a secure fashion	1. High risk of data leakage upon decrypting the transferred data.
McGhin et al. (2019)	Blockchain technology	Presents a clear roadmap on the application of blockchain technology to handle topmost concerns of healthcare applications, including transfer of medical records, data sharing, interoperability, and authentication	1. Scalability, privacy concerns, decentralized storage, and lack of standardization 2. Key management and IoT overhead
Shickel et al. (2018)	DL framework	Provides a detailed account of current research efforts on adopting DL techniques for clinical tasks based on EHR data	1. Deep EHR interpretability and lack of universal benchmarks. 2. Data heterogeneity and unified representation
Chenthara et al. (2020)	Ethereum, POA, Smart contracts, IPFS	Provides a foundational background on blockchain-based security solutions to greatly enhance security, privacy, integrity, and scalability of healthcare records. Model is built on hyperledger fabric (permissioned distributed ledger), IPFS	1. Model does not extend to multiple nodes. 2. Data leakage on purpose or accidentally by patients upon decryption of updated records

(Continued)

Table 9.1 Existing solutions to solve electronic health records concerns. *Continued*

Reference	Technologies used	Main contributions	Limitations
		(data storage), and a cryptographic public-key encryption (data encryption)	
Nguyen et al. (2019)	IPFS and blockchain technology	Considers the beneficial combination of the decentralized IPFS and blockchain to facilitate flexible and secure data exchanges on a mobile cloud platform	1. Data uploading privacy and low network latency 2. Scalability
Rifi et al. (2017)	Smart contract role-based access control	Proposed to solve the topmost concerns of EHRs by exploiting distinctive benefits of smart contracts and security solutions of blockchain technology. More so, users can manage their personal health data and share it securely.	1. Bottleneck may appear in terms of privacy preservation 2. Implementation environment is limited 3. Related cost
Liu et al. (2018)	Cloud storage	Proposes a blockchain-based privacy preserving technique to improve the shareability of healthcare data. Data access is stored securely in the cloud, thus reducing storage burden of blockchain	1. High risk of collision attacks from requestors and cloud servers 2. Building fullytrusted third parties remain a prime challenge

BSN, *Body sensor network;* DL, *deep learning;* IPFS, *interplanetary file system;* IoT, *Internet of Things;* POA, *proof of authority;* SKE, *symmetric key encryption.*

9.4 Federated learning-based electronic health records

Over the years, healthcare systems worldwide have been faced with the challenge of advancing patient care delivery, patient outcomes, and health data research while working within limited healthcare budgets (Oyebiyi et al., 2023). The advent of EHRs has revolutionized the healthcare sector by improving care coordination, data accessibility, information management, patient safety, data-driven decision-making, evidence-based practices and health research, and cost-effectiveness. With EHRs, the errors associated with manual record-keeping, such as misplaced files, duplication of efforts, or illegible handwriting, are significantly reduced. Moreover, healthcare providers can track medical histories, access critical data in real time, and share patient records, leading to personalized care plans, proactive interventions, and seamless collaboration among healthcare organizations (Cowie, 2017). However, the sensitive and distributed nature of EHR in real time applications poses significant challenges in analyzing and aggregating the data for meaningful insights. Specifically, upholding the necessary data privacy, obtaining large and diverse datasets, learning from data residents in health-related organizations, and securing data sharing protocols present a prime challenge for EHR systems (Pilares et al., 2022).

Although, recent breakthroughs in ML-based healthcare solutions hold immense promise and innovation in a variety of medical data processes (disease diagnosis, segmentation, and biomedical data analysis) by learning digital healthcare information obtained from EHRs, conducting ML on EHRs faces several ethical and legal hurdles, given the sensitive nature of healthcare data (Li, 2021). It is worth mentioning that ML technologies and modern data mining typically require a large amount of high-quality medical data covering a full range of possible anatomies and meaningful patterns (Kavitha, 2022). In practice, small healthcare providers may lack sufficient training data to learn more informative patterns, which may skew their prediction performance when applied to a different setting. Linkage across multiple sources and central data repositories has proven to be quite expensive and vulnerable to user information leakage (Rieke, 2020). Interestingly, ML on EHRs across multiple centers can become significantly more feasible if a decentralized approach to collaborative learning (e.g., FL), which eliminates the need for aggregating all data in a central location, is adopted (Joshi et al., 2022).

FL is becoming an innovative candidate technology to mitigate the shortcomings of centralized data storage by sharing only aggregated mathematical parameters and metadata across multiple institutions without exchanging the actual data, thereby enhancing the model's external validity and preserving patient privacy (Pfitzner et al., 2021). In this context, the models are trained locally in each healthcare ecosystem using their own EHR data. The training process involves iteratively updating the model using local datasets, which are then shared with a central server, where aggregation occurs to create a global model. Summarily, the implementation of FL involves three basic steps: initial model updates, local model training, and global model aggregation (Nguyen, 2022). As opposed to its centralized counterpart, FL is a prominent approach to bringing algorithms closer to the data. The fundamental distinction between centralized learning (CL) and FL is illustrated in Eq. (9.1) (Majeed et al., 2022):

$$\text{Case (CLFL)} = \begin{cases} \text{data} \rightarrow \text{algorithms, CL} \\ \text{algorithm} \rightarrow \text{data, FL} \end{cases} \tag{9.1}$$

FL is built upon the federated averaging (FedAvg) technique, which forms the basis of its operation. In each round of FedAvg, the objective is to minimize the global model's parameter 'w', which is the sum of the weighted average of the losses from the local device. A simplified overview of the FedAvg algorithm's operation is given in Eq. (9.2) (Qammar et al., 2022):

$$w_t + 1 = w_t + \eta \cdot \frac{\sum_{i=1}^{kN} d_i \Delta w_{t+1}^i}{\sum_{i=1}^{kN} d_i} \tag{9.2}$$

In this context, the set of data owners, denoted by N and d_i, represents the particular dataset of each owner. Moreover, $w_t + 1$ accounts for the updated model at time $t + 1$, w_t accounts for the previous model updates at time t, η represents the learning rate of the global FL model, and Δw_{t+1}^i accounts for the updated parameters of each global model. At time t, the FL server randomly chooses a subset of clients, denoted as k, from the entire client pool $N(0 < K \leq 1)$. Each participant is fed a global model with weights w_t after which local training for E epochs of gradient descent is performed. Then, these newly trained parameters or gradients are sent back to a central server via an encrypted communication link, which aggregates the local models to obtain a new global model, denoted as $w_t + 1$. This training process is iterative and continues for an additional d rounds of communication.

Given the distinctive characteristics of FL in supporting healthcare services while adequately ensuring data privacy, the state-of-the-art has gained attention for providing adaptable as well as confidential EHR systems. Healthcare institutions can potentially derive much more insights and reliable solutions for data harmonization, analysis, and processing in EHR systems. Moreover, FL for EHRs offers several benefits in terms of collaborative opportunities, increased data diversity, data sovereignty, and privacy preservation (Xu et al., 2021). A figure representing the application of FL in a hospital setting is presented in Fig. 9.6.

Several studies have characterized the effectiveness of solving healthcare challenges using the FL paradigm. Mondrejevski et al. (2022) characterized the performance of an FL model aimed at predicting in-hospital mortality among intensive care unit (ICU) patients. The findings revealed that models trained within the FL setup achieved comparable performance to those trained centrally. An innovative federated reinforcement learning system that incorporates the double deep Q network to offer personalized clinical decision support leveraging data from EMRs and smart devices at the edge is proposed by Xue (2021). Lu et al. (2020) presented an innovative FL framework for extracting latent features from patient data. Lee (2021) presented a comparative analysis between the performance of conventional DL techniques and FL regarding ultrasound image analysis to ascertain whether thyroid nodules were benign or malignant. Kim et al. (2017) proposed an innovative federated tensor factorization model to enable computational phenotyping of massive EHRs. Going forward, Salim and Park (2023) presented an insightful exposition on the performance of an FL-based EHR sharing model designed to secure valuable hospital biomedical data. Available results validate the superiority of the proposed FL-based convolutional neural network (CNN) as opposed to the traditional centralized approach.

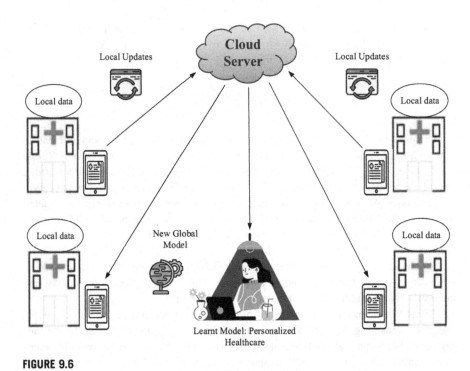

FIGURE 9.6

Federated learning architecture applied in a hospital setting to share sensitive data across multiple hospitals.

9.4.1 Learning algorithms

9.4.1.1 FedAvg

FedAvg, proposed by Google (Brendan et al., 2017), was presented as a communication-efficient and privacy-centric mechanism for FL and is the most widely adopted FL algorithm. The state-of-the-art assumes uniform distribution across participating devices with the aim of reducing the global model's objective 'w.' Unfortunately, FedAvg is best suited for cases with statistical homogeneity (Ahmad et al., 2023). In the event that all local datasets are non-IID, the convergence rate performs poorly. Therefore the problems posed by data heterogeneity remain a prime challenge for FedAvg. A typical illustration of how the FedAvg algorithm operates is expressed in Eq. (9.3) (Shaheen et al., 2022):

$$minf(w) = \sum_{k=1}^{N} p_k F_k(w) \tag{9.3}$$

where p_k denotes the weight of participant k; N denotes the number of participants; and $F_k(w)$ accounts for the loss on device k.

9.4.1.2 Federated proximal

Federated proximal (Fedprox) addresses the challenge of heterogeneity in federated systems by correcting client drifts, which arise due to variations among clients. In this case, a proximal term in conjunction with partial information from stragglers is adopted to constrain local updates to remain closer to the most recent global update, enhancing the convergence rate compared with FedAvg (Li et al., 2020). Therefore the local objective with the added proximal term is optimized, leading to improved convergence behavior.

9.4.1.3 Stochastic controlled averaging algorithm

In Karimireddy et al. (2020), a novel algorithm called stochastic controlled averaging algorithm (SCAFFOLD) is proposed to address the challenges of data heterogeneity and slow convergence using control variates. SCAFFOLD works by estimating the extent of drift between the server model and each client and making appropriate adjustments to their local updates, leading to a substantially reduced effect of data heterogeneity and fewer communication rounds required for convergence.

9.4.1.4 Sparse ternary compression

Building upon the top-k gradient technique by incorporating downstream compression, Sattler et al. (2020) introduced an insightful sparse ternary compression (STC) algorithm as an innovative solution to fulfill the overall requirement of FL. A comparative analysis between the proposed algorithm and the FedAvg algorithm in both IID and non-IID scenarios is performed using the CIFAR-10 and MNIST datasets. The results indicate that the STC method outperforms the FedAvg algorithm considerably, particularly in non-IID environments, offering a significant improvement in FL performance.

9.4.1.5 Federated mediation

Federated mediation (FedMed) is proposed by Yin et al. (2018) as a distributed optimization approach aimed at minimizing the uncertainty in FL settings, comparable with a Byzantine failure, while achieving statistical optimality. The algorithm accomplishes this by utilizing a coordinate-wise median to aggregate local solutions, requiring only a single communication round. It is important to note that the proposed FedMed algorithm effectively ensures optimal statistical performance while providing resilience against Byzantine failure.

9.4.1.6 Federated primal-dual algorithm

A federated primal-dual (FedPD) algorithm is a novel meta-algorithm introduced by Zhang et al. (2020) aimed at addressing general nonconvex objective functions in FL. The FedPD algorithm builds upon the primal-dual optimization strategy and is targeted at maximizing the communication and optimization complexities, especially in scenarios where the environment is non-IID.

9.4.1.7 Federated dropout (Fed-Dropout)

Asad et al. (2021) characterized a lossy-based compression and a federated dropout (Fed-Dropout) model to minimize expenses resulting from server-to-participant interactions. In this case, a submodel is generated using activation functions during a predetermined set of iterations within the fully connected framework. This submodel is then shared with the participants for their local training. Subsequently, the updated submodel is transmitted to the global model, resulting in an absolute deep neural network paradigm. Simulation results validate the superiority of the proposed Fed-Dropout algorithm in effectively minimizing the communication costs from the server to participants as well as decreasing the size of updates sent from participants to the server.

9.4.2 Statistical concerns in federated learning-based electronic health records

Conventional ML models are often built under the assumption that data is IID. In real-world situations, EHR systems are intrinsically heterogeneous as medical data is generated, collected, and distributed in an inconsistent manner. Moreover, with the synergistic interconnection of multiple healthcare institutions involving different numbers of data points, data samples, device availability, and device capabilities, the problem of statistical heterogeneity in FL systems presents a prime challenge when training ML models. As such, federated algorithms must be heterogeneity-aware, robust, and capable of handling faults (Tahir & Ali, 2022). This section discusses solutions to tolerate statistical heterogeneity across medical data providers. Specifically, an insightful exposition of the performance of consensus and pluralistic solutions is presented.

9.4.2.1 Consensus solution

The conventional centralized models are optimized to minimize losses concerning uniform distribution. These models are trained using aggregated training data derived from locally drawn samples across clients and can be represented by the following equation:

$$\overline{\mathcal{U}} = \sum_{k=1}^{K} \frac{n_k}{n} \mathcal{U}_k,$$

where $\overline{\mathcal{U}}$ specifies the target data distribution for the model (Smith et al., 2017).

This uniform distribution is often insufficient to handle the issue of data diversity. The consensus solution has been explored in recent times to alleviate this bottleneck (Zhu et al., 2021). This involves modeling the target distribution or enforcing data adaptation toward a uniform distribution. One such approach is the minimax optimization scheme known as agnostic FL (AFL), proposed by Mohri et al. (2019). AFL optimizes the centralized model for any target distribution formed by a mixture of client distributions. Going forward, Du et al., (2021) characterized the performance of a fairness-aware AFL model under data shift scenarios, taking into consideration reweighing functions on each training sample for fairness constraint and loss function. Results from experimental studies demonstrate the effectiveness of the proposed centralized model on unknown testing data. Li et al. (2020) introduced a novel q-Fair FL optimization scheme to reduce variance in the distribution of accuracy within the network. The optimization objective, which is aimed at assigning higher weights to participants with poorer performance, is effectively solved using a communication-efficient q-FedAvg algorithm. Compared with existing baselines, the proposed technique shows significant improvement in terms of flexibility, fairness, and efficiency.

9.4.2.2 Pluralistic solution

Although the consensus solution holds promising potential for modeling the target distribution or enforcing data adaptation toward a uniform distribution, finding a consensus solution that benefits all components is generally challenging. In this regard, heterogeneity-aware federated solutions are gaining momentum in the research community. Multitask learning (MTL) comes naturally into this realm, and it has certainly found application in handling data from various distributions. MTL exploits the distinctive interplay between unbalanced and non-IID data by establishing their interconnectedness to learn a single global model (Li et al., 2020). Consequently, graph-based relationships, sharing low-rank structure and sparsity, may be adapted to capture the relatedness among unbalanced and non-IID data. Specifically, Smith et al. (2017) characterized the performance of a novel system-aware optimization paradigm, MOCHA, to address the challenge of distributed MTL. The MTL problem considers the issues of fault tolerance, stragglers, and high communication overhead. Simulation results show that the proposed technique achieves significant speedups in comparison with existing baselines. As a step further, Liu et al. (2022) introduced a novel federated multitask graph learning model to solve the problem of MTL resulting from multiple analysis tasks on decentralized graphs. The results obtained validate the superiority of the proposed framework in terms of effectiveness and scalability. Regardless of the potential presented by MTL, its success is dependent on the validity of the chosen relatedness assumptions. In contrast, the pluralistic solution holds great promise in handling statistical and system heterogeneity without requiring the relatedness between diverse datasets as in MTL (Xu et al., 2021). Konečný and Richtárik (2017) provided insightful analysis of the performance of a semistochastic gradient descent framework to minimize the average of a large number of smooth convex loss functions. Moreover, Eichner et al. (2019) demonstrated the effectiveness of a pluralistic approach in the presence of block-cyclic data. Remarkably simple, the pluralistic approach provides a modest solution to the challenge of data heterogeneity, even outperforming the ideal IID baseline when the component distributions are distinct.

9.4.3 Communication efficiency of federated learning-based electronic health records

In FL settings, the server and participating devices, which can sometimes be in the millions, engage in multiple communication rounds until the desired accuracy is achieved. During each communication round, the central server shares a lightweight version of the global model across devices within the federated environment, after which their local models are sent back to the central server for aggregation toward the fully trained global model. The presence of a substantial client base, combined with statistical heterogeneity and limited edge node computation and network bandwidth, introduces challenges that contribute to sluggish communication rounds (Asad et al., 2021). It is crucial to optimize the communication overhead to improve efficiency. While certain algorithms, such as STC and Fed-Dropout, implicitly improve the communication bandwidth by enhancing the training process during each communication round, several other communication-efficient solutions have been proposed and are being evaluated (Pouriyeh, 2022). This section introduces insightful propositions to improve the communication efficiency of FL-based EHR systems.

9.4.3.1 Client selection

Client selection is one of the most straightforward techniques to overcome the challenge of communication overhead in FL environments. This is achieved by restricting the number of participating clients or choosing which parameter could be updated at each communication round. Therefore this approach determines which clients should be involved in the training process and what fraction of parameters should be aggregated for the local updates (Xu et al., 2021). Specifically, Nishio and Yonetani (2019) introduced a novel FL protocol, federated learning with client selection (FedCS), that solves the problem of client selection with resource constraints. In this regard, the central server is tasked with managing the resources of multiple clients while actively aggregating as many client updates as possible based on various resource information, such as computation capabilities and wireless channel states. Results from experimental evaluation reveal that FedCS outperforms the original FL protocol in terms of bandwidth cost and training latency. Xu and Wang (2021) posited that communication rounds are not only temporally interdependent but also have varying relevance to the final learning outcome. As such, the study proposes an algorithm to provide desirable client selection patterns that are adaptive to wireless FL environments. Data-driven experiments validate the efficiency of the proposed model as opposed to benchmark schemes. Anh et al. (2019) introduced a mobile crowd-ML (MCML) algorithm that allows the server to learn and make optimal decisions on mobile device resource management. Their proposed MCML algorithm is shown to reduce training latency and energy consumption considerably. Moreover, Zhang et al. (2021) jointly characterized the performance of a load balancing and computation offloading model to optimize the training and computation overhead of mobile-edge computing networks. This technique also holds great promise for minimizing communication delays in FL settings.

9.4.3.2 Model update reduction

One possible solution to potentially lower the communication cost is to have fewer but more efficient model updates. Therefore effectively ascertaining how these updates are computed can further minimize the required communication rounds, resulting in improved communication overhead (Xu et al., 2021). Kamp et al. (2019) introduced a novel and dynamic model averaging approach for training that adjusts according to the effectiveness of communication without compromising the predictive performance of the model. The results manifest a modest trade-off between communication and model performance. Wu and Wang (2021) presented an investigative analysis of the performance of a federated adaptive weighting algorithm designed to dynamically reinforce positive node contributions for updating the global model per round of communication. Their approach proves valuable in non-IID environments, resulting in greater communication efficiency. Besides, Li et al. (2021) proposed to extend the perspective on one-shot FL by presenting an innovative one-shot algorithm named FedKT. This algorithm, applied to FL settings using only a single round of communication between the central server and participating devices, is shown to be better suited for one-shot communication. As a step further, Wang et al. (2019) considered the synergistic interconnection between the FL framework and deep reinforcement learning techniques with edge systems. The proposed "In-Edge AI" architecture is designed to provide improved training and inference of the models while preserving communication bandwidth. Although some unnecessary system communication burdens are minimized, the overhead of learning is relatively low.

9.4.3.3 Model compression

FL requires continuous downlink and uplink communication during each round, and the number of communication rounds incurs heavy communication overhead. Model compression seeks to mitigate this roadblock by compressing server-to-client exchanges. This is achieved through two main approaches: structured updates and lossy compression (Xu et al., 2021). A technique where the update is learned directly from a parameter-restricted space represented using sparse variables is referred to as structured updates (KPMG LLP, 2018). Specific examples include model distillation, weight quantization, minimizing the least useful connections in a system, and sparse or low-rank representations (Chen et al., 2020; Zhu & Jin, 2020). Moreover, a technique where a full model update is initially learned and subsequently compressed via the interplay of subsampling, random rotations, and quantization before being sent to the server is referred to as lossy compression (Agarwal et al., 2018). The Fed-Dropout algorithm delineated above is one such approach that minimizes both the size of uplink updates and the downlink communication cost while lowering local computational costs as well (Asad et al., 2021). Su et al. (2023) proposed to optimize the performance of an FL framework with lossy communications using model compression, retransmission, and forward error correction. Compared with existing baselines, the proposed system is shown to provide a significant reduction in communication traffic. Similarly, Jiang (2023) presented an adaptive model pruning framework to jointly optimize the communication and computation efficiency of FL networks over heterogeneous scenarios. Chang et al. (2023) adopted the model pruning technique to design a communication-efficient FL model. Available results indicate a robust improvement in communication efficiency with acceptable accuracy.

9.4.3.4 Peer-to-peer learning

As previously mentioned, within an FL setup, a central server interacts with several participating devices, which can often reach into millions, conducting multiple communication rounds in a decentralized manner. Despite the promising potential presented by this approach, the issue of communication overhead remains a major bottleneck. Moreover, relying on a central server to coordinate the training process may prove costly, as its failure could lead to training interruptions. Fully decentralized or peer-to-peer learning has appeared as an alternative approach to enable clients to learn from their neighbor's knowledge without the need for a central server (Tang et al., 2018). The participating node communicates only with its one-hop neighbors and each node updates its model on its dataset and then aggregates information from its neighbors in the network (Chen et al., 2022). He et al. (2019) considered the performance of a central server-free FL model algorithm aimed at a generic social network setting. The results from a rigorous regret analysis reveal that the proposed online push-sum algorithm effectively manages the intricate network while achieving optimal convergence rates. In addition, Lalitha et al. (2019) introduced a distributed learning algorithm designed to train ML models in a fully decentralized manner. Each node updates its local model via aggregated information from its immediate neighbors. In contrast to node learning without cooperation, the proposed system indicates a robust improvement in the accuracy of learning.

9.4.4 Data security and privacy

In the following sections, the security and privacy challenges associated with FL networks are presented. Interestingly, FL inherently possesses a privacy-preserving characteristic and holds a crucial role in data-driven industries. It is also worth mentioning that a typical FL setup involves an excess supply of participating clients, such as medical organizations and phones, potentially numbering in the thousands or millions, increasing the risk of clients being malicious. Although the technique of FL is aimed at training models locally without explicitly centralizing the training data, clients may infer from data across multiple entities. As a result, the risk of user information leakage and data breaches remains a major roadblock across various aspects of the FL system. To gain a comprehensive understanding, a summarized discussion of common vulnerabilities as well as potential solutions is presented (Long et al., 2022).

9.4.4.1 Common vulnerabilities associated with federated learning
9.4.4.1.1 Client data manipulations
Unfortunately, there is no constant verification of local devices or clients in the context of FL frameworks, which implies that a compromised device or client could deliberately or unintentionally learn from malicious data. As a result, incorrect gradients may be uploaded to the server.

9.4.4.1.2 Communication protocol
Although the data stored on a device remains local and is not directly transmitted, it is essential for the client to communicate with the central server. This communication plays a pivotal role in transferring global parameters from the central server to the local user and sending newly acquired local updates from the user's device back to the central server. However, this network-based interaction heightens the susceptibility to eavesdropping, exposing the system to potential security breaches.

9.4.4.1.3 Weaker aggregation algorithm
The central server employs an aggregation algorithm to incorporate updates obtained from various local clients into the global model. This algorithm is designed to ensure the system's integrity by detecting abnormal client updates and discarding updates from skeptical clients. Nonetheless, weaker algorithms such as FedAvg lack this essential structure, increasing the system's vulnerability to data manipulation. Fig. 9.7 presents a diagrammatic illustration of possible threats to FL.

9.4.4.2 Potential solutions
The following FL-based solutions can be leveraged to eliminate or minimize the vulnerabilities associated with FL frameworks:

9.4.4.2.1 Data sanitization and pruning
These are widely adopted approaches for safeguarding against backdoor and poisoning attacks in FL systems. While data sanitization is utilized in anomaly detection to eliminate questionable training scenarios, data pruning finds application in assessing whether a unit is expected to remain dormant during clean data but activated during updates (Koh et al., 2022; Liu et al., 2019). Unfortunately, these defense techniques raise serious questions as regards efficiency in the event of

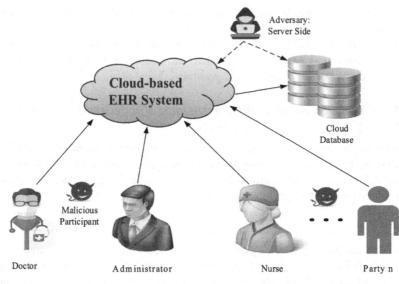

FIGURE 9.7

Possible threats to FL. *FL*, Federated learning.

stronger data poisoning attacks and violation of the privacy principle as it relates to accessing clean holdout data.

9.4.4.2.2 Secure multiparty computation

Recently, some researchers explored the distinctive combination of differential privacy (DP) and secure multiparty computation (SMC) to protect against data leakage (Owusu-Agyemeng et al., 2023). In SMC settings, the concept of homomorphic encryption, which is a public key system that enables parties to encrypt their data using a known public key, is exploited to preserve probable privacy (Fang & Qian, 2021). In this case, individual participants utilize oblivious transfer as well as cryptographic techniques to collaboratively compute a function of their private data. Thus, more participants can collaborate as the parties can run computations on encrypted data across multiple institutions (Bonawitz, 2017).

In the context of FL and ML, SMC can be utilized to bolster data security and privacy through various means. SMC ensures the confidentiality of individual data contributions from each participant and allows the computation of aggregated updates or model parameters without the necessity for any party to directly disclose its private data.

SMC contributes to resilience against malicious participants or nodes in FL. Even if some parties act maliciously or provide inaccurate information, the computation can still yield correct results, provided a sufficient number of participants behave honestly. Additionally, SMC facilitates the computation of functions over encrypted data, particularly advantageous in FL scenarios where data resides on diverse devices with varying security levels. Ongoing advancements in cryptography and optimizations are making SMC and FL increasingly practical for real-world applications.

It is crucial to note that, while SMC provides robust privacy and security enhancements to FL and ML, it introduces challenges related to computational overhead, scalability, and algorithmic complexity. It is also less resistant to mitigating adversaries from gaining certain individual insights. Additionally, due to the need for recurrent communication between clients regarding certain encrypted results and iterated encryption or decryption, SMCs tend to incur high communication overhead and are vulnerable to attacks from colluding parties (Ohata, 2020). Careful consideration of the trade-offs between privacy, security, and performance is necessary when deciding to implement SMC for FL-based EHR systems.

9.4.4.2.3 Differential privacy

DP, a widely adopted method for safeguarding the privacy of individual data, introduces random noise into client privacy data to prevent the leakage of personal information (McMahan et al., 2018). This technique also finds widespread deployment in DL analysis, support vector machine, and principal component analysis (Zheng et al., 2020). DP has become a widely studied concept in FL research because it ensures that the inclusion or exclusion of any particular data point does not significantly affect the analysis outcome. However, due to its lossy nature and the introduction of random noise into the gradient updates, DP may lead to reduced prediction accuracy and model degeneration. This technique is also accompanied by a certain degree of loss in statistical data quality and difficulty for the central server to validate the uploaded models from clients as the number of parties increases (Cheng, 2021). It is also interesting to note that

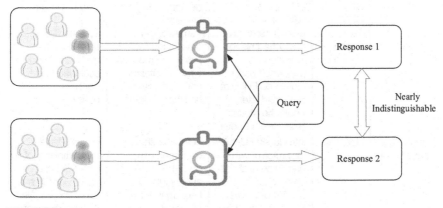

FIGURE 9.8

Differential privacy.

disruptive defense strategies to mitigate against gradient-based attacks exist. This is achieved by increasing the difficulty of the attacks without compromising model performance to the same degree. The following is a proposition by Wei et al. (2020) and Ouadrhiri and Abdelhadi (2022) to realize this objective: altering the activation function in the local model, using higher-resolution training images or videos in local clients, and lifting the batch size of mini-batch stochastic gradient descent in clients. Fig. 9.8 is a diagrammatic illustration of a DP privacy analysis designed to guarantee a unanimous conclusion among viewers. In other words, response 1 and response 2 are nearly indistinguishable.

9.5 Industrial use cases of federated learning implementation in electronic health record systems

As a disruptive technique to preserve patient-level information while ensuring improved solutions for intelligent data analytics in EHR management, FL has shown tremendous potential in smart healthcare applications. More precisely, the integration of FL paradigms into EHR systems is one promising source for an amalgamation of critical biomedical research while adequately preserving the privacy of users. A comprehensive summary of several representative industrial use cases of the state-of-the-art in EHR systems is presented in Table 9.2.

Table 9.2 Industrial use cases of federated learning implementation in electronic health record systems.

Reference	FL clients	Key contributions	Results obtained	Challenges observed
Vaid (2021)	Mount Sinai COVID-19 Health Center	Mortality prediction of hospitalized patients with COVID-19 within 7 days of admission. Training was performed using real-world EHRs at 5 different sites within the Mount Sinai Health Center	Available results demonstrate the efficacy of FL framework in developing robust predictive models while maintaining confidentiality of healthcare data	Issues related to load balancing, model accessibility, additional data modalities, and scalability have not been considered
Boughorbel et al. (2019)	Hospitals	Introduced FUALA for preterm-birth prediction analytics on a cohort of 87,574 deliveries spread across top 50 hospitals with the most delivery episodes	FUALA is shown to outperform both the FedAvg and weighted FedAvg in terms of model uncertainty prediction from distributed EHR	Issues related to the complexity of local training at each hospital have been ignored. Also, the uncertainty of the model when making mistakes is quite significant

(Continued)

Table 9.2 Industrial use cases of federated learning implementation in electronic health record systems. *Continued*

Reference	FL clients	Key contributions	Results obtained	Challenges observed
Choudhury (2020)	Hospitals	Characterized the performance of a differential privacy-based FL framework capable of learning from distributed health data. The evaluation process is performed leveraging real-world EHRs and administrative claims data of 1 million patients held locally at different sites	Results from experimental and comparative analysis indicate the feasibility and effectiveness of the proposed model in mitigating the drawbacks associated with CL	Although an elevated level of privacy is achieved, the predictive capability of the global model considerably deteriorated due to introduction of excessive amount of noise resulting from differential privacy scheme
Brisimi et al. (2018)	Boston Medical Center	Hospitalization prediction during a target year for cardiac events spread among various data sources. Adopts an iterative cPDS algorithm for different agents, $m \in \{5, 10\}$, to solve the sparse support vector machine problem	cPDS algorithm is shown to achieve similar prediction accuracy and faster convergence rate compared with centralized barrier scheme and alternative distributed algorithms	Improved convergence rate is at the expense of some communication overhead between agents. Analysis of cPDS under time-varying conditions has not been considered
Choudhury et al. (2019)	Data clients	Presented an FL-based approach to mitigate the challenges associated with existing ADR prediction models. Training data was procured from the EHR of over 1 million commercially insured patients held locally at different sites for a 5-year period	The predictive capability of the FL model is shown to outperform localized learning models while achieving comparable performance with CL models. Moreover, for varying data distributions, results obtained manifest a robust improvement in terms of accuracy, recall, and precision	Model is simple with a lack of detailed analysis on algorithms with gradient boosting, decision trees, as well as large-scale distributed datasets. Also, privacy preservation against adversarial attacks while training the models was not effectively verified.

ADR, *Adverse drug reaction;* cPDS, *cluster primal-dual splitting;* EHR, *electronic health record;* FedAvg, *federated averaging;* FUALA, *federated uncertainty-aware learning algorithm.*

9.6 Open research issues

FL-based EHR systems are still in a very infant stage, and certainly not all technical concerns have been solved. This section provides a useful resource to guide future research by summarizing the various open challenges associated with this framework.

9.6.1 Security and privacy

Although FL holds great promise in upholding the security and privacy of sensitive health-related data and has been among the initial objectives for exploring the state-of-the-art, several participants on the client side and external adversaries on the server side may pose serious concerns classified as unintentional information leakage, membership inference attacks, and generative adversarial systems. Therefore, a participant may try to poison local data and shared updates, while an external adversary may try to access sensitive information about the aggregated global model. Although more recent security solutions, such as collaborative training, SMC, and DP, have been largely characterized, this is an ongoing and interesting direction requiring further exploration to mitigate malicious attacks and minimize performance trade-offs.

9.6.2 Resource management

In addition to data accumulated in medical centers, EHRs encompass not only data from mobile phones but a broader range of sources such as wearable devices, Internet of Medical Things devices, ML model providers, and the cloud. Given the heavy communication overhead and computation requirements as well as the dynamic and rapid variations of wireless channels during the federated model training process, proper resource management can provide considerable improvement in the learning performance. Thus, well-designed and robust resource management schemes are required to facilitate efficient FL-enabled EHR systems.

9.6.3 Unsupervised learning

In current FL research, supervised or semisupervised techniques are predominantly adopted. However, in real-world scenarios, the issue of label deficiency is evident, prompting the exploration of alternative methods. Although unsupervised learning is intrinsically more difficult than supervised learning as it requires a large training set to produce outcomes, it will undoubtedly be a focus of study in the coming years.

9.6.4 Heterogeneity

Heterogeneity (statistical and system) in FL networks can introduce numerous bottlenecks, such as training divergence, quality degradation, or even long convergence times during the training process. Although consensus and pluralistic solutions have been characterized in this study, there is a need for more sophisticated solutions to overcome the non-IID challenge in FL-based EHR systems.

9.6.5 Secure aggregation

An important process in FL is aggregating newly learned local gradients or parameters to obtain an updated global model. In this regard, a potential concern for medical data providers is ensuring secure and trustworthy aggregation of learning models at the aggregation server, as the presence of malicious users or computationally expensive verification protocols may slow down the convergence rate of FL. Consequently, secure and trustful aggregation protocols for FL using low-complexity schemes constitute an attractive future research direction.

9.6.6 Data quality

Thanks to the rapid development of the FL setup, isolated hospitals, and health organizations can share experiences while maintaining privacy. However, the heterogeneity of data standards, data quality, and computational capabilities across multiple medical institutions can significantly degrade the training quality in FL settings. Therefore well-designed incentive mechanisms are critical to realizing high-quality data training and aggregation requirements.

9.7 Conclusion

FL is an emerging collaborative paradigm that holds promising potential for eliminating personal data orchestration at a central location while preserving provable privacy guarantees. This new learning approach has found its way into digital healthcare networks and applications. Given the infancy of this framework, this chapter has provided an extensive exposition on the integration of FL paradigms into EHR systems. A summarized introduction to the foundational background, key motivations, technical requirements, and types of FL for smart healthcare is presented. An insightful exposition on EHR concerns, existing proposals to solve issues associated with EHR systems, and FL algorithms to maximize the integrity of valuable hospital biomedical data are also presented. Moreover, innovative solutions to statistical concerns, communication efficiency, data security, and privacy issues are delineated. Finally, various open challenges that exist in FL-based EHR systems are presented. It is worthy of note that the application of federated architecture in collaborative healthcare systems is still in its infancy and is envisioned to provide more efficient healthcare services.

References

Abikoye, O. C., Oladipupo, E. T., Imoize, A. L., Awotunde, J. B., Lee, C. C., & Li, C. T. (2023). Securing critical user information over the internet of medical things platforms using a hybrid cryptography scheme. *Future Internet*, *15*(3), 99. Available from https://doi.org/10.3390/fi15030099.

Agarwal, N., Suresh, A. T., Yu, F., Kumar, S., & Brendan McMahan, H. (2018). *CPSGD: Communication-efficient and differentially-private distributed SGD. Advances in neural information processing systems*.

Ahmad, A., Luo, W., & Robles-Kelly, A. (2023). Robust federated learning under statistical heterogeneity via hessian-weighted aggregation. *Machine Learning*, *112*(2). Available from https://doi.org/10.1007/s10994-022-06292-8.

Al Omar, A., Rahman, M.S., Basu, A., & Kiyomoto, S. (2017). MediBchain: A blockchain based privacy preserving platform for healthcare data, in *Lecture Notes in Computer Science (including subseries Lecture Notes in Artificial Intelligence and Lecture Notes in Bioinformatics)*. Available from https://doi.org/10.1007/978-3-319-72395-2_49.

Amalraj, J. R., & Lourdusamy, R. (2022). Security and privacy issues in federated healthcare − An overview. *Open Computer Science*, *12*(1). Available from https://doi.org/10.1515/comp-2022-0230.

Anh, T. T., Luong, N. C., Niyato, D., Kim, D. I., & Wang, L. C. (2019). Efficient training management for mobile crowd-machine learning: A deep reinforcement learning approach. *IEEE Wireless Communications Letters*, *8*(5). Available from https://doi.org/10.1109/LWC.2019.2917133.

Antunes, R. S., Da Costa, C. A., Küderle, A., Yari, I. A., & Eskofier, B. (2022). Federated learning for healthcare: Systematic review and architecture proposal. *ACM Transactions on Intelligent Systems and Technology*, *13*(4). Available from https://doi.org/10.1145/3501813.

Asad, M., Moustafa, A., Ito, T., & Aslam, M. (2021). Evaluating the communication efficiency in federated learning algorithms. *Proceedings of the 2021 IEEE 24th International Conference on Computer Supported Cooperative Work in Design, CSCWD 2021*. Available from https://doi.org/10.1109/CSCWD49262.2021.9437738.

Batko, K., & Ślęzak, A. (2022). The use of Big Data Analytics in healthcare. *Journal of Big Data*, *9*(1). Available from https://doi.org/10.1186/s40537-021-00553-4.

Bonawitz, K., et al. (2017). Practical secure aggregation for privacy-preserving machine learning. *Proceedings of the ACM Conference on Computer and Communications Security*. Available from https://doi.org/10.1145/3133956.3133982.

Boobalan, P., et al. (2022). Fusion of federated learning and industrial Internet of Things: A survey. *Computer Networks*, *212*. Available from https://doi.org/10.1016/j.comnet.2022.109048.

Boughorbel, S., Jarray, F., Venugopal, N., Moosa, S., Elhadi, H., & Makhlouf, M. (2019). Federated uncertainty-aware learning for distributed hospital EHR data. *arXiv Prepr*, arXiv:1910.12191.

Brendan McMahan, H., Moore, E., Ramage, D., Hampson, S., & Agüera y Arcas, B. "Communication-efficient learning of deep networks from decentralized data," in Proceedings of the 20th International Conference on Artificial Intelligence and Statistics, AISTATS 2017, 2017.

Brisimi, T. S., Chen, R., Mela, T., Olshevsky, A., Paschalidis, I. C., & Shi, W. (2018). Federated learning of predictive models from federated Electronic Health Records.

International Journal of Medical Informatics, 112. Available from https://doi.org/10.1016/j.ijmedinf.2018.010.007.

Chang, M. K., Chan, Y. W., & Wu, T. E. (2023). Communication-efficient federated learning with model pruning," in Lecture notes in electrical engineering. Available from https://doi.org/10.1007/978-981-99-1428-9_8.

Chen, Y., Sun, X., & Jin, Y. (2020). Communication-efficient federated deep learning with layerwise asynchronous model update and temporally weighted aggregation. *IEEE Transactions on Neural Networks and Learning Systems, 31*(10). Available from https://doi.org/10.1109/TNNLS.2019.2953131.

Chen, Z., Liao, W., Tian, P., Wang, Q., & Yu, W. (2022). A fairness-aware peer-to-peer decentralized learning framework with heterogeneous devices. *Future Internet, 14*(5). Available from https://doi.org/10.3390/fi14050138.

Cheng, K., et al. (2021). SecureBoost: A lossless federated learning framework. *IEEE Intelligent Systems, 36*(6). Available from https://doi.org/10.1109/MIS.2021.3082561.

Chenthara, S., Ahmed, K., Wang, H., Whittaker, F., & Chen, Z. (2020). Healthchain: A novel framework on privacy preservation of electronic health records using blockchain technology. *PLoS One, 15*(12). Available from https://doi.org/10.1371/journal.pone.0243043, December.

Choudhury, O., Park, Y., Salonidis, T., Gkoulalas-Divanis, A., Sylla, I., & Das, A. K. (2019). Predicting adverse drug reactions on distributed health data using federated learning. *AMIA Annu. Symp. proceedings. AMIA Symp, 2019*, 313−322.

Choudhury, O., et al. (2020). Differential privacy-enabled federated learning for sensitive health data. *arXiv Prepr*, arXiv:1910.02578.

Coorevits, P., et al. (2013). Electronic health records: New opportunities for clinical research. *Journal of Internal Medicine, 274*(6). Available from https://doi.org/10.1111/joim.12119.

Cowie, M. R., et al. (2017). Electronic health records to facilitate clinical research. *Clinical Research in Cardiology, 106*(1). Available from https://doi.org/10.1007/s00392-016-1025-6.

Dang, T. K., Lan, X., Weng, J., & Feng, M. (2022). Federated learning for electronic health records. *ACM Transactions on Intelligent Systems and Technology, 13*(5). Available from https://doi.org/10.1145/3514500.

Dasaradharami Reddy, K., & Gadekallu, T. R. (2023). A comprehensive survey on federated learning techniques for healthcare informatics. *Computational Intelligence and Neuroscience, 2023*. Available from https://doi.org/10.1155/2023/8393990.

Du, W., Xu, D., Wu, X., & Tong, H. (2021). Fairness-aware agnostic federated learning. *SIAM International Conference on Data Mining, SDM 2021*. Available from https://doi.org/10.1137/1.9781611976700.21.

Eichner, H., Koren, T., McMahan, H. B., Srebro, N., & Talwar, K. (2019). Semi-cyclic stochastic gradient descent. In 36th International Conference on Machine Learning, ICML 2019.

Fang, H., & Qian, Q. (2021). Privacy preserving machine learning with homomorphic encryption and federated learning. *Future Internet, 13*(4). Available from https://doi.org/10.3390/fi13040094.

Garg, S. K., & Chari, S. T. (2020). Early detection of pancreatic cancer. *Current Opinion in Gastroenterology, 36*(5). Available from https://doi.org/10.1097/MOG.0000000000000663.

Gosselin, R., Vieu, L., Loukil, F., & Benoit, A. (2022). Privacy and security in federated learning: A survey. *Applied Sciences, 12*(19). Available from https://doi.org/10.3390/app12199901.

Guo, R., Shi, H., Zhao, Q., & Zheng, D. (2018). Secure attribute-based signature scheme with multiple authorities for blockchain in electronic health records systems. *IEEE Access, 6*. Available from https://doi.org/10.1109/ACCESS.2018.2801266, vol.

He, C., Tan, C., Tang, H., Qiu, S., & Liu, J. (2019). Central server free federated learning over single-sided trust social networks. *arXiv*, arXiv:1910.04956.

Jawadur Rahman, K. M., et al. (2021). Challenges, applications and design aspects of federated learning: A survey. *IEEE Access, 9*. Available from https://doi.org/10.1109/ACCESS.2021.3111118, vol.

Jiang, Z., et al. (2023). Computation and communication efficient federated learning with adaptive model pruning. *IEEE Transactions on Mobile Computing*. Available from https://doi.org/10.1109/TMC.2023.3247798.

Joshi, M., Pal, A., & Sankarasubbu, M. (2022). Federated learning for healthcare domain — Pipeline, applications and challenges. *ACM Transactions on Computing for Healthcare, 3*(4). Available from https://doi.org/10.1145/3533708.

Kamp, M., et al. (2019). Efficient decentralized deep learning by dynamic model averaging, in Lecture notes in computer science (including subseries lecture notes in artificial intelligence and lecture notes in bioinformatics). Available from https://doi.org/10.1007/978-3-030-10925-7_24.

Karimireddy, S. P., Kale, S., Mohri, M., Reddi, S. J., Stich, S. U., & Suresh, A. T. (2020). SCAFFOLD: Stochastic controlled averaging for federated learning, in 37th international conference on machine learning, ICML 2020.

Kavitha, K. V. N., et al. (2022). On the use of wavelet domain and machine learning for the analysis of epileptic seizure detection from EEG signals. *Journal of Healthcare Engineering, 2022*, 1−16. Available from https://doi.org/10.1155/2022/8928021.

Khan, L. U., Saad, W., Han, Z., Hossain, E., & Hong, C. S. (2021). Federated learning for internet of things: Recent advances, taxonomy, and open challenges. *IEEE Communications Surveys and Tutorials, 23*(3). Available from https://doi.org/10.1109/COMST.2021.3090430.

Kim, Y., Sun, J., Yu, H., & Jiang, X. (2017). Federated tensor factorization for computational phenotyping, in Proceedings of the ACM SIGKDD international conference on knowledge discovery and data mining. Available from https://doi.org/10.1145/3097983.3098118.

Koh, P. W., Steinhardt, J., & Liang, P. (2022). Stronger data poisoning attacks break data sanitization defenses. *Machine Learning, 111*(1). Available from https://doi.org/10.1007/s10994-021-06119-y.

Konečný, J., & Richtárik, P. (2017). Semi-stochastic gradient descent methods. *Frontiers in Applied Mathematics and Statistics, 3*. Available from https://doi.org/10.3389/fams.2017.00009, May.

KPMG LLP, "Federated learning: Strategies for improving communication efficiency," Iclr, vol. 2008. 2018.

Kruse, C. S., Stein, A., Thomas, H., & Kaur, H. (2018). The use of electronic health records to support population health: A systematic review of the literature. *Journal of Medical Systems, 42*(11). Available from https://doi.org/10.1007/s10916-018-1075-6.

Lalitha, A., Kilinc, O. C., Javidi, T., & Koushanfar, F. (2019). Peer-to-peer federated learning on graphs. *arXiv*, arXiv:1901.11173.

Lee, H., et al. (2021). Federated learning for thyroid ultrasound image analysis to protect personal information: Validation study in a real health care environment. *JMIR Medical Informatics, 9*(5). Available from https://doi.org/10.2196/25869.

Lee, S. H., & Yang, C. S. (2018). Fingernail analysis management system using microscopy sensor and blockchain technology. *International Journal of Distributed Sensor Networks*, *14*(3). Available from https://doi.org/10.1177/1550147718767044.

Li, Q., He, B., & Song, D. (2021). Practical one-shot federated learning for cross-silo setting. *IJCAI International Joint Conference on Artificial Intelligence*. Available from https://doi.org/10.24963/ijcai.2021/205.

Li, T., Sanjabi, M., Beirami, A., & Smith, V. (2020). Fair resource allocation in federated learning, in 8th International Conference on Learning Representations, ICLR 2020.

Li, T., Sahu, A. K., Zaheer, M., Sanjabi, M., Talwalkar, A., & Smith, V. (2020). Federated optimization in heterogeneous networks tian. *MLSys*.

Li, T., Hu, S., Beirami, A., & Smith, V. (2020). Federated multi-task learning for competing constraints. *arXiv*, no. Section 4.

Li, W., et al. (2021). A comprehensive survey on machine learning-based big data analytics for IoT-enabled smart healthcare system. *Mobile Networks and Applications*, *26*(1). Available from https://doi.org/10.1007/s11036-020-01700-6.

Lim, W. Y. B., et al. (2020). Federated learning in mobile edge networks: A comprehensive survey. *IEEE Communications Surveys and Tutorials*, *22*(3). Available from https://doi.org/10.1109/COMST.2020.2986024.

Liu, B., Wang, L., & Liu, M. (2019). Lifelong federated reinforcement learning: A learning architecture for navigation in cloud robotic systems. *IEEE Robotics and Automation Letters*, *4*(4). Available from https://doi.org/10.1109/LRA.2019.2931179.

Liu, J., Li, X., Ye, L., Zhang, H., Du, X., & Guizani, M. (2018). BPDS: A blockchain based privacy-preserving data sharing for electronic medical records, in 2018 IEEE global communications conference, GLOBECOM 2018 - Proceedings. Available from https://doi.org/10.1109/GLOCOM.2018.8647713.

Liu, Y., Han, D., Zhang, J., Zhu, H., Xu, M., & Chen, W. (2022). Federated multi-task graph learning. *ACM Transactions on Intelligent Systems and Technology*, *13*(5). Available from https://doi.org/10.1145/3527622.

Long, G., Shen, T., Tan, Y., Gerrard, L., Clarke, A., & Jiang, A. (2022). Federated learning for privacy-preserving open innovation future on digital health, in Humanity driven AI. Available from https://doi.org/10.1007/978-3-030-72188-6_6.

Lu, S., Zhang, Y., & Wang, Y. (2020). Decentralized federated learning for electronic health records, in 2020 54th Annual Conference on Information Sciences and Systems, CISS 2020. Available from https://doi.org/10.1109/CISS48834.2020.1570617414.

Majeed, A., Zhang, X., & Hwang, S. O. (2022). Applications and challenges of federated learning paradigm in the big data era with special emphasis on COVID-19. *Big Data and Cognitive Computing*, *6*(4). Available from https://doi.org/10.3390/bdcc6040127.

McGhin, T., Choo, K. K. R., Liu, C. Z., & He, D. (2019). Blockchain in healthcare applications: Research challenges and opportunities. *Journal of Network and Computer Applications*, *135*, 62−75. Available from https://doi.org/10.1016/j.jnca.2019.020.027.

McMahan, H. B., Ramage, D., Talwar, K., & Zhang, L. (2018) Learning differentially private recurrent language models, in 6th international conference on learning representations, ICLR 2018 - Conference track proceedings.

Menachemi, N., & Collum, T. H. (2011). Benefits and drawbacks of electronic health record systems. *Risk Management and Healthcare Policy*, *4*, 47−55. Available from https://doi.org/10.2147/RMHP.S12985.

Mohri, M., Gary, S. & Ananda, T. S. (2019). Agnostic federated learning, in 36th international conference on machine learning, proceedings of machine learning research, pp. 4615–4625.

Mondrejevski, L., Miliou, I., Montanino, A., Pitts, D., Hollmen, J., & Papapetrou, P. (2022). FLICU: A federated learning workflow for intensive care unit mortality prediction, in Proceedings - IEEE symposium on computer-based medical systems. Available from https://doi.org/10.1109/CBMS55023.2022.00013.

Nguyen, D. C., Pathirana, P. N., Ding, M., & Seneviratne, A. (2019). Blockchain for secure EHRs sharing of mobile cloud based E-health systems. *IEEE Access*, *7*. Available from https://doi.org/10.1109/ACCESS.2019.2917555, vol.

Nguyen, D. C., et al. (2022). Federated learning for smart healthcare: A survey. *ACM Computing Surveys*, *55*(3). Available from https://doi.org/10.1145/3501296.

Nishio, T., & Yonetani, R. (2019). Client selection for federated learning with heterogeneous resources in mobile edge," in IEEE International Conference on Communications. Available from https://doi.org/10.1109/ICC.2019.8761315.

Ohata, S. (2020). Recent advances in practical secure multi-party computation. *IEICE Transactions on Fundamentals of Electronics, Communications and Computer Sciences*, *E103A*(10). Available from https://doi.org/10.1587/transfun.2019DMI0001.

Ouadrhiri, A. El, & Abdelhadi, A. (2022). Differential privacy for deep and federated learning: A survey. *IEEE Access*, *10*. Available from https://doi.org/10.1109/ACCESS.2022.3151670.

Owusu-Agyemeng, K., Qin, Z., Xiong, H., Liu, Y., Zhuang, T., & Qin, Z. (2023). MSDP: multi-scheme privacy-preserving deep learning via differential privacy. *Personal and Ubiquitous Computing*, *27*(2). Available from https://doi.org/10.1007/s00779-021-01545-0.

Oyebiyi, O. G., Abayomi-Alli, A., 'Tale Arogundade, O., Qazi, A., Imoize, A. L., & Awotunde, J. B. (2023). A systematic literature review on human ear biometrics: Approaches, algorithms, and trend in the last decade. *Information*, *14*(3), 192. Available from https://doi.org/10.3390/info14030192.

Pfitzner, B., Steckhan, N., & Arnrich, B. (2021). Federated learning in a medical context: A systematic literature review. *ACM Transactions on Internet Technology*, *21*(2). Available from https://doi.org/10.1145/3412357.

Pilares, I. C. A., Azam, S., Akbulut, S., Jonkman, M., & Shanmugam, B. (2022). Addressing the challenges of electronic health records using blockchain and IPFS. *Sensors*, *22*(11). Available from https://doi.org/10.3390/s22114032.

Poongodi, T., & Imoize, A. L. (2021). Internet of things, artificial intelligence and blockchain technology. Cham: Springer International Publishing. Available from https://doi.org/10.1007/978-3-030-74150-1.

Pouriyeh, S., et al. (2022). Secure smart communication efficiency in federated learning: Achievements and challenges. *Applied Sciences*, *12*(18). Available from https://doi.org/10.3390/app12188980.

Qammar, A., Ding, J., & Ning, H. (2022). Federated learning attack surface: Taxonomy, cyber defences, challenges, and future directions. *Artificial Intelligence Review*, *55*(5). Available from https://doi.org/10.1007/s10462-021-10098-w.

Rieke, N., et al. (2020). The future of digital health with federated learning. *npj Digital Medicine*, *3*(1). Available from https://doi.org/10.1038/s41746-020-00323-1.

Rifi, N., Rachkidi, E., Agoulmine, N., & Taher, N. C. (2017). Towards using blockchain technology for eHealth data access management, in International conference on advances in biomedical engineering, ICABME. Available from https://doi.org/10.1109/ICABME.2017.8167555.

Salim, M. M., & Park, J. H. (2023). Federated learning-based secure electronic health record sharing scheme in medical informatics. *IEEE Journal of Biomedical and Health Informatics, 27*(2). Available from https://doi.org/10.1109/JBHI.2022.3174823.

Sattler, F., Wiedemann, S., Muller, K. R., & Samek, W. (2020). Robust and communication-efficient federated learning from non-i.i.d. data. *IEEE Transactions on Neural Networks and Learning Systems, 31*(9). Available from https://doi.org/10.1109/TNNLS.2019.2944481.

Shaheen, M., Farooq, M. S., Umer, T., & Kim, B. S. (2022). Applications of federated learning; taxonomy, challenges, and research trends. *Electron, 11*(4). Available from https://doi.org/10.3390/electronics11040670.

Shahnaz, A., Qamar, U., & Khalid, A. (2019). Using blockchain for electronic health records. *IEEE Access, 7.* Available from https://doi.org/10.1109/ACCESS.2019.2946373, vol.

Shickel, B., Tighe, P. J., Bihorac, A., & Rashidi, P. (2018). Deep EHR: A survey of recent advances in deep learning techniques for electronic health record (EHR) analysis. *IEEE Journal of Biomedical and Health Informatics, 22*(5). Available from https://doi.org/10.1109/JBHI.2017.2767063.

Smith, V., Chiang, C.K., Sanjabi, M. & Talwalkar, A. (2017). Federated multi-task learning, in Advances in neural information processing systems.

Su, X., Zhou, Y., Cui, L., & Liu, J. (2023). On model transmission strategies in federated learning with lossy communications. *IEEE Transactions on Parallel and Distributed Systems, 34*(4). Available from https://doi.org/10.1109/TPDS.2023.3240883.

Tahir, M., & Ali, M. I. (2022). On the performance of federated learning algorithms for IoT. *Internet of Things, 3*(2). Available from https://doi.org/10.3390/iot3020016.

Tang, H., Gan, S., Zhang, C., Zhang, T., & Liu, J. (2018). Communication compression for decentralized training, in Advances in neural information processing systems.

Vaid, A., et al. (2021). Federated learning of electronic health records to improve mortality prediction in hospitalized patients with COVID-19: Machine learning approach. *JMIR Medical Informatics, 9*(1). Available from https://doi.org/10.2196/24207.

Wang, X., Han, Y., Wang, C., Zhao, Q., Chen, X., & Chen, M. (2019). In-edge AI: Intelligentizing mobile edge computing, caching and communication by federated learning. *IEEE Network, 33*(5). Available from https://doi.org/10.1109/MNET.2019.1800286.

Wei, W., et al. (2020). A framework for evaluating gradient leakage. *arXiv e-prints.*

Wu, H., & Wang, P. (2021). Fast-Convergent Federated Learning with Adaptive Weighting. *IEEE Transactions on Cognitive Communications and Networking, 7*(4). Available from https://doi.org/10.1109/TCCN.2021.3084406.

Xu, J., & Wang, H. (2021). Client selection and bandwidth allocation in wireless federated learning networks: A long-term perspective. *IEEE Transactions on Wireless Communications, 20*(2). Available from https://doi.org/10.1109/TWC.2020.3031503.

Xu, J., Glicksberg, B. S., Su, C., Walker, P., Bian, J., & Wang, F. (2021). Federated learning for healthcare informatics. *Journal of Healthcare Informatics Research, 5*(1). Available from https://doi.org/10.1007/s41666-020-00082-4.

Xue, Z., et al. (2021). A resource-constrained and privacy-preserving edge-computing-enabled clinical decision system: A federated reinforcement learning approach. *IEEE Internet of Things Journal, 8*(11). Available from https://doi.org/10.1109/JIOT.2021.3057653.

Yang, Q., Liu, Y., Chen, T., & Tong, Y. (2019). Federated machine learning: Concept and applications. *ACM Transactions on Intelligent Systems and Technology, 10*(2). Available from https://doi.org/10.1145/3298981.

Yin, D., Chen, Y., Ramchandran, K., & Bartlett, P. (2018). Byzantine-robust distributed learning: Towards optimal statistical rates, in 35th international conference on machine learning, ICML 2018.

Zhang, W. Z., et al. (2021). Secure and optimized load balancing for multitier IoT and edge-cloud computing systems. *IEEE Internet of Things Journal, 8*(10). Available from https://doi.org/10.1109/JIOT.2020.3042433.

Zhang, X., Hong, M., Dhople, S., Yin, W., & Liu, Y. (2020). FedPD: A federated learning framework with optimal rates and adaptivity to non-IID data. *arXiv.*

Zhao, H., Zhang, Y., Peng, Y., & Xu, R. (2017). Lightweight backup and efficient recovery scheme for health blockchain keys. In *Proceedings — 2017 IEEE 13th international symposium on autonomous decentralized systems, ISADS 2017.* Available from https://doi.org/10.1109/ISADS.2017.22.

Zheng, H., Hu, H., & Han, Z. (2020). Preserving user privacy for machine learning: Local differential privacy or federated machine learning? *IEEE Intelligent Systems, 35*(4). Available from https://doi.org/10.1109/MIS.2020.3010335.

Zheng, X., Mukkamala, R.R., Vatrapu, R., & Ordieres-Mere, J. (2018). Blockchain-based personal health data sharing system using cloud storage, in 2018 IEEE 20th international conference on e-Health networking, applications and services, Healthcom 2018. Available from https://doi.org/10.1109/HealthCom.2018.8531125.

Zhu, H., & Jin, Y. (2020). Multi-objective evolutionary federated learning. *IEEE Transactions on Neural Networks and Learning Systems, 31*(4). Available from https://doi.org/10.1109/TNNLS.2019.2919699.

Zhu, H., Xu, J., Liu, S., & Jin, Y. (2021). Federated learning on non-IID data: A survey. *Neurocomputing, 465*, 371−390. Available from https://doi.org/10.1016/j.neucom.2021.07.098.

Technical considerations of federated learning in digital healthcare systems

10

Emmanuel Alozie[1], Hawau I. Olagunju[1], Nasir Faruk[1,2] and Salisu Garba[3]

[1]*Department of Information Technology, Sule Lamido University, Kafin Hausa, Jigawa, Nigeria*
[2]*Directorate of Information and Communication Technology, Sule Lamido University, Kafin Hausa, Jigawa, Nigeria*
[3]*Department of Software Engineering, Sule Lamido University, Kafin Hausa, Jigawa, Nigeria*

10.1 Introduction

The technological advancement and the exponential growth of generated data have resulted in the development of several data-driven machine learning (ML) and deep learning (DL) models. These algorithms offer efficient strategies for deriving valuable insights from the massive amounts of medical data generated within modern healthcare systems (Rieke et al., 2020). Radiology, pathology, genomics, and other sciences have benefited significantly from the progress of artificial intelligence (AI), particularly the advancement in ML and DL algorithms (Iguoba & Imoize, 2022; Rieke et al., 2020). Furthermore, DL has demonstrated remarkable results in smart healthcare applications aimed at improving healthcare. It has enabled data-driven medical diagnosis and treatment, recognizing texts in medical lab reports, tumor segmentation and categorization in magnetic resonance imaging (MRI) scans, cancer detection and prognosis, electroencephalogram (EEG) signal classification, and the analysis of biomedical images for early detection of severe ailments, thus enhancing healthcare efficacy (Kavitha et al., 2022; Nwaneri et al., 2022; Prayitno et al., 2021; Rahman et al., 2022). The effectiveness of these algorithms in the digital healthcare system is heavily reliant on the availability of a variety and a significant number of input datasets derived from various sources, including biomedical sensors, distinct patients, clinical associations, healthcare facilities, drug firms, and health insurance firms (Crowson et al., 2022; Prayitno et al., 2021). However, obtaining enough datasets to train and verify the effectiveness of the developed ML and DL algorithms has proven to be a significant issue due to the limited number of patients and ailments that have minimal prevalence rates present in a hospital (Prayitno et al., 2021). Additionally, due to concerns about data protection and management, transferring sensitive biological and medical records between facilities may be challenging.

Federated learning (FL) is an ML-based approach that can be utilized to address data confidentiality and management concerns when developing ML/DL

Federated Learning for Digital Healthcare Systems. DOI: https://doi.org/10.1016/B978-0-443-13897-3.00009-6

models. FL enables the development of ML/DL models using datasets dispersed across data centers, such as medical facilities, clinical research facilities, and mobile devices, while maintaining the privacy of data by developing the algorithms collectively without the need to share any dataset (Joshi et al., 2022; Rieke et al., 2020). With FL, medical data proprietors, such as hospitals and other healthcare facilities, can prevent transferring clinical data. Also healthcare personnel may develop the model locally and share the model parameters with the aggregator for data aggregation. However, despite the numerous benefits of FL, there remain challenges with data standards and quality, as well as implementation impediments. Therefore this chapter provides an overview of the technical considerations of FL algorithms in digital healthcare systems, the technical challenges in executing FL models in real-world applications, and proposed solutions.

10.1.1 Key contributions

This section outlines the contributions of the chapter:

1. An overview of FL, including its various applications in the healthcare domain in conjunction with novel technologies.
2. A critical examination and discussion on the technical considerations of FL algorithms in healthcare systems.
3. An evaluation of the technical challenges of executing the FL model in real-world applications.
4. A review of the proposed solutions to these technical considerations and challenges in FL.

10.1.2 Chapter outline

The remainder of this chapter is outlined as follows: Section 10.2 provides a general overview of FL models, different scenarios and features, as well as the various applications of FL in conjunction with other novel technologies. The technical considerations of FL are provided in Section 10.3. Section 10.4 presents the technical challenges of executing FL models in real-world applications, and Section 10.5 presents the proposed solutions for the technical considerations and challenges in FL. The key lessons learned from the review are presented in Section 10.6, and finally, a conclusion and recommendation are provided in Section 10.7.

10.2 Overview of federated learning

This section provides a background understanding of FL based on the different scenarios, the different types of FL, as well as its various applications in conjunction with novel technologies.

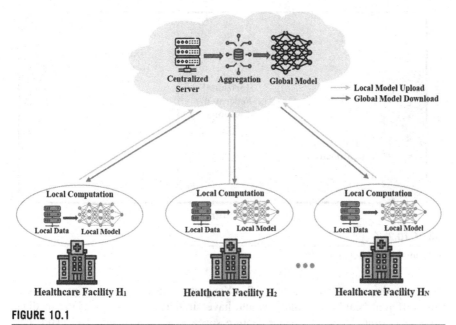

FIGURE 10.1

Federated learning architecture for healthcare.

The conventional ML model approach, which requires a large volume of data to be obtained and sent to a centralized server for easy access to use for developing and evaluating the efficacy of the algorithm, has proven to be inefficient and can be prone to data privacy violations, particularly when it pertains to healthcare or medical data where data privacy and management are very crucial concerns. This challenge has hindered the development of ML models, limiting the impact of ML in some of these sectors.

FL, originally conceptualized by Google, represents a decentralized learning paradigm in which ML/DL models can be developed from datasets distributed across numerous data servers, eliminating data loss, as depicted in Fig. 10.1 (Shaheen et al., 2022; Yang et al., 2019). In contrast to the conventional ML model approach, FL presents a systematic solution to the existing data privacy concerns associated with ML algorithm development. Notably, it eliminates the necessity of sending data to a centralized server. FL can be categorized into three distinct types, each based on specific aspects of the dataset, including sample, label, and feature spaces. These categories are known as horizontal, vertical, and transfer learning.

10.2.1 Horizontal federated learning

Horizontal FL (HFL), also known as sample-based FL, employs datasets with identical feature and label spaces but different sample spaces across participating

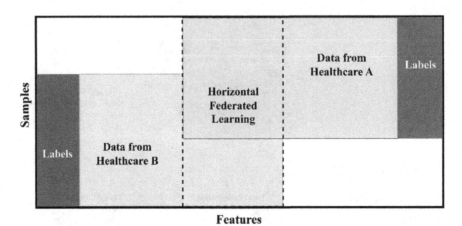

FIGURE 10.2

Horizontal federated learning.

clients (Rodríguez-Barroso et al., 2023; Yang et al., 2019). For example, datasets from different healthcare centers would have similarities in terms of the medical parameters, thus having the same feature space as shown in Fig. 10.2. HFL can be expressed as follows:

$$F_i = F_j, l_i = l_j, S_i \neq S_j \forall D_i, D_j, i \neq j \tag{10.1}$$

where F denotes feature space, l denotes label space, S denotes sample space, D is the dataset of collaborators (i,j).

10.2.2 Vertical federated learning

Vertical FL (VFL), also known as feature-based FL, employs datasets with identical sample spaces but distinct feature and label spaces (Rodríguez-Barroso et al., 2023; Yang et al., 2019). For example, datasets from a local bank and a local healthcare center would contain some similarities in terms of the samples, such as information about most of the residents in that area. However, the features would be different as the bank would retain data about the user's revenue and net worth, while the healthcare center would retain data about the user's ailment, prescribed drugs, etc., as shown in Fig. 10.3.

VFL is the method of collecting several distinct attributes and confidentially estimating learning loss and gradients to develop a model collaboratively, utilizing data from both sides (Yang et al., 2019). It can be expressed as follows:

$$F_i \neq F_j, l_i \neq l_j, S_i = S_j \forall D_i, D_j, i \neq j \tag{10.2}$$

where F denotes feature space, l denotes label space, S denotes sample space, D is the dataset of collaborators (i,j).

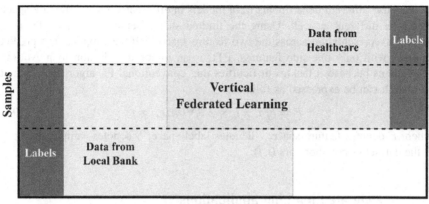

FIGURE 10.3

Vertical federated learning.

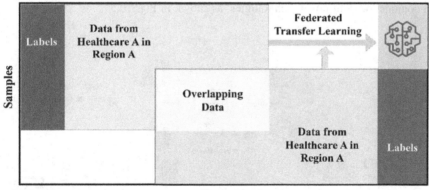

FIGURE 10.4

Federated transfer learning.

10.2.3 **Federated transfer learning**

Federal transfer learning (FTL) shares some similarities with VFL but is typically employed when the datasets have no similarities based on feature, sample, or label spaces (Rodríguez-Barroso et al., 2023; Yang et al., 2019). For example, if a dataset is obtained from a department in a healthcare facility in country A and another dataset is obtained from another department at a healthcare facility in country B, the sample spaces will differ since the patients from the two healthcare centers are different, as shown in Fig. 10.4.

Because different departments train models in different ways, the feature space would be different as well. Using the limited shared sample sets, FTL learns a common representation across the two feature spaces and then applies it to predict samples with only one-side features. FTL is an important addition to prevailing FL systems because it tackles difficulties that conventional FL algorithms cannot resolve. It can be expressed as follows:

$$F_i \neq F_j, l_i \neq l_j, S_i \neq S_j \forall D_i, D_j, i \neq j \qquad (10.3)$$

where F denotes feature space; l denotes label space; S denotes sample space; D is the dataset of collaborators (i, j).

10.2.4 Federated learning applications

FL methods have been the most commonly employed way of maintaining confidentiality in corporate and medical artificial intelligence applications. As a result, recent studies from the ML community have shown that using FL techniques may significantly improve efficiency and aid in solving problems in healthcare systems worldwide. This section discusses how FL has been used in the medical domain in conjunction with novel technologies, as shown in Fig. 10.5.

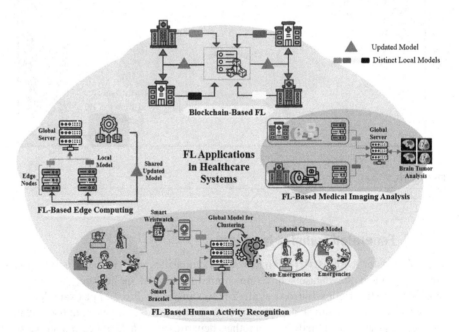

FIGURE 10.5

Federated learning applications in healthcare systems.

10.2.4.1 Blockchain-based federated learning for healthcare

Blockchain technology is a distributed, immutable database that simplifies resource management and transaction documentation within enterprise networks. It comprises an ever-expanding collection of encrypted and interconnected documents, termed blocks. Each block houses transaction data, a timestamp, and an encrypted hash of the preceding block. The timestamp certifies that the transaction data existed at the time of block creation. These interconnected blocks form an unchangeable chain; thus, once a transaction is recorded, it becomes irreversible without erasing all subsequent blocks, ensuring the immutability of blockchain transactions (Ghosh et al., 2023; Raja, 2021; Ramasamy et al., 2021). Numerous authors have conducted comprehensive reviews of the potential applications of blockchain technology in healthcare, encompassing health data analytics, data sharing, remote patient monitoring, data management, and protection (Ghosh et al., 2023; Haleem et al., 2021; Javed et al., 2021; Khezr et al., 2019; Saeed et al., 2022). However, there are still challenges posed by this technology.

As mentioned earlier, the major aim of FL is to preserve the confidentiality of sensitive datasets, such as healthcare-related datasets, which can be used in developing and evaluating new ML algorithms. To accomplish this, one of the most promising methods for effectively deploying FL applications in the healthcare sector and securely storing health records is to use blockchain technology (Rauniyar et al., 2022). The conjunction of blockchain technology with FL has been extensively reviewed (Hou et al., 2021; Li, Luo, et al., 2022; Qammar et al., 2023), where it was noted that blockchain can improve the performance of FL while addressing some of its challenges. Furthermore, blockchain-based FL for healthcare has been extensively discussed (Chang et al., 2021; Myrzashova et al., 2023; Ngan Van et al., 2022; Rehman et al., 2022), where findings showed that FL algorithms are vulnerable to inference attacks, of which homomorphic encryption and differential privacy can be utilized to protect against these attacks.

10.2.4.2 Federated learning based on edge computing for healthcare

Edge computing (EC), also known as fog computing, is a smart gateway that operates as an intermediary between devices and cloud data servers. It specializes in sending data between nodes at the edge and those in proximity to the location of the user applications, rather than to a centralized server (Brecko et al., 2022). Similar to blockchain technology, edge computing has been applied to the smart healthcare system. For example (Tripathy et al., 2022), a new edge computing-based framework was developed for accurate diagnosis and therapy recommendations for heart illnesses using an advanced DL method in an i-fogsim simulation environment, with results demonstrating excellent precision in heart disease diagnosis with relatively reduced network bandwidth and minimum energy use compared with equivalent methods. Similarly, in a study by Hartmann et al. (2022), edge computing applications were said to focus primarily on health data classification involving monitoring vital signs and recognizing falls. The study included an in-depth review of edge computing data activities, which include transmitting, classifying, reducing, encrypting, authenticating, and predicting. Some challenges were also identified, one of which included patient data privacy. This challenge resulted in the integration of EC and FL, notably for healthcare systems. FL is a well-suited EC application because it can leverage the processing capability of edge servers as well as data acquired from a variety of edge nodes (Xia et al., 2021). Abreha et al. (2022), Brecko et al. (2022), and Xia et al. (2021) conducted an extensive review of FL for EC, where three major applications were identified, which include computation offloading and content caching, malware and anomaly detection, and finally, task scheduling and resource allocation. Other applications include healthcare systems, vehicular networks, intelligent recommendation, unmanned aerial vehicle (UAV) networks, etc. Several challenges were also identified, which include, but are not limited to, hardware requirements, communication overhead, limited resources, scalability, and susceptibility to several security and privacy attacks, such as Byzantine attacks.

10.2.4.3 Federated learning-based medical imaging analysis

Medical imaging analysis is the practice of using multiple imaging modalities and digital image analysis techniques to diagnose and evaluate human body conditions, diseases, and different types of malignancies. Its major application involves the differentiation, categorization, and detection of abnormalities using images obtained from the different clinical image modalities, which include ultrasound, MRI, X-ray, computed tomography (CT), diffusion tensor imaging, positron emission tomography (PET), etc. (Anwar et al., 2018; van der Velden et al., 2022). Several works on medical image analysis utilized DL models, particularly convolutional neural networks (CNNs) and explainable artificial intelligence (XAI), which were extensively reviewed by Anwar et al. (2018) and van der Velden et al. (2022), where both techniques were seen to have a significant impact in all subfields of medical imaging analysis, such as categorization, identification, and differentiation. However, the major challenge faced by these techniques is the lack of learning data and constrained labels, which can be resolved using FL, especially the FTL among other techniques. FL applications in medical imaging analysis with deep neural networks (DNN) were extensively reviewed in (Nazir & Kaleem, 2023) where it was noted that FL improved the performance of the DNN, and addressed labeling issues through model-constructive FL and incremental learning approaches. Other noteworthy reviews on FL for medical imaging are provided in Kamble and Phophalia (2022), Mouhni et al. (2022), and Sohan and Basalamah (2023).

10.2.4.4 Federated learning-based advanced human activity recognition

Advanced human activity recognition (HAR) is an approach to automatically identifying and evaluating human activities, particularly elderly persons or patients, based on data gathered from a variety of sensors included in smartphones and wearables, including accelerometer and gyroscope sensors, time and location sensors, and a variety of other environmental sensors (Hayat et al., 2022; Schrader et al., 2020). Several ML and DL models, such as k-nearest neighbors (KNN), random forest (RF), support vector machine (SVM), artificial neural networks (ANN), and long short-term memory (LSTM), are utilized for human activity recognition, of which LSTM is said to be the most suitable with the best accuracy compared with other algorithms (Hayat et al., 2022). A comprehensive review was presented (Demrozi et al., 2020) on the significant contribution of ML in developing HAR applications using inertial sensor technologies associated with other physical and environmental sensors. However, there is still the challenge of utilizing a large dataset for both training and testing the efficacy of the algorithm, which, due to privacy issues, is not readily available. Thus, this led to the integration of FL algorithms. FL was utilized (Sozinov et al., 2018) to develop a HAR classifier, and then its performance was compared with two centralized models, such as a DNN and a SoftMax regression model, developed on both synthesized and real-world datasets. Results obtained from the work showed that FL is sufficiently reliable under an array of workloads and develops models with reasonable precision, equivalent to centralized learning models. According to Gad and Fadlullah (2022), the conventional FL method could not train heterogeneous model architectures and thus proposed an FL via augmented knowledge distillation (FedAKD) for decentralized training of heterogeneous models using two HAR datasets. The results obtained showed that the proposed FedAKD not only outperformed the conventional FL methods, but it was more robust in statistically heterogeneous settings.

10.2.4.5 Federated learning-based COVID-19 related studies in healthcare

Since its emergence as an epidemic in 2019, COVID-19 has proven to be one of the most devastating health crises that has significantly affected and is still affecting millions of people worldwide. Infected individuals may show symptoms such as a wheezing cough, a high body temperature, digestive problems, muscle pain, and loss of smell, which pose a significant threat to global health security (Malik, Naeem, et al., 2023). When an infected individual sneezes, coughs, or speaks, respiratory particulates are discharged into the air, allowing the COVID-19 virus to spread quickly (Malik, Naeem, et al., 2023). Early identification of COVID-19 is crucial for controlling the spread of infection. ML technology, as a fundamental element of AI, has seen significant application by scientists and other stakeholders in the healthcare sector, for example in the scenario of health monitoring, and it is also expected to be a promising approach for the effective identification of COVID-19 and chest disorders, specifically using DL techniques on healthcare diagnostic images such as chest X-ray imaging (Chowdhury et al., 2023;

Malik, Anees, et al., 2023; Malik, Naeem, et al., 2023). However, conventional ML is based on manual feature extraction, which is not only error-prone but also tedious and challenging to develop, especially in COVID-19 scenarios with highly sensitive and widely distributed data. Instead of relying on manual extraction, DL learns hierarchical depictions from data, allowing for improved scalability with additional data. COVID-19 data, however, may be limited for DL investigations. To address the issue, FL has been proposed, where instead of sending data from individual healthcare centers to a centralized server, the FL algorithm is implemented on nodes for local development using raw data. The model parameters are then shared to a centralized server for global development, thus preserving sensitive healthcare information (Samuel et al., 2023). An extensive review of FL for COVID-19 identification was provided by Naz et al. (2022). The study reviewed several research works on FL-based COVID-19 experiments to show the application of FL in addressing health research challenges. Furthermore, different impediments to FL deployments in the healthcare sector were also identified and elaborated.

10.2.4.6 *Federated learning integrated cloud computing for Internet of Things-based healthcare monitoring*

The Internet of Things (IoT) represents a cutting-edge technological trend that describes the networking of uniquely identifiable smart devices and gadgets. The IoT encompasses numerous objects that are covertly integrated into our surroundings (Verma et al., 2018). The Internet of Medical Things (IoMT), on the other hand, is an interconnected set of healthcare technology and software that reduces visits to hospitals and enables physicians to remotely monitor patients (Awotunde et al., 2022; Imoize et al., 2022; Kashyap et al., 2022). The motivations for the spread of IoMT are its high reliability, affordability, and minimal lag in providing medical services. Recent improvements in IoMT enabled real-time monitoring of basic medical parameters such as blood testing, diabetes management, and blood pressure measurement in the comfort of the user's home. As a result, healthcare is transitioning away from hospitals toward home-based services. Furthermore, advancements in communication technologies, body sensor networks, fog computing, and cloud computing have enabled monitoring and identification, medical evaluations, and pharmaceutical prescriptions to be performed at the patient's door (Awotunde et al., 2022; Kashyap et al., 2022). Several health monitoring systems have been proposed and developed based on IoT and cloud computing (Ijaz et al., 2021; Siam et al., 2019, 2021). However, concerns such as the sharing of data over IoMT and stored electronic health records (EHRs) have been raised due to privacy regulations. As a result, FL was proposed for decentralized or edge-learning paradigms that allow the training of algorithms on local data without explicitly sharing data. A comprehensive survey on FL for IoMT has been provided by Prasad et al. (2023) where FL-IoMT architecture has been presented with a focus on privacy and connectivity requirements, backed up with actual use-case scenarios.

10.3 Technical considerations of federated learning algorithms in healthcare systems

This section presents the different technical considerations of FL algorithms in digital healthcare systems, as presented in Fig. 10.6.

10.3.1 Bias

In dispersed networks, bias is a common concern. Bias occurs when a neural network has a preference toward the distribution of one client over the distribution of other clients. As a result, the model outperforms on a particular client while underperforming on others. Bias can be caused by differences in the magnitude or distribution of client data. However, the FL model itself could be biased (Darzidehkalani et al., 2022). As a result, effectively identifying the source of bias in an FL system is crucial to adequately mitigating it without losing model quality. Several studies have been undertaken to reduce bias in an FL system:

1. Preprocessing Bias Mitigation Method: This method involves modifying the training data to obtain a less-biased dataset, such as selecting and reassessing training samples. This approach is general and can be used irrespective of the ML model.
2. Inprocessing Bias Mitigation Method: In this method, the optimization objective of the underlying classifier is adjusted by incorporating regularizations that are sensitive to discrimination or fairness constraints in the optimization formulation. However, this approach is limited to certain ML models and training methods.

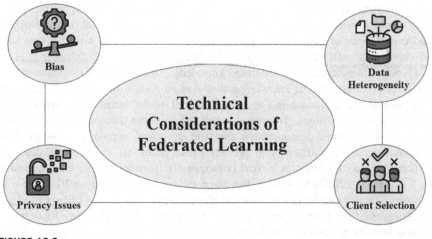

FIGURE 10.6

Technical considerations of federated learning algorithms in healthcare systems.

3. Postprocessing Bias Mitigation Method: This method considers the trained model to be a black box in which neither the training data nor the learning process may be changed. These strategies aid trained classifiers in making more accurate predictions for a given dataset.

The challenge of bias in FL was discussed by Abay et al. (2020), and three pre- and inprocessing methods were proposed to mitigate bias without compromising data privacy. Experimental analysis was conducted on the proposed methods across various data distributions based on model performance, bias learning patterns, and fairness metrics. The results demonstrated the effectiveness of the proposed methods. Similarly, the subject of bias in FL was discussed, and a comparative evaluation of current mitigation methods was conducted (Ferraguig et al., 2021), where findings suggested that the majority of existing works on FL bias reduction concentrate on cross-silo scenarios.

10.3.2 Data and system heterogeneity

Data heterogeneity describes the context in which data held by FL participants has a diverse distribution of data or features, often known as non-independently and identically distributed (non-IID) data (Yoo et al., 2022). In addition to data heterogeneity, there is also system heterogeneity that can be caused by differences in devices in the FL environment based on their hardware configuration, computing power, communication cost, and internet access (Yoo et al., 2022). Several techniques have been developed to address the problem of data and system diversity in FL. For instance, Li et al. (2022) proposed FL via group client selection known as federated group selection (FedGS), which is a hierarchical cloud-edge-end FL framework for 5G-enabled sectors, to improve the performance of industrial-based FL on non-IID data. The results obtained from experiments and other analyses showed that the proposed model outperformed FedAvg in terms of convergence performance. Furthermore, it was effective and efficient for non-IID data. In Zhou et al. (2022), the issue of data heterogeneity in FL was addressed using a federated knowledge fusion (FedKF) scheme where the server returned global knowledge that was merged with local knowledge in each learning cycle, enabling the local model to be normalized toward the global optimal conditions. The results obtained from the theoretical analysis and intensive experiment demonstrated that the proposed approach accomplished excellent accuracy, fairness, and privacy preservation all at the same time. A different approach known as virtual homogeneity learning (VHL) was proposed by Tang et al. (2022) to resolve the issue of diversity in data in FL, where a synthetic dataset can be produced from shared pure noise across clients to align the features of the heterogeneous clients. The empirical analysis results demonstrated that the proposed method improved the FL's efficiency of convergence and ability to generalize. Distributional transformation (DisTrans), a novel framework, was proposed by Yuan et al. (2022) to resolve the challenge of diversity in data in FL and improve performance using training and testing-time distributional changes, as well as a model structure with two input channels. The proposed framework worked by modifying the distributional deviations and models for each FL client to change their data distribution and then aggregating these deviations at the centralized server to significantly improve performance in events of distributional diversity. Evaluation of the proposed framework on multiple benchmark datasets showed that the framework outperformed other conventional FL and data augmentation methods in a variety of scenarios and with varying degrees of client distributional diversity. Other methods such as divide and conquer (Chandran et al., 2021), FedAlign (Mendieta et al., 2021), classifier retraining with federated features (CReFF) (Shang et al., 2022), and FedNH (Dai et al., 2022) have also been proposed to resolve the challenge of diversity in data in FL.

10.3.3 Client selection and management

The key responsibilities of clients are to train local models, sign updates to local parameters, and upload parameter updates together with the signature (Wang et al., 2023). Unlike centralized ML models that aggregate and analyze user data on a single server, FL requires each client to contribute their data to the system. As a result, for effective training, a client management system must be implemented to discourage clients who are searching for advantages (free-riding) while not contributing to model training. The server is required to identify the list of involved clients according to limited data and authenticate their specific contributions (Yoo et al., 2022). Various approaches for selecting clients have been proposed in the literature. Rai et al. (2022) proposed a novel sampling method known as the irrelevance sampling technique, which is a privacy-preserving, computationally inexpensive, and intuitive method for selecting a subset of clients according to the reliability and volume of data on edge devices. The proposed model's performance evaluation revealed that it performed well on actual application datasets. Federated client selection (FedCS) is a novel FL protocol to solve the problem of client selection with resource constraints while actively managing clients (Nishio & Yonetani, 2019). Experiments using freely accessible huge-scale image datasets showed that the proposed protocol required a shorter time to complete its training process compared with the conventional FL protocols. Other schemes that can be utilized for client management in FL have been highlighted (Fu et al., 2022).

10.3.4 Privacy and security

Although FL was designed to protect the confidentiality of user data during model development, it perpetually has some security and privacy flaws due to the sharing of model parameters, a higher number of training iterations, and communications between client devices and the centralized server, which opens the federated environment to new risks and loopholes for attackers and malware to further manipulate the FL models (Khan et al., 2021). FL is prone to various security and privacy attacks some of which are discussed below.

10.3.4.1 Poisoning attack

A poisoning attack involves an attacker modifying the variables of the target learning model during the training stage by manipulating a portion of the training data with misleading labels. This manipulation results in a corrupted model that misclassifies specific inputs according to the attacker's objectives during the inference phase (Zhang et al., 2019). Poisoning attack is classified into two types: data poisoning and model poisoning. Fig. 10.7 illustrates the poisoning attack in FL.

Data poisoning is an attack that occurs when an adversary tampers with the model by introducing low-quality data into the training process to generate erroneous model variables that are then sent to the centralized server. Data injection and data modification are examples of data poisoning attacks, where an attacker injects harmful data into a client's local model processing or alters the local model variables to poison the global model with malicious data (Mothukuri et al., 2021). Model poisoning, on the other hand, is an attack where the attacker targets the global model directly without necessarily using malicious data, as in a data poisoning attack. Model poisoning happens when an attacker modifies an updated model before transmitting it to the centralized server, poisoning the global model (Mothukuri et al., 2021). An extensive review of poisoning attacks in FL was conducted by Xia et al. (2023), presenting several taxonomies based on attack methods, attack goals, and defense strategies. The defense strategies were categorized

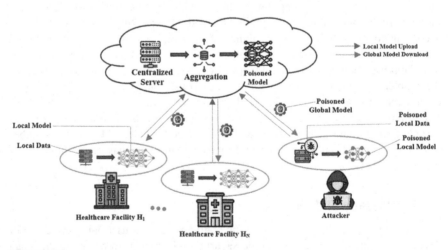

FIGURE 10.7

Poisoning attack in FL. *FL*, Federated learning.

into three main approaches, which include model analysis, Byzantine robust aggregation, and verification-based methods.

10.3.4.2 Byzantine attack

FL is susceptible to Byzantine attacks, in which an attacker introduces counterfeit models or gradients to disrupt the learning process or actively corrupts the training data, causing the global model to learn inaccurate information (Shi et al., 2022). Fig. 10.8 illustrates the Byzantine attack in FL.

Several works have been conducted to mitigate byzantine attacks in FL. For instance, FLTrust, introduced by Cao et al. (2020), offers a novel Byzantine-proof FL technique where the server ensures trust rather than solely depending on the client's local model updates. The proposed model was evaluated, and the results obtained showed its effectiveness in defending against both existing and adaptive attacks. Additionally, DiverseFL was proposed by Prakash & Avestimehr (2020) to mitigate the attack of Byzantine in FL. The model was demonstrated to be effective not only in mitigating the attack but also in identifying the Byzantine clients. Other proposed models include the Byzantine-resilient secure aggregation (BREA) framework (So et al., 2021), privacy-preserving Byzantine-robust FL (PBFL) scheme (Ma, Zhou, et al., 2022), DisBezant (Ma, Jiang, et al., 2022), dynamic federated aggregation operator (DFAO) (Rodríguez-Barroso et al., 2022), BytoChain (Li et al., 2021), SecureFL (Hao et al., 2021), etc., where a different level of efficiency and effectiveness was achieved based on the performance evaluation. However, these schemes are not without limitations and challenges, especially when evaluated against a new Byzantine attack such as the Weight attack (Shi et al., 2022), warranting further improvements.

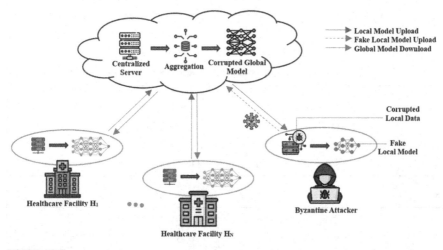

FIGURE 10.8

Byzantine attack on FL. *FL,* Federated learning.

Table 10.1 Limitations of existing research works.

References	Technical considerations					Technical challenges			
	Bias	Data and system heterogeneity	Client selection and management	Privacy and security	Communication challenge	Statistical challenge	Computational challenge	Hardware design and cost implications
Rieke et al. (2020)	✗	✓	✗	✓	✗	✗	✗	✗
Joshi et al. (2022)	✓	✗	✗	✓	✓	✗	✓	✗
Shaheen et al. (2022)	✓	✓	✗	✓	✓	✓	✗	✗
Rauniyar et al. (2022)	✗	✗	✗	✓	✗	✗	✓	✗
Darzidehkalani et al. (2022)	✓	✓	✗	✓	✓	✗	✗	✗
Yoo et al. (2022)	✗	✓	✓	✓	✗	✗	✗	✗
Majeed et al. (2022)	✗	✗	✓	✓	✓	✗	✗	✗
Li et al. (2019)	✗	✓	✗	✓	✓	✓	✗	✗
Mammen (2021)	✗	✓	✗	✓	✓	✗	✗	✗
Wen et al. (2023)	✗	✓	✓	✓	✓	✗	✗	✗
Iqbal and Chan (2021)	✗	✗	✗	✓	✓	✓	✗	✗
Imteaj et al. (2022)	✓	✓	✗	✓	✓	✗	✓	✗
Yu et al. (2022)	✗	✗	✗	✓	✓	✓	✓	✓
Ali, Alam, et al. (2022)	✗	✓	✗	✓	✓	✗	✗	✗
Chettri (2022)	✗	✓	✗	✓	✓	✗	✗	✗
Ding et al. (2022)	✓	✗	✗	✓	✗	✓	✓	✗
Zhang, Song, et al. (2022)	✗	✓	✗	✓	✓	✓	✗	✗
This chapter	✓	✓	✓	✓	✓	✓	✓	✓

10.3.4.3 Inference attack

This is a privacy attack that focuses primarily on mining data, where the attacker leverages techniques associated with data mining to evaluate the data and extract usable information from it (Ali, Naeem, et al., 2022). Membership inference and generative adversarial networks (GAN)-based reconstructive attacks are a few examples of inference attacks that allow the system to unintentionally disclose information about the training data (Yoo et al., 2022). Several inference attacks have been presented in the literature (Hu et al., 2021) to assess the privacy concerns in FL. A novel inference attack known as the source inference attack (SIA) was presented to acquire an optimal evaluation of the source of an FL client. The efficacy of the proposed attack was proved by results acquired from experiments on both synthesized and real datasets. Similarly, another novel inference attack was proposed by Zari et al. (2021) known as the passive membership inference attack, which utilized lower computational capacity and memory compared with conventional methods. Empirical results obtained showed that the attack achieved high accuracy. However, to defend FL from inference attacks, Boutet et al. (2021) proposed MixNN, a proxy-based privacy-preserving system to protect clients' privacy from hostile aggregation servers attempting to deduce confidential information, such as membership and attribute inferences. The performance of the proposed scheme was evaluated with different datasets and compared with other baselines using local differential privacy where results obtained showed that the proposed method provides a greater degree of protection than the baseline methods while preserving the same model accuracy. Similarly, a digestive neural network (DNN) was proposed by Lee et al. (2021) as a novel protection scheme against inference attacks in FL.

FL is vulnerable to several other security and privacy attacks including evasion attacks, man-in-the-middle (MITM) attacks, free-riding attacks, backdoor attacks, watermark attacks, model inversion attacks, etc., which have been extensively reviewed (Bouacida & Mohapatra, 2021; Gosselin et al., 2022; Lyu et al., 2020; Mothukuri et al., 2021; Rodríguez-Barroso et al., 2023; Zhang, Zhu, et al., 2022) while considering some of the novel schemes that can be utilized to defend against some of these attacks. Table 10.1 presents the limitations of existing works on the technical considerations and challenges of FL in several systems, such as healthcare, vehicular networks, etc.

From Table 10.1, it can be seen that all research works that considered the challenges of FL in various fields highlighted privacy and security issues as one of the major issues, whereas issues such as bias, client selection, and management, as well as hardware designs and cost implications, etc., were not discussed.

10.4 Technical challenges in executing federated learning models in real-world applications

Regarding the implementation of FL in actual healthcare facilities, several challenges have been discovered. This section presents and describes these associated challenges. These challenges have been categorized into four major categories (Rauniyar et al., 2022), which include communication, statistical, computational, and security and privacy challenges. However, other challenges exist beyond these categories.

10.4.1 Communication challenge

In an FL environment, communication between client devices and the centralized server is critical. Communication cost is one of the main issues in FL due to its decentralized nature, where participating clients or devices need to constantly communicate with their FL model parameters and other updates to train the global model, which may result in a communication bottleneck (Pouriyeh et al., 2022). This communication bottleneck can be a result of a large number of participants, limited edge node computational capability, and, most importantly, insufficient communication bandwidth. Furthermore, the geographical distribution of healthcare centers can also be a hindrance to communication, as some facilities may have unstable or unreliable network connections, which can hinder their participation in the overall FL process (Pouriyeh et al., 2022; Rauniyar et al., 2022). Although this challenge persists, several techniques and frameworks have been proposed in the literature. Some of these include the FL method with periodic averaging and quantization (FedPAQ) (Reisizadeh et al., 2019), communication-mitigated FL (CMFL) (Wang et al., 2019), FetchSGD (Rothchild et al., 2020), communication-efficient FL (CEFL) framework (Chen et al., 2021), FedCAMS (Wang et al., 2022), and LotteryFL (Li et al., 2020), all aimed at resolving the challenges associated with communication and scalability in FL. Additionally, other techniques for improving communication efficiency in FL have been explored and reviewed (Shahid et al., 2021).

10.4.2 Statistical challenge

In contrast to the conventional data center where data is believed to be IID among clients, data distribution in FL may be non-IID, leading to statistical challenges in FL. This challenge stems mostly from the discrepancy in the dissemination of data between the several clients collaborating in the FL process. To address this challenge, Zhu et al. (2021) proposed the federated one-versus-all (FedOVA) method, an efficient FL algorithm that demonstrated the ability to accommodate a significant number of clients while achieving greater reliability and rapid convergence compared with FedAvg and data-sharing methods. Similarly, Zeng et al. (2022) proposed the FedDynamic scheme to address the statistical issues of FL due to non-IID, which when evaluated against other schemes, such as FedProx, FedAvg, and Scaffold, performed better in terms of accuracy and convergence based on experiments conducted on different datasets. Furthermore, Zhao et al. (2018) addressed the statistical issues of FL when local data is non-IID, proposing a data-sharing technique for enhancing FedAvg with non-IID data by distributing only a tiny percentage of globally shared data, comprising samples from each class, to all edge devices. Experimental results obtained showed that accuracy can be increased by 30% if only 5% of the data is centralized and distributed.

10.4.3 Privacy and security challenge

FL is vulnerable to several privacy and security attacks, as highlighted and discussed in Section 10.3.4, despite its initial intent to safeguard the privacy of highly confidential data, such as medical data, during the development of ML/DL algorithms. Since FL operates a decentralized system, distributed participating clients or nodes can collaborate and develop an ML/DL model by exchanging model variables to a centralized server. This setup introduces vulnerabilities, as a malicious client posing as a legitimate participant could exploit backdoor functionalities to manipulate inputs and compromise the global model on the central server, potentially violating data protection regulations such as the General Data Protection Regulation. Backdoor attacks have drawn a lot of attention because of their ability to disrupt real-world deep-learning applications. Although some mitigation mechanisms against these attacks have been proposed, such as FedGrad (Nguyen et al., 2023), these schemes are not without challenges. Nguyen et al. (2024) provided an extensive review of the existing backdoor attack techniques and defenses in FL and also provided an exhaustive analysis of several methods, including the issues. Furthermore, instead of limiting the review to only backdoor attacks, Sikandar et al. (2023) provided an extensive review of the numerous attacks in FL and also introduced a layered protection system that can be employed to protect against these attacks.

10.4.4 Computation challenge

The processing capability, memory, and power usage of different network devices may differ due to the variety of hardware components. Some participating healthcare devices might be unreliable or stop working due to certain energy constraints during the FL process, which ordinarily involves several communication cycles with the centralized server, affecting the data and model quality (Rauniyar et al., 2022). Apart from the hardware configuration of the edge device of the participating clients, the data disparity among participating clients might result in an increased number of FL iterations to attain the required accuracy level, thus increasing the computational time. To address this issue, Zhan et al. (2020) proposed an experience-driven computational algorithm that leverages deep reinforcement learning (DRL) to dynamically allocate computational resources in FL, achieving near-optimal solutions without relying on knowledge of network quality. Similarly, Wang et al. (2021) proposed an SVM-based FL model for local training tasks across devices or nodes in a wireless high-altitude balloon network. The approach can help optimize task and resource allocation within the network by utilizing SVMs as the underlying learning model for federated training.

These pioneering approaches provide valuable insights and potential solutions to combat the computational challenge in FL, ultimately promoting the effective and efficient deployment of FL within the healthcare sector.

10.4.5 Hardware design issues and cost implications

Considerable costs are incurred in executing FL in terms of learning time and energy consumption. These costs can arise from various factors, such as extensive hardware requirements, the large number of selected clients involved, and the number of local iterations in each training round (Luo et al., 2021). Since FL requires a variety of distributed devices to collaboratively develop and share models, concerns regarding hardware design heterogeneity have been raised. The issue pertains to variations in CPU capabilities, RAM sizes, and battery lifespans, which can affect the FL development process in terms of the training time, reduced accuracy levels, and rounds of communications (Farcas et al., 2022; Yang et al., 2022). Schemes and frameworks, such as model elasticity (Farcas et al., 2022) and FLASH (Yang et al., 2022), have been proposed, which not only increase the training efficiency but also allow the different hardware to learn different model architectures. A cost-effective sampling-based algorithm is presented by Luo et al. (2021) to minimize the total cost based on the learning time and energy consumption while ensuring convergence. Results obtained from experiments validated the theoretical analysis and showed the effectiveness and efficiency of the proposed algorithm. Although these schemes and frameworks yielded positive results in terms of enhancing training efficiency and lowering costs, the challenge remains because the financial expense of providing dedicated hardware devices to develop and share FL models was not considered.

10.5 Proposed solutions to federated learning technical considerations and challenges

This section presents the proposed solutions to the different challenges in FL that have been identified and discussed in the earlier sections of this chapter.

In addition to these challenges discussed, several other challenges have been highlighted in the literature, some of which include traceability and accountability issues, explainability of FL models, lack of standard data, system architecture, model performance, data skewness, level of trust, etc. (Aledhari et al., 2020; Darzidehkalani et al., 2022; Reddy & Gadekallu, 2023). From Table 10.2, it is evident that numerous models and frameworks have been proposed to tackle specific challenges within FL. These solutions encompass various techniques and strategies, each with its unique approach and characteristics. However, an essential observation is that the performance and efficiency of these models can vary significantly when assessed in real-world scenarios. Therefore, there is a critical

Table 10.2 Summary of technical considerations and challenges with proposed solutions.

Technical considerations and challenges	Proposed solutions	Description	References
Bias	Pre- and inprocessing	Preprocessing bias mitigation techniques involve adjusting data before training, while inprocessing techniques make bias corrections during training process in FL.	Abay et al. (2020)
	FairFL	FairFL integrates deep multiagent reinforcement learning and a secure data aggregation protocol to optimize model accuracy and fairness within client privacy constraints.	Zhang et al. (2020)
	Bias-free FedGAN	Bias-Free FedGAN employs generative adversarial networks to generate synthetic data that can be used to reduce bias in the federated dataset.	Mugunthan et al. (2021)
	Astraea	Astraea mitigates imbalance by augmenting runtime data and adjusting client training schedules based on data distribution divergence using a mediator.	Duan et al. (2019)
	FELICIA	FELICIA enhances data sharing and synthetic data generation in FL, particularly benefiting biased sites through secure assistance from others.	Rajotte et al. (2021)
Data and system heterogeneity	FedGS	FedGS addresses global model convergence in heterogeneous FL systems by adapting the FL process to accommodate varying local data distributions and system characteristics.	Li, He, et al. (2022)
	FedKF	FedKF utilizes global knowledge of the server to regularize local models, mitigating client model drift.	Zhou et al. (2022)

(Continued)

Table 10.2 Summary of technical considerations and challenges with proposed solutions. *Continued*

Technical considerations and challenges	Proposed solutions	Description	References
	VHL	Tackles data heterogeneity in FL by training on a simulated virtual dataset that is more homogeneous, devoid of sensitive information, and easily separable, streamlining learning across diverse client devices.	Tang, Zhang, et al. (2022)
	DisTrans	Optimizes distributional offsets for FL clients, adjusting data distributions and aggregating these offsets at the server to enhance performance in cases of distributional heterogeneity.	Yuan et al. (2022)
	Divide and Conquer	Partitions global model into segments and allocates them to clients based on their capabilities and data size.	Chandran et al. (2021)
	FedAlign	Provides methods to align data distributions across devices, reducing impact of data and system heterogeneity.	Mendieta et al. (2021)
	CReFF	Improves FL adaptability to diverse data and long-tail challenges using federated features and classifier retraining.	Shang et al. (2022)
	FedNH	Focuses on addressing data and system heterogeneity by enabling neighborhood-based learning among devices.	Dai et al. (2022)
Client selection and management	Irrelevance Sampling Technique	Enhances federated model training efficiency and fairness by prioritizing clients with pertinent data and significant contributions while diminishing the impact of less relevant or noisy clients.	Rai et al. (2022)

(Continued)

Table 10.2 Summary of technical considerations and challenges with proposed solutions. *Continued*

Technical considerations and challenges	Proposed solutions	Description	References
	FedCS	Optimizes client selection to efficiently aggregate client updates and enhance ML model performance while considering resource constraints.	Nishio and Yonetani (2019)
	FedSS	Enhances resource utilization and model performance in decentralized settings.	Tahir et al. (2022)
	FedCor	Leverages correlation-based selection to select most suitable clients for participation in FL process	Tang, Ning, et al. (2022)
	HACCS	Leverages statistical heterogeneity by encapsulating distinguishable data distributions rather than individual devices during training procedure.	Wolfrath et al. (2022)
	Power-of-Choice	It is a versatile client selection framework that balances convergence speed and solution bias efficiently in terms of communication and computation.	Jee Cho et al. (2022)
	FedDCS	Employs a parameter estimation algorithm to select the most suitable clients to participate in collaboration, ultimately leading to enhancement of global ML model.	Zou et al. (2021)
	GraphCS	Utilizes a graph-based approach to address both data and hardware heterogeneity within an FL system concurrently.	Chang et al. (2023)
	FedChoice	Employs loss function optimization for client selection, aiming to enhance global model convergence by identifying suitable local models.	Zeng et al. (2023)

(Continued)

Table 10.2 Summary of technical considerations and challenges with proposed solutions. *Continued*

Technical considerations and challenges	Proposed solutions	Description	References
	CBE3	Focuses on balancing exploration and exploitation by considering contributions of clients, enabling more informed and effective client selection strategies to improve global model's performance.	Lin et al. (2022)
	PyramidFL	Not only considers divergence between selected and nonselected participants for client selection but also maximizes utilization of data and system heterogeneity within chosen clients to enhance their utility profiling more effectively.	Li, Zeng, et al. (2022)
	WCL	Combines both model weight divergence and client training loss to select most suitable clients for participation, ultimately enhancing performance of global model.	Guo et al. (2021)
	PFedRe	PFedRe customizes client selection within FL by incorporating individual relevance, quantified through Wasserstein distances among dummy datasets.	Ma et al. (2023)
	AUCTION	Evaluates client performance and selects them based on quality awareness for an FL task within a set budget.	Deng et al. (2022)
Privacy and security	FLTrust	Enhances security by establishing trust among devices within FL network through a bootstrapping mechanism, making the system resilient to Byzantine attacks.	Cao et al. (2020)
	BREA	BREA integrates stochastic quantization, verifiable outlier detection, and secure model aggregation to ensure Byzantine resilience, privacy, and convergence.	So et al. (2021)

(Continued)

Table 10.2 Summary of technical considerations and challenges with proposed solutions. *Continued*

Technical considerations and challenges	Proposed solutions	Description	References
	PBFL	PBFL combines privacy protection techniques with Byzantine fault-tolerant mechanisms to ensure the privacy of individual data while maintaining the integrity and security of the collaborative learning process, even in presence of malicious clients	Ma, Zhou, et al. (2022)
	DisBezant	Demonstrates strong resistance against Byzantine attacks while simultaneously ensuring efficient and privacy-preserving model training.	Ma, Jiang, et al. (2022)
	DFAO	Adapts aggregation methods in FL based on characteristics of participating clients and their data, optimizing model aggregation for improved FL performance	Rodríguez-Barroso et al. (2022)
	BytoChain	Boosts model verification efficiency with parallel verifiers and uses a Byzantine-resistant PoA for detecting attacks.	Li et al. (2021)
	SecureFL	Enhances FL security using TEE and optimizes TEE efficiency across cloud and edge workers through partitioning and aggregation methods.	Hao et al. (2021)
	DNN	Alters input data to enhance classification accuracy in FL while simultaneously minimizing accuracy of potential inference attacks	Lee et al. (2021)
	FedGrad	FedGrad is a robust defense against advanced backdoor attacks, including edge cases, and performs well with diverse client data and numerous compromised clients.	Nguyen, Nguyen, Wong, et al. (2023)

(Continued)

Table 10.2 Summary of technical considerations and challenges with proposed solutions. *Continued*

Technical considerations and challenges	Proposed solutions	Description	References
Communication challenge	FedPAQ	Is a communication-efficient FL approach that reduces communication overhead by periodically averaging models and quantizing updates, striking a balance between model accuracy and communication cost.	Reisizadeh et al. (2019)
	CMFL	Reduces communication overhead in FL by dynamically identifying and excluding irrelevant client updates before uploading them to the central server.	Wang et al. (2019)
	FetchSGD	Tackles communication bottleneck in FL by utilizing Count Sketch-based compression of model updates, enhancing its practicality and scalability for large-scale distributed scenarios.	Rothchild et al. (2020)
	CEFL	Prioritizes device selection, employs quantization methods, and efficient wireless resource allocation to significantly enhance convergence speed and training loss	Chen et al. (2021)
	FedCAMS	Significantly decreases communication expenses through implementation of error feedback and compression strategies	Wang et al. (2022)
	LotteryFL	Leverages lottery ticket hypothesis to identify crucial subnetworks within client models, reducing communication costs and enhancing personalization in FL on non-IID datasets.	Li et al. (2020)
Statistical challenge	FedOVA	Tackles non-IID data by employing a one-versus-all strategy, enhancing learning from diverse data sources in federated settings.	Zhu et al. (2021)

(Continued)

Table 10.2 Summary of technical considerations and challenges with proposed solutions. *Continued*

Technical considerations and challenges	Proposed solutions	Description	References
	FedDynamic	Optimizes aggregation based on device-specific data distributions, prioritizing devices with more relevant data to enhance global model performance and convergence in FL.	Zeng et al. (2022)
	DWFed	Assigns dynamic weights to individual clients during model aggregation process, improving overall performance of FL systems.	Chen et al. (2022)
Computation challenge	SVM-Based FL	Employs SVM as machine learning model for local tasks across network nodes, optimizing task and resource allocation in federated training within a wireless high-altitude balloon network.	Wang et al. (2021)
	DRL-Based FL	Dynamically allocates computational resources in FL, achieving near-optimal solutions without relying on knowledge of network quality.	Zhan et al. (2020)
	Joint Device Scheduling and Resource Allocation Scheme	Optimizes device and resource allocation while considering delay and energy constraints to improve FL efficiency.	Shi et al. (2021)
	Heterogeneous Computation and Resource Allocation Framework	Optimizes computational resource allocation in wireless-powered federated edge learning, efficiently managing device capabilities for distributed machine learning tasks.	Feng et al. (2022)
Hardware design issues and cost implications	Model Elasticity	Model elasticity enables local devices to train distinct models from same architectural family, each tailored to match resource constraints of different edge devices.	Farcas et al. (2022)
	FLASH	Addresses hardware heterogeneity and cost implications by optimizing client selection and resource allocation for improved scalability and efficiency.	Yang et al. (2022)

FL, *Federated learning.*

need to develop a comprehensive mitigation model that not only addresses all major FL challenges but also offers high accuracy in the development and implementation of FL models. Such a holistic approach would potentially offer more robust solutions that can adapt to the complexities and nuances of practical FL deployments.

10.5.1 Bias

Addressing bias in FL is essential to ensuring fair and equitable model training. Several techniques and models have been proposed to address this challenge. For instance, three preprocessing and inprocessing bias mitigation methods were proposed by Abay et al. (2020). The preprocessing bias mitigation technique aims to preprocess the data at each client to reduce bias before aggregation, while the inprocessing technique focuses on modifying the learning process to mitigate bias during model training and aims to preprocess federated datasets before or during FL training, removing sources of bias. FairFL, proposed by Zhang et al. (2020), comprises a well-structured deep multiagent reinforcement learning framework and a secure data aggregation protocol that aims to enhance both the accuracy and fairness of the trained model, all while adhering to the stringent privacy requirements of the clients. Bias-free federated generative adversarial network (FedGAN) explores generative adversarial networks to generate synthetic data to reduce bias in both the FL data and models (Mugunthan et al., 2021). Astraea addresses global imbalance by dynamically augmenting runtime data and mitigates local imbalance during averaging by introducing a mediator that adjusts client training schedules based on the Kullback–Leibler divergence of their data distributions (Duan et al., 2019). FELICIA (Rajotte et al., 2021) enables more secure data sharing, facilitating the creation of synthetic data within the context of FL. In this framework, a biased site can receive secure assistance from another site through FELICIA, with greater benefits realized as the level of bias increases. However, it is important to note that, while these solutions are effective, complete bias elimination can be challenging, especially when bias is deeply ingrained in the data. Additionally, these methods may require well-annotated data, which can be costly and time-consuming to obtain, limiting their practical applicability to some scenarios.

10.5.2 Data and system heterogeneity

Heterogeneous data and systems are inherent in federated environments, making it challenging to ensure model convergence and collaboration. Solutions such as FedGS (Li, He, et al., 2022) employ a gradient-based binary permutation algorithm to choose specific devices from each factory, creating consistent super nodes that actively engage in FL training. FedKF involves the server providing the global knowledge to combine with local knowledge in each training round, allowing for the regularization of local models toward the global optimum and addressing the problem of client model drift (Zhou et al., 2022). VHL addresses data heterogeneity by training on a simulated, homogeneous, nonsensitive, and easily separable virtual dataset to facilitate learning across diverse client devices. Meanwhile, DisTrans optimizes distributional offsets for FL clients, enhancing performance in cases of distributional heterogeneity by adjusting data distributions and aggregating these offsets at the server (Tang, Zhang, et al., 2022; Yuan et al., 2022). The divide and conquer algorithm (Chandran et al., 2021) partitions the global model into segments and allocates them to clients based on their capabilities and data size, thus addressing data and system heterogeneity in FL. FedAlign utilizes distillation-based regularization to enhance local learning generality, effectively managing resource usage and addressing the issue of heterogeneity within FL (Mendieta et al., 2021). CReFF leverages federated features, making it more adaptable to diverse data distributions and is capable of addressing long-tail data challenges, while FedNH enhances the performance of local models, focusing on both personalization and generalization, by integrating class prototype uniformity and semantics (Dai et al., 2022; Shang et al., 2022). However, these solutions may introduce computational complexity, potentially slowing down the FL process, especially in scenarios with numerous participating devices.

10.5.3 Client selection and management

Efficient client selection and management are crucial in FL for faster and more accurate model development. Rai et al. (2022). proposed an irrelevance sampling technique that aims to improve the overall efficiency and fairness of the federated model training process by giving more importance to clients that have relevant data or higher contributions to the global model while reducing the influence of less relevant or noisy clients. FedCS (Nishio & Yonetani, 2019) addresses the challenge of client selection under resource constraints, enabling efficient aggregation of client updates by the server and accelerating performance enhancement in ML models. FL with a smart selection of clients (FedSS) intelligently chooses clients for participation in the FL process, optimizing the use of resources and improving overall model performance in decentralized settings, while federated correlation (FedCor) leverages correlation-based selection to select the most suitable clients for participation in the FL process (Tahir et al., 2022; Tang, Ning, et al., 2022). These models aim to select clients that contribute most effectively to the global model, thus improving efficiency. Heterogeneity-aware clustered client selection (HACCS) and power-of-choice provide different strategies for selecting clients based on their capabilities and contributions; that is, they consider client capabilities, making informed choices based on device resources and data quality (Jee Cho et al., 2022; Wolfrath et al., 2022). Federated dynamic client selection (FedDCS), graph-based client selection (GraphCS), FedChoice, contribution-based exponential-weight algorithm for exploration and exploitation (CBE3), PyramidFL, WCL, personalized FL with relevance (PFedRe), and

AUCTION collectively provide a rich array of client selection and management techniques. These encompass dynamic client selection strategies, graph-based methodologies, weighted client prioritization, and the utilization of auction-based methods. These comprehensive techniques effectively balance device contributions and optimize network resources. However, these methods depend on specific FL scenarios and may require accurate client information, posing potential challenges to suboptimal decision-making.

10.5.4 Privacy and security

Protecting privacy and ensuring security are paramount in FL, particularly when dealing with highly sensitive data, such as healthcare data. To address this issue, Cao et al. (2020) proposed FLTrust, which enhances security by establishing trust among devices within the FL network through a bootstrapping mechanism, making the system resilient to Byzantine attacks. Similarly, So et al. (2021) proposed BREA, which utilizes a combined strategy involving stochastic quantization, verifiable outlier detection, and secure model aggregation to ensure Byzantine resilience, privacy preservation, and convergence in a unified framework. PBFL combines privacy protection techniques with Byzantine fault-tolerant mechanisms to ensure the privacy of individual data while maintaining the integrity and security of the collaborative learning process, even in the presence of malicious clients (Ma, Zhou, et al., 2022). DisBezant (Ma, Jiang, et al., 2022) ensures efficient and privacy-preserving model training during model aggregation to effectively defend against Byzantine attacks over non-iid datasets. DFAO adapts and applies aggregation techniques to balance privacy preservation and model quality while accommodating the decentralized and heterogeneous nature of client devices and data (Rodríguez-Barroso et al., 2022). BytoChain enhances model verification efficiency by introducing verifiers that can concurrently execute resource-intensive verification processes while also detecting Byzantine attacks through the implementation of a Byzantine-resistant consensus mechanism known as proof-of-accuracy (PoA) (Li et al., 2021). SecureFL ensures comprehensive security for FL by incorporating readily accessible trusted execution environments (TEE) to safeguard against privacy breaches, and it further enhances TEE efficiency across both cloud and edge workers through meticulous partitioning and aggregation methods (Hao et al., 2021). DNN alters input data to enhance the classification accuracy in FL while simultaneously minimizing the accuracy of potential inference attacks (Lee et al., 2021). FedGrad represents an innovative defense mechanism in FL, demonstrating resilience against sophisticated backdoor attacks, including challenging edge-case attacks, while maintaining high performance even when faced with heterogeneous client data and a substantial number of compromised clients (Nguyen et al., 2023). While the reviewed solutions have demonstrated their effectiveness in preserving security and privacy in FL, it is evident that the primary emphasis has been on addressing Byzantine and backdoor attacks. FL is vulnerable to a variety of other potential threats, necessitating the development of countermeasures for a more comprehensive range of attacks, including MITM attacks, free-riding attacks, watermark attacks, model inversion attacks, etc. Also, these privacy and security solutions may introduce additional computational overhead, potentially making them less suitable for resource-constrained client devices. Furthermore, the effectiveness of these techniques can vary depending on the nature of the FL environment and the sophistication of potential attackers.

10.5.5 Communication challenge

Efficient communication is essential in FL, particularly over bandwidth-limited networks. Different solutions have been proposed to reduce the amount of data transmitted during FL updates, thus ensuring communication efficiency in FL. For instance, Reisizadeh et al. (2019) proposed FedPAQ, a communication-efficient FL method that addresses the issue of high communication overhead in FL based on periodic model averaging and quantization techniques, thus striking a balance between model accuracy and communication cost. Wang et al. (2019) also tried to mitigate communication overhead in FL by proposing a communication-mitigated FL (CMFL), which utilizes dynamically identifying and excluding irrelevant client updates before uploading them to the central server. FetchSGD (Rothchild et al., 2020) resolves the issue of communication bottle-necks in FL by compressing model updates using a Count Sketch, making it more practical and scalable for large-scale distributed settings. Chen et al. (2021) also attempted to mitigate communication delays in FL by proposing a CEFL framework that prioritizes device selection, employs quantization methods, and utilizes efficient wireless resource allocation to significantly enhance convergence speed and training loss. FedCAMS and LotteryFL were proposed by Wang et al. (2022) and Li et al. (2020), respectively, to reduce the communication cost in FL. While FedCAMS utilizes error feedback and compression strategies to reduce this communication cost, LotteryFL utilizes a lottery ticket network. Both models, evaluated based on experiments, show great promise in terms of personalization and communication costs. While these communication-efficient methods show potential for reducing bandwidth usage, their effectiveness may vary depending on the specific FL scenario and network conditions, and further research is needed to fine-tune their practical implementation in various contexts.

10.5.6 Statistical challenge

Addressing statistical challenges is crucial for a stable FL. This has attracted the attention of authors who have proposed and developed different models to address this challenge. FedOVA (Zhu et al., 2021) addresses the challenge of non-IID data by training models using a one-versus-all strategy, enabling more effective learning from heterogeneous data sources in a federated setting. FedDynamic (Zeng et al., 2022) also aims to address the statistical challenge in

FL caused by non-IID by adapting its aggregation process to the data distribution on each device, ensuring that devices with more relevant data contribute more to the global model, improving overall performance and convergence in FL. Chen et al. (2022) also attempted to address the statistical heterogeneity in FL by proposing a dynamic weighted model aggregation algorithm called DWFed, which assigns dynamic weights to individual clients during the model aggregation process, improving the overall performance of FL systems. While these solutions are said to be effective, considering the complexity of data variations in FL, these models may not comprehensively address all possible scenarios. They may also introduce computational overhead and necessitate adaptability while not completely eradicating statistical challenges. Therefore further research in this area is required.

10.5.7 Computation challenge

Computation challenges can impact the scalability of FL, therefore there is a need to address this challenge. Solutions such as the SVM-based FL model, proposed by Wang et al. (2021), involve using SVM as the ML model for local training tasks across devices or nodes in a wireless high-altitude balloon network. The approach can help optimize task and resource allocation within the network by utilizing SVMs as the underlying learning model for federated training. Similarly, Zhan et al. (2020) attempted to address the computation challenge in FL and proposed an experience-driven computational algorithm that leverages DRL to dynamically allocate computational resources in FL, achieving near-optimal solutions without relying on knowledge of network quality. The joint device scheduling and resource allocation scheme (Shi et al., 2021) focuses on optimizing the allocation of devices and resources while considering constraints related to delay and energy consumption, ultimately enhancing the efficiency and effectiveness of FL processes. The heterogeneous computation and resource allocation framework (Feng et al., 2022) was designed to optimize the allocation of computational resources in wireless-powered federated edge learning systems. This framework effectively manages the varying computation capabilities of edge devices while ensuring efficient resource utilization for distributed ML tasks. A notable limitation of these approaches is that some may require specialized hardware or resource allocation strategies, potentially limiting their universal applicability. Additionally, the efficacy of these solutions can be influenced by the scale, topology, and dynamics of the network, which may result in reduced accuracy when deployed in large-scale and complex environments.

Consideration of hardware design and cost implications is essential for practical FL deployment. Farcas et al. (2022) tackled the challenge of hardware diversity among edge devices in FL. They introduced a novel concept called "model elasticity" and proposed a framework that combines model design with architecture, allowing edge devices to train distinct models from the same architecture family. These models are tailored to fit the resource constraints of different edge devices, addressing the issue of hardware heterogeneity in FL. FLASH (Yang et al., 2022) focuses on mitigating the challenges associated with hardware heterogeneity and cost implications. It achieves this by intelligently selecting clients with similar hardware configurations, dynamically allocating resources to optimize efficiency, and ensuring cost-effective model training. This approach enhances the scalability and overall efficiency of FL in diverse and resource-constrained environments. However, a common limitation is that hardware constraints and cost considerations can vary significantly across different FL deployments, making it challenging to develop a one-size-fits-all solution.

In summary, these proposed solutions effectively tackle distinct technical challenges in FL. However, they are not without limitations and may include potential computational overhead and the necessity for adaptation to specific FL scenarios and network conditions. Consequently, their universal applicability across all FL scenarios remains a challenge. Table 10.2 presents a summary of the technical considerations and challenges with their respective proposed solutions in the literature.

10.6 Lessons learned

The key points from the review are identified and discussed in this section, which can be used to guide further studies on FL in digital healthcare.

10.6.1 Lesson 1: Federated learning can guarantee data confidentiality and management

FL was proposed to address the security and privacy flaws of conventional ML model development, especially when dealing with sensitive datasets such as patient records. The ability of FL to deliver on these critical aspects is intricately linked to its distributed architecture, as depicted in Fig. 10.1. In the FL paradigm, participating healthcare centers, or simply clients, conduct localized model training using their respective sensitive datasets, after which they only share the local model parameters or gradients with a central server for aggregation and global learning. Subsequently, the central server disseminates the processed global model to all participating healthcare centers or clients. This inherently robust process guarantees the preservation of data confidentiality at each local healthcare center, as the actual datasets remain securely on-premises without the necessity of uploading sensitive information. Nevertheless, despite the numerous benefits of FL, it is not without challenges. Therefore there is a need for continuous refinement and optimization of FL to overcome the challenges associated with its implementation in real-world healthcare settings.

10.6.2 Lesson 2: Federated learning can be applied in healthcare system in conjunction with novel technologies

The chapter provides an in-depth exploration of the versatile applications of FL within healthcare systems, leveraging various cutting-edge technologies, including blockchain, edge computing, cloud computing, IoT, and medical imaging. The review demonstrates that integrating these innovative technologies into FL models holds great potential for enhancing computational efficiency, fortifying security and privacy measures, and enabling the early detection of critical medical conditions. However, this chapter revealed some of the intricate challenges associated with the integration of FL with these novel technologies, including hardware requisites, communication overhead, resource constraints, and the imperative need for scalability. As such, while promising, the harmonization of FL and these advanced technologies within healthcare systems requires meticulous considerations and innovative solutions to these challenges to fully harness their transformative potential.

10.6.3 Lesson 3: Federated learning has security and privacy flaws

Despite its design to enhance data security and privacy during model development through its decentralized nature, FL is susceptible to several security and privacy threats. Among the primary threats are poisoning attacks, where malicious participants inject biased or false data into the training process; inference attacks that extract sensitive information from model updates; and Byzantine attacks that involve clients behaving maliciously or inaccurately. In addition to these major concerns, FL also faces other security and privacy risks, including evasion attacks, MITM attacks, free-riding attacks by participants who contribute inadequately, backdoor attacks that introduce hidden vulnerabilities, watermark attacks that can identify the source of data, model inversion attacks that reverse-engineer sensitive data, etc. While these challenges have drawn the attention of researchers, as shown in Table 10.1, who have proposed various mitigation models and techniques to address these threats in FL, the review showed that these solutions are not without their own set of challenges. These challenges may encompass issues such as computational overhead, adaptability to specific FL scenarios, and the need for further optimization. Hence, there remains a critical need for ongoing research and development to reinforce the security and privacy aspects of FL and ensure its robustness against evolving threats.

10.6.4 Lesson 4: Model development with federated learning can be expensive

Another crucial insight gained from the review is the substantial cost associated with model development in FL. This expense encompasses various factors, including the high costs of dedicated hardware components and the significant energy resources required for computationally intensive model training and development. Additionally, costs may arise from extended training times and communication rounds, which may demand high bandwidth for efficient parameter exchange between participating clients and the global server, thus amplifying expenses. While some researchers have proposed cost-reduction models and methods, further research efforts are needed to strike a balance between computational and hardware costs in FL model development.

10.6.5 Lesson 5: Proposed solutions to federated learning challenges are still in the theoretical phase

This chapter has extensively explored the technical challenges faced by FL, encompassing biases, heterogeneity, security and privacy, communication, statistical, and cost-related hardware issues. Remarkably, researchers have proactively proposed various models and techniques to alleviate these challenges in FL. However, a prevalent limitation is that a significant portion of these proposals remain in a theoretical phase, lacking real-life validation. This holds particularly true for models aiming to execute FL in real-world applications, contributing to their relatively low adoption in critical sectors such as healthcare. Hence, there is a need to validate and evaluate these theoretical models in real medical environments. Moreover, to enhance the adoption and precision of FL models across diverse sectors, it is imperative to provide comprehensive training, especially to those tasked with local data modeling.

10.6.6 Lesson 6: Ongoing research and personnel education are vital

The review emphasizes the dynamic nature of FL and the need for continuous research and education. FL is a rapidly evolving field with ongoing advancements in technology, security, and privacy. Researchers, healthcare practitioners, and policymakers must stay abreast of these developments to harness the full potential of FL while addressing its challenges effectively. Moreover, comprehensive training and education initiatives are crucial to equipping professionals with the knowledge and skills required for responsible FL implementation, particularly in the area of local model training using sensitive data. This commitment to ongoing research and education is essential to driving innovation, fostering ethical practices, and ensuring the long-term success of FL in healthcare and other sectors.

10.7 Conclusion and recommendations

FL has several applications in different sectors, particularly in digital healthcare systems, due to its decentralized manner of developing ML/DL models and the privacy preservation of sensitive datasets. Nonetheless, it faces several challenges.

This chapter presented and discussed these numerous challenges. Specifically, the chapter discussed the technical considerations of FL in healthcare systems, which include bias, data heterogeneity, client selection, and management, as well as privacy and security issues. In addition, five significant technical challenges of implementing FL in real-world applications have been discussed, including their mitigation techniques or schemes. Although several research works on FL challenges have been conducted and numerous models and frameworks have been proposed to mitigate these challenges, findings show that these works largely focus on FL privacy and security issues. The majority of the proposed models have not been evaluated in real-world scenarios, particularly models associated with privacy and security challenges, which is one of the most critical issues in FL. Therefore, for future work, the development of a robust privacy-preserved and secure FL system with high accuracy is recommended while integrating other developed models to address challenges such as bias and data heterogeneity.

Acknowledgment

This work is funded by the Federal Republic of Nigeria under the National Research Fund (NRF) of the Tertiary Education Trust Fund (TETFund), grant no. TETF/ES/DR&D-CE/NRF-2021/SETI/ICT/00112/VOL.1.

References

Abay, A., Zhou, Y., Baracaldo, N., Rajamoni, S., Chuba, E., & Ludwig, H. (2020). Mitigating Bias in Federated Learning. arXiv Prepr. arXiv2012.02447.

Abreha, H. G., Hayajneh, M., & Serhani, M. A. (2022). Federated learning in edge computing: A systematic survey. *Sensors*, *22*(2), 450. Available from https://doi.org/10.3390/s22020450.

Aledhari, M., Razzak, R., Parizi, R. M., & Saeed, F. (2020). Federated learning: A survey on enabling technologies, protocols, and applications. *IEEE Access*, *8*, 140699—140725. Available from https://doi.org/10.1109/ACCESS.2020.3013541.

Ali, H., Alam, T., Househ, M., & Shah, Z. (2022). *Federated learning and internet of medical things — Opportunities and challenges. Studies in health technology and informatics* (295, pp. 201—204). IOS Press, ISBN 9781643682907.

Ali, M., Naeem, F., Tariq, M., & Kaddoum, G. (2022). Federated learning for privacy preservation in smart healthcare systems: A comprehensive survey. *IEEE Journal of Biomedical and Health Informatics*, *27*(2), 778—789. Available from https://doi.org/10.1109/JBHI.2022.3181823.

Anwar, S. M., Majid, M., Qayyum, A., Awais, M., Alnowami, M., & Khan, M. K. (2018). Medical image analysis using convolutional neural networks: A review. *Journal of Medical Systems*, *42*, 226. Available from https://doi.org/10.1007/s10916-018-1088-1.

Awotunde, J. B., Imoize, A. L., Ayoade, O. B., Abiodun, M. K., Do, D.-T., Silva, A., & Sur, S. N. (2022). An enhanced hyper-parameter optimization of a convolutional neural network model for leukemia cancer diagnosis in a smart healthcare system. *Sensors*, *22*, 9689. Available from https://doi.org/10.3390/s22249689.

Bouacida, N., & Mohapatra, P. (2021). Vulnerabilities in federated learning. *IEEE Access*, *9*, 63229−63249. Available from https://doi.org/10.1109/ACCESS.2021.3075203.

Boutet, A., Lebrun, T., Aalmoes, J., & Baud, A. (2021). MixNN: Protection of federated learning against inference attacks by mixing neural network layers. arXiv Prepr. arXiv2109.12550.

Brecko, A., Kajati, E., Koziorek, J., & Zolotova, I. (2022). Federated learning for edge computing: A survey. *Applied Sciences*, *12*, 9124. Available from https://doi.org/10.3390/app12189124.

Cao, X., Fang, M., Liu, J., & Gong, N.Z. (2020). FLTrust: Byzantine-robust federated learning via trust bootstrapping. arXiv Prepr. arXiv2012.13995.

Chandran, P., Bhat, R., Chakravarthy, A., & Chandar, S. (2021). *Divide-and-conquer federated learning under data heterogeneity. Proceedings of the AI, machine learning and applications* (Vol. 11, pp. 21−33). Academy and Industry Research Collaboration Center (AIRCC).

Chang, Y., Fang, C., & Sun, W. (2021). A blockchain-based federated learning method for smart healthcare. *Computational Intelligence and Neuroscience*, *2021*, 1−12. Available from https://doi.org/10.1155/2021/4376418.

Chang, T., Li, L., Wu, M., Yu, W., Wang, X., & Xu, C. (2023). GraphCS: Graph-based client selection for heterogeneity in federated learning. *Journal of Parallel and Distributed Computing*, *177*, 131−143. Available from https://doi.org/10.1016/j.jpdc.2023.030.003.

Chen, A., Fu, Y., Wang, L., & Duan, G. (2022). DWFed: A statistical- heterogeneity-based dynamic weighted model aggregation algorithm for federated learning. *Frontiers in Neurorobotics*, *16*, 1041553. Available from https://doi.org/10.3389/fnbot.2022.1041553.

Chen, M., Shlezinger, N., Poor, H. V., Eldar, Y. C., & Cui, S. (2021). Communication-efficient federated learning. *Proceedings of the National Academy of Sciences of the United States of America*, *118*, e2024789118. Available from https://doi.org/10.1073/pnas.2024789118.

Chettri, S. (2022). Federated learning: Approaches, possibilities, and challenges. *SSRN Electronic Journal*, 1−9. Available from https://doi.org/10.2139/ssrn.4209578.

Chowdhury, D., Banerjee, S., Sannigrahi, M., Chakraborty, A., Das, A., Dey, A., & Dwivedi, A. D. (2023). Federated learning based Covid-19 detection. *Expert Systems*, *40*. Available from https://doi.org/10.1111/exsy.13173.

Crowson, M. G., Moukheiber, D., Arévalo, A. R., Lam, B. D., Mantena, S., Rana, A., Goss, D., Bates, D. W., & Celi, L. A. (2022). A systematic review of federated learning applications for biomedical data. *PLOS Digital Health*, *1*, e0000033. Available from https://doi.org/10.1371/journal.pdig.0000033.

Dai, Y., Chen, Z., Li, J., Heinecke, S., Sun, L., & Xu, R. (2022). *Tackling data heterogeneity in federated learning with class prototypes*. arXiv pre-print Serv.

Darzidehkalani, E., Ghasemi-rad, M., & van Ooijen, P. M. A. (2022). Federated learning in medical imaging: Part II: Methods, challenges, and considerations. *Journal of the American College of Radiology: JACR*, *19*, 975−982. Available from https://doi.org/10.1016/j.jacr.2022.030.016.

Demrozi, F., Pravadelli, G., Bihorac, A., & Rashidi, P. (2020). Human activity recognition using inertial, physiological and environmental sensors: A comprehensive survey. *IEEE Access*, *8*, 210816−210836. Available from https://doi.org/10.1109/ACCESS.2020.3037715.

Deng, Y., Lyu, F., Ren, J., Wu, H., Zhou, Y., Zhang, Y., & Shen, X. (2022). AUCTION: Automated and quality-aware client selection framework for efficient federated learning. *IEEE Transactions on Parallel and Distributed Systems*, *33*, 1996−2009. Available from https://doi.org/10.1109/TPDS.2021.3134647.

Ding, J., Tramel, E., Sahu, A.K., Wu, S., Avestimehr, S., & Zhang, T. (2022). Federated learning challenges and opportunities: an outlook. In Proceedings of the ICASSP 2022 - 2022 IEEE International Conference on Acoustics, Speech and Signal Processing (ICASSP); IEEE, May 23 2022; Vol. 2022-May, pp. 8752−8756.

Duan, M., Liu, D., Chen, X., Tan, Y., Ren, J., Qiao, L., & Liang, L. (2019). Astraea: Self-Balancing Federated Learning for Improving Classification Accuracy of Mobile Deep Learning Applications. In Proceedings of the 2019 IEEE 37th International Conference on Computer Design (ICCD); IEEE, November 2019; pp. 246−254.

Farcas, A.-J., Chen, X., Wang, Z., & Marculescu, R. (2022). Model Elasticity for Hardware Heterogeneity in Federated Learning Systems. In Proceedings of the Proceedings of the 1st ACM Workshop on Data Privacy and Federated Learning Technologies for Mobile Edge Network; ACM: New York, NY, USA, October 17 2022; pp. 19−24.

Feng, J., Zhang, W., Pei, Q., Wu, J., & Lin, X. (2022). Heterogeneous computation and resource allocation for wireless powered federated edge learning systems. *IEEE Transactions on Communications*, *70*, 3220−3233. Available from https://doi.org/10.1109/TCOMM.2022.3163439.

Ferraguig, L., Djebrouni, Y., Bouchenak, S., & Marangozova, V. (2021). Survey of bias mitigation in federated learning. In *Proceedings of the Conférence francophone d'informatique en Parallélisme, Architecture et Système*.

Fu, L., Zhang, H., Gao, G., Wang, H., Zhang, M., & Liu, X. (2022). Client selection in federated learning: Principles, challenges, and opportunities. arXiv Prepr. arXiv2211.01549.

Gad, G., & Fadlullah, Z. (2022). Federated learning via augmented knowledge distillation for heterogenous deep human activity recognition systems. *Sensors*, *23*, 6. Available from https://doi.org/10.3390/s23010006.

Ghosh, P. K., Chakraborty, A., Hasan, M., Rashid, K., & Siddique, A. H. (2023). Blockchain application in healthcare systems: A review. *Systems*, *11*, 38. Available from https://doi.org/10.3390/systems11010038.

Gosselin, R., Vieu, L., Loukil, F., & Benoit, A. (2022). Privacy and security in federated learning: A survey. *Applied Sciences*, *12*, 9901. Available from https://doi.org/10.3390/app12199901.

Guo, Y., Huang, K., & Chen, J. (2021). WCL: Client selection in federated learning with a combination of model weight divergence and client training loss for internet traffic classification. *Wireless Communications and Mobile Computing*, *2021*, 1−10. Available from https://doi.org/10.1155/2021/3381998.

Haleem, A., Javaid, M., Singh, R. P., Suman, R., & Rab, S. (2021). Blockchain technology applications in healthcare: An overview. *International Journal of Intelligent Networks*, *2*, 130−139. Available from https://doi.org/10.1016/j.ijin.2021.090.005.

Hao, M., Li, H., Xu, G., Chen, H., & Zhang, T. Efficient, private and robust federated learning. In *Proceedings of the annual computer security applications conference*; ACM: New York, NY, USA, December 6 2021; pp. 45−60.

Hartmann, M., Hashmi, U. S., & Imran, A. (2022). Edge computing in smart health care systems: Review, challenges, and research directions. *Transactions on Emerging Telecommunications Technologies*, *33*, e3710. Available from https://doi.org/10.1002/ett.3710.

Hayat, A., Morgado-Dias, F., Bhuyan, B., & Tomar, R. (2022). Human activity recognition for elderly people using machine and deep learning approaches. *Information*, *13*, 275. Available from https://doi.org/10.3390/info13060275.

Hou, D., Zhang, J., Man, K.L., Ma, J., Peng, Z. A systematic literature review of blockchain-based federated learning: Architectures, applications and issues. In *Proceedings of the 2021 2nd information communication technologies conference (ICTC)*; IEEE, May 7 2021; pp. 302−307.

Hu, H., Salcic, Z., Sun, L., Dobbie, G., & Zhang, X. Source inference attacks in federated learning. In *Proceedings of the 2021 IEEE International Conference on Data Mining (ICDM)*; IEEE, December 2021; Vol. 2021-Decem, pp. 1102−1107.

Iguoba, V. A., & Imoize, A. L. (2022). *The psychology of explanation in medical decision support systems. Explainable artificial intelligence in medical decision support systems* (pp. 489−506). Institution of Engineering and Technology.

Ijaz, M., Li, G., Lin, L., Cheikhrouhou, O., Hamam, H., & Noor, A. (2021). Integration and applications of fog computing and cloud computing based on the internet of things for provision of healthcare services at home. *Electronics*, *10*, 1077. Available from https://doi.org/10.3390/electronics10091077.

Imoize, A. L., Gbadega, P. A., Obakhena, H. I., Irabor, D. O., Kavitha, K. V. N., & Chakraborty, C. (2022). *Artificial intelligence-enabled internet of medical things for COVID-19 pandemic data management*. Explainable artificial intelligence in medical decision support systems (pp. 357−380). Institution of Engineering and Technology.

Imteaj, A., Thakker, U., Wang, S., Li, J., & Amini, M. H. (2022). A survey on federated learning for resource-constrained IoT devices. *IEEE Internet of Things Journal*, *9*, 1−24. Available from https://doi.org/10.1109/JIOT.2021.3095077.

Iqbal, Z., & Chan, H. Y. (2021). Concepts, key challenges and open problems of federated learning. *International Journal of Engineering*, *34*, 1667−1683. Available from https://doi.org/10.5829/ije.2021.34.07a.11.

Javed, W., Aabid, F., Danish, M., Tahir, H., & Zainab, R. Role of blockchain technology in healthcare: A systematic review. In *Proceedings of the 2021 International Conference on Innovative Computing (ICIC)*; IEEE, November 9 2021; pp. 1−8.

Jee Cho, Y., Wang, J., & Joshi, G. (2022). Towards understanding biased client selection in federated learning. In *Proceedings of the Proceedings of The 25th International Conference on Artificial Intelligence and Statistics*; PMLR; Vol. 151, pp. 10351−10375.

Joshi, M., Pal, A., & Sankarasubbu, M. (2022). Federated learning for healthcare domain − Pipeline, applications and challenges. *ACM Transactions on Computing for Healthcare*, *3*, 1−36. Available from https://doi.org/10.1145/3533708.

Kamble, V., & Phophalia, A. Medical image analysis using federated learning frameworks: Technical review. In *Proceedings of the 2022 IEEE 10th Region 10 Humanitarian Technology Conference (R10-HTC)*; IEEE, September 16 2022; Vol. 2022. pp. 44−48.

Kashyap, V., Kumar, A., Kumar, A., & Hu, Y.-C. (2022). A systematic survey on fog and IoT driven healthcare: Open challenges and research issues. *Electronics*, *11*, 2668. Available from https://doi.org/10.3390/electronics11172668.

Kavitha, K. V. N., Ashok, S., Imoize, A. L., Ojo, S., Selvan, K. S., Ahanger, T. A., & Alhassan, M. (2022). On the use of wavelet domain and machine learning for the analysis of epileptic seizure detection from EEG signals. *Journal of Healthcare Engineering*, *2022*, 1−16. Available from https://doi.org/10.1155/2022/8928021.

Khan, L. U., Saad, W., Han, Z., Hossain, E., & Hong, C. S. (2021). Federated learning for internet of things: Recent advances, taxonomy, and open challenges. *IEEE Communications Surveys & Tutorials, 23*, 1759—1799.

Khezr, S., Moniruzzaman, M., Yassine, A., & Benlamri, R. (2019). Blockchain technology in healthcare: A comprehensive review and directions for future research. *Applied Sciences, 9*, 1736. Available from https://doi.org/10.3390/app9091736.

Lee, H., Kim, J., Ahn, S., Hussain, R., Cho, S., & Son, J. (2021). Digestive neural networks: A novel defense strategy against inference attacks in federated learning. *Computers & Security, 109*, 102378. Available from https://doi.org/10.1016/j.cose.2021.102378.

Lin, W., Xu, Y., Liu, B., Li, D., Huang, T., & Shi, F. (2022). Contribution-based federated learning client selection. *International Journal of Intelligent Systems, 37*, 7235—7260. Available from https://doi.org/10.1002/int.22879.

Li, Z., He, Y., Yu, H., Kang, J., Li, X., Xu, Z., & Niyato, D. (2022). Data heterogeneity-robust federated learning via group client selection in industrial IoT. *IEEE Internet of Things Journal, 9*, 17844—17857. Available from https://doi.org/10.1109/JIOT.2022.3161943.

Li, D., Luo, Z., & Cao, B. (2022). Blockchain-based federated learning methodologies in smart environments. *Cluster Computing, 25*, 2585—2599. Available from https://doi.org/10.1007/s10586-021-03424-y.

Li, T., Sahu, A. K., Talwalkar, A., & Smith, V. (2019). Federated learning: Challenges, methods, and future directions. *IEEE Signal Processing Magazine, 37*, 50—60. Available from https://doi.org/10.1109/MSP.2020.2975749.

Li, A., Sun, J., Wang, B., Duan, L., Li, S., Chen, Y., & Li, H. (2020). LotteryFL: Personalized and communication-efficient federated learning with lottery ticket hypothesis on non-IID datasets. arXiv Prepr. arXiv2008.03371.

Li, Z., Yu, H., Zhou, T., Luo, L., Fan, M., Xu, Z., & Sun, G. (2021). Byzantine resistant secure blockchained federated learning at the edge. *IEEE Network, 35*, 295—301. Available from https://doi.org/10.1109/MNET.011.2000604.

Li, C., Zeng, X., Zhang, M., & Cao, Z. (2022). *PyramidFL: A fine-grained client selection framework for efficient federated learning. Proceedings of the proceedings of the 28th annual international conference on mobile computing and networking* (pp. 158—171). New York, NY, USA: ACM.

Luo, B., Li, X., Wang, S., Huang, J., & Tassiulas, L. Cost-effective federated learning design. In Proceedings of the IEEE INFOCOM 2021 - IEEE Conference on Computer Communications; IEEE, May 10 2021; Vol. 2021-May, pp. 1—10.

Lyu, L., Yu, H., Zhao, J., & Yang, Q. (2020). *Threats to federated learning. Lecture notes in computer science (including subseries Lecture Notes in Artificial Intelligence and Lecture Notes in Bioinformatics)* (Vol. 12500, pp. 3—16). LNCS.

Majeed, A., Zhang, X., & Hwang, S. O. (2022). Applications and challenges of federated learning paradigm in the big data era with special emphasis on COVID-19. *Big Data and Cognitive Computing, 6*, 127. Available from https://doi.org/10.3390/bdcc6040127.

Malik, H., Anees, T., Naeem, A., Naqvi, R. A., & Loh, W.-K. (2023). Blockchain-federated and deep-learning-based ensembling of capsule network with incremental extreme learning machines for classification of COVID-19 using CT scans. *Bioengineering, 10*, 203. Available from https://doi.org/10.3390/bioengineering10020203.

Malik, H., Naeem, A., Naqvi, R. A., & Loh, W.-K. (2023). DMFL_Net: A federated learning-based framework for the classification of COVID-19 from multiple chest diseases using X-rays. *Sensors, 23*, 743. Available from https://doi.org/10.3390/s23020743.

Mammen, P.M. (2021). Federated learning: Opportunities and challenges. arXiv Prepr. arXiv2101.05428.

Ma, X., Jiang, Q., Shojafar, M., Alazab, M., Kumar, S., & Kumari, S. (2022). DisBezant: Secure and robust federated learning against byzantine attack in IoT-enabled MTS. *IEEE Transactions on Intelligent Transportation Systems*, *24*, 1−11. Available from https://doi.org/10.1109/TITS.2022.3152156.

Ma, Z., Lu, Y., Li, W., & Cui, S. (2023). *Beyond random selection: A perspective from model inversion in personalized federated learning. Joint european conference on machine learning and knowledge discovery in databases* (pp. 572−586). Springer, ISBN 9783031264115.

Ma, X., Zhou, Y., Wang, L., & Miao, M. (2022). Privacy-preserving byzantine-robust federated learning. *Computer Standards & Interfaces*, *80*, 103561. Available from https://doi.org/10.1016/j.csi.2021.103561.

Mendieta, M., Yang, T., Wang, P., Lee, M., Ding, Z., & Chen, C. Local learning matters: Rethinking data heterogeneity in federated learning. Proc. IEEE Comput. Soc. Conf. Comput. Vis. Pattern Recognit. 2021, 2022-June, 8387−8396, Available from https://doi.org/10.1109/CVPR52688.2022.00821.

Mothukuri, V., Parizi, R. M., Pouriyeh, S., Huang, Y., Dehghantanha, A., & Srivastava, G. (2021). A Survey on security and privacy of federated learning. *Future Generation Computer Systems*, *115*, 619−640. Available from https://doi.org/10.1016/j.future.2020.100.007.

Mouhni, N., Elkalay, A., Chakraoui, M., Abdali, A., Ammoumou, A., & Amalou, I. (2022). Federated learning for medical imaging: An updated state of the art. *Ingénierie des systèmes d Inf*, *27*, 143−150. Available from https://doi.org/10.18280/isi.270117.

Mugunthan, V., Gokul, V., Kagal, L., & Dubnov, S. (2021). Bias-free FedGAN: A federated approach to generate bias-free datasets. arXiv Prepr. arXiv2103.09876.

Myrzashova, R., Alsamhi, S. H., Shvetsov, A. V., Hawbani, A., & Wei, X. (2023). Blockchain meets federated learning in healthcare: A systematic review with challenges and opportunities. *IEEE Internet of Things Journal*. Available from https://doi.org/10.1109/JIOT.2023.3263598.

Nazir, S., & Kaleem, M. (2023). Federated learning for medical image analysis with deep neural networks. *Diagnostics*, *13*, 1532. Available from https://doi.org/10.3390/diagnostics13091532.

Naz, S., Phan, K. T., & Chen, Y. P. P. (2022). A comprehensive review of federated learning for COVID-19 detection. *International Journal of Intelligent Systems*, *37*, 2371−2392. Available from https://doi.org/10.1002/int.22777.

Ngan Van, L., Hoang Tuan, A., Phan The, D., Vo, T.-K., & Pham, V.-H. A privacy-preserving approach for building learning models in smart healthcare using blockchain and federated learning. In *Proceedings of the The 11th International Symposium on Information and Communication Technology*; ACM: New York, NY, USA, December 2022; pp. 435−441.

Nguyen, T.D., Nguyen, A.D., Wong, K.-S., Pham, H.H., Nguyen, T.H., Nguyen, P.Le, & Nguyen, T.T. (2023). FedGrad: Mitigating backdoor attacks in federated learning through local ultimate gradients inspection. arXiv Prepr.

Nishio, T., & Yonetani, R. Client selection for federated learning with heterogeneous resources in mobile edge. In Proceedings of the IEEE International Conference on Communications; IEEE, May 2019; Vol. 2019-May, pp. 1−7.

Nguyen, T. D., Nguyen, T., Le Nguyen, P., Pham, H. H., Doan, K. D., & Wong, K. S. (2024). Backdoor attacks and defenses in federated learning: Survey, challenges and future research directions. *Engineering Applications of Artificial Intelligence, 127*, 107166.

Nwaneri, S. C., Yinka-Banjo, C., Uregbulam, U. C., Odukoya, O. O., & Imoize, A. L. (2022). *Explainable neural networks in diabetes mellitus prediction. Explainable artificial intelligence in medical decision support systems* (pp. 313–334). Institution of Engineering and Technology.

Pouriyeh, S., Shahid, O., Parizi, R. M., Sheng, Q. Z., Srivastava, G., Zhao, L., & Nasajpour, M. (2022). Secure smart communication efficiency in federated learning: Achievements and challenges. *Applied Sciences, 12*, 8980. Available from https://doi.org/10.3390/app12188980.

Prakash, S., & Avestimehr, A. S. (2020). Mitigating byzantine attacks in federated learning. *arXiv preprint arXiv:2010.07541*.

Prasad, V. K., Bhattacharya, P., Maru, D., Tanwar, S., Verma, A., Singh, A., Tiwari, A. K., Sharma, R., Alkhayyat, A., Ţurcanu, F. E., et al. (2023). Federated learning for the internet-of-medical-things: A survey. *Mathematics, 11*, 1–47. Available from https://doi.org/10.3390/math11010151.

Prayitno., Shyu, C.-R., Putra, K. T., Chen, H.-C., Tsai, Y.-Y., Hossain, K. S. M. T., Jiang, W., & Shae, Z.-Y. (2021). A systematic review of federated learning in the healthcare area: From the perspective of data properties and applications. *Applied Sciences, 11*, 11191. Available from https://doi.org/10.3390/app112311191.

Qammar, A., Karim, A., Ning, H., & Ding, J. (2023). Securing federated learning with blockchain: A systematic literature review. *Artificial Intelligence Review, 56*, 3951–3985. Available from https://doi.org/10.1007/s10462-022-10271-9.

Rahman, A., Hossain, M. S., Muhammad, G., Kundu, D., Debnath, T., Rahman, M., Khan, M. S. I., Tiwari, P., & Band, S. S. (2022). Federated learning-based AI approaches in smart healthcare: Concepts, taxonomies, challenges and open issues. *Cluster Computing.* Available from https://doi.org/10.1007/s10586-022-03658-4.

Rai, S., Kumari, A., & Prasad, D. K. (2022). Client selection in federated learning under imperfections in environment. *AI, 3*, 124–145. Available from https://doi.org/10.3390/ai3010008.

Raja, G. B. (2021). Impact of internet of things, artificial intelligence, and blockchain technology in Industry 4.0. In R. L. Kumar, Y. Wang, T. Poongodi, & A. L. Imoize (Eds.), *Internet of Things, artificial intelligence and blockchain technology* (pp. 157–178). Cham: Springer International Publishing, ISBN 9783030741501.

Rajotte, J.-F., Mukherjee, S., Robinson, C., Ortiz, A., West, C., Ferres, J. M. L., & Ng, R. T. (2021). *Reducing bias and increasing utility by federated generative modeling of medical images using a centralized adversary. Proceedings of the proceedings of the conference on information technology for social good* (pp. 79–84). New York, NY, USA: ACM.

Ramasamy, L. K., Khan, K. P. F., Imoize, A. L., Ogbebor, J. O., Kadry, S., & Rho, S. (2021). Blockchain-based wireless sensor networks for malicious node detection: A survey. *IEEE Access, 9*, 128765–128785. Available from https://doi.org/10.1109/ACCESS.2021.3111923.

Rauniyar, A., Hagos, D. H., Jha, D., Håkegård, J. E., Bagci, U., Rawat, D. B., & Vlassov, V. (2022). Federated learning for medical applications: A taxonomy, current trends, challenges, and future research directions. arXiv Prepr. arXiv2208.03392.

Reddy, K. D., & Gadekallu, T. R. (2023). A comprehensive survey on federated learning techniques for healthcare informatics. *Computational Intelligence and Neuroscience*, *2023*, 266. Available from https://doi.org/10.1155/2023/8393990.

Rehman, A., Abbas, S., Khan, M. A., Ghazal, T. M., Adnan, K. M., & Mosavi, A. (2022). A secure Healthcare 5.0 system based on blockchain technology entangled with federated learning technique. *Computers in Biology and Medicine*, *150*, 106019. Available from https://doi.org/10.1016/j.compbiomed.2022.106019.

Reisizadeh, A., Mokhtari, A., Hassani, H., Jadbabaie, A., & Pedarsani, R. (2019). FedPAQ: A communication-efficient federated learning method with periodic averaging and quantization. In *Proceedings of the International Conference on Artificial Intelligence and Statistics*; PMLR; pp. 2021−2031.

Rieke, N., Hancox, J., Li, W., Milletarì, F., Roth, H. R., Albarqouni, S., Bakas, S., Galtier, M. N., Landman, B. A., Maier-Hein, K., et al. (2020). The future of digital health with federated learning. *npj Digital Medicine.*, *3*, 119. Available from https://doi.org/10.1038/s41746-020-00323-1.

Rodríguez-Barroso, N., Jiménez-López, D., Luzón, M. V., Herrera, F., & Martínez-Cámara, E. (2023). Survey on federated learning threats: Concepts, taxonomy on attacks and defences, experimental study and challenges. *Information Fusion*, *90*, 148−173. Available from https://doi.org/10.1016/j.inffus.2022.090.011.

Rodríguez-Barroso, N., Martínez-Cámara, E., Luzón, M. V., & Herrera, F. (2022). Dynamic defense against byzantine poisoning attacks in federated learning. *Future Generation Computer Systems*, *133*, 1−9. Available from https://doi.org/10.1016/j.future.2022.030.003.

Rothchild, D., Panda, A., Ullah, E., Ivkin, N., Stoica, I., Braverman, V., Gonzalez, J., & Arora, R. Fetchsgd: Communication-Efficient Federated Learning with Sketching. In Proceedings of the 37th International Conference on Machine Learning, ICML 2020; PMLR, 2020; Vol. PartF16814, pp. 8223−8235.

Saeed, H., Malik, H., Bashir, U., Ahmad, A., Riaz, S., Ilyas, M., Bukhari, W. A., & Khan, M. I. A. (2022). Blockchain technology in healthcare: A systematic review. *PLoS One*, *17*, e0266462. Available from https://doi.org/10.1371/journal.pone.0266462.

Samuel, O., Omojo, A. B., Onuja, A. M., Sunday, Y., Tiwari, P., Gupta, D., Hafeez, G., Yahaya, A. S., Fatoba, O. J., & Shamshirband, S. (2023). IoMT: A COVID-19 healthcare system driven by federated learning and blockchain. *IEEE Journal of Biomedical and Health Informatics*, *27*, 823−834. Available from https://doi.org/10.1109/JBHI.2022.3143576.

Schrader, L., Vargas Toro, A., Konietzny, S., Rüping, S., Schäpers, B., Steinböck, M., Krewer, C., Müller, F., Güttler, J., & Bock, T. (2020). Advanced sensing and human activity recognition in early intervention and rehabilitation of elderly people. *Journal of Population Ageing*, *13*, 139−165. Available from https://doi.org/10.1007/s12062-020-09260-z.

Shaheen, M., Farooq, M. S., Umer, T., & Kim, B.-S. (2022). Applications of federated learning; taxonomy, challenges, and research trends. *Electronics*, *11*, 670. Available from https://doi.org/10.3390/electronics11040670.

Shahid, O., Pouriyeh, S., Parizi, R.M., Sheng, Q.Z., Srivastava, G., & Zhao, L. (2021). Communication efficiency in federated learning: Achievements and challenges. arXiv Prepr. arXiv2107.10996.

Shang, X., Lu, Y., Huang, G., & Wang, H. (2022). Federated learning on heterogeneous and long-tailed data via classifier re-training with federated features. *Proc.*

Thirty-First Int. Jt. Conf. Artif. Intell., 2218−2224. Available from https://doi.org/10.24963/ijcai.2022/308.

Shi, W., Sun, Y., Zhou, S., & Niu, Z. (2021). Device scheduling and resource allocation for federated learning under delay and energy constraints. In Proceedings of the 2021 IEEE 22nd International Workshop on Signal Processing Advances in Wireless Communications (SPAWC); IEEE, September 27 2021; Vol. 2021-Septe, pp. 596−600.

Shi, J., Wan, W., Hu, S., Lu, J., & Yu Zhang, L. (2022). Challenges and approaches for mitigating byzantine attacks in federated learning. In Proceedings of the 2022 IEEE International Conference on Trust, Security and Privacy in Computing and Communications (TrustCom); IEEE, December 2022; pp. 139−146.

Siam, A. I., Abou Elazm, A., El-Bahnasawy, N. A., El Banby, G., & Abd El-Samie, F. E. (2019). F.E.F.E.A.E.-S. Smart health monitoring system based on IoT and cloud computing. *Menoufia Journal of Electronic Engineering Research*, 28, 37−42. Available from https://doi.org/10.21608/mjeer.2019.76711.

Siam, A. I., Almaiah, M. A., Al-Zahrani, A., Elazm, A. A., El Banby, G. M., El-Shafai, W., El-Samie, F. E. A., & El-Bahnasawy, N. A. (2021). Secure health monitoring communication systems based on IoT and cloud computing for medical emergency applications. *Computational Intelligence and Neuroscience*, 2021, 1−23. Available from https://doi.org/10.1155/2021/8016525.

Sikandar, H. S., Waheed, H., Tahir, S., Malik, S. U. R., & Rafique, W. (2023). A detailed survey on federated learning attacks and defenses. *Electronics*, 12, 260. Available from https://doi.org/10.3390/electronics12020260.

Sohan, M. F., & Basalamah, A. (2023). A systematic review on federated learning in medical image analysis. *IEEE Access*, 11, 28628−28644. Available from https://doi.org/10.1109/ACCESS.2023.3260027.

Sozinov, K., Vlassov, V., & Girdzijauskas, S. Human activity recognition using federated learning. In Proceedings of the 2018 IEEE Intl Conf on Parallel & Distributed Processing with Applications, Ubiquitous Computing & Communications, Big Data & Cloud Computing, Social Computing & Networking, Sustainable Computing & Communications (ISPA/IUCC/BDCloud/SocialCom/SustainCom); IEEE, December 1 2018; pp. 1103−1111.

So, J., Guler, B., & Salman Avestimehr, A. (2021). Byzantine-resilient secure federated learning. *IEEE Journal on Selected Areas in Communications*, 39, 2168−2181. Available from https://doi.org/10.1109/JSAC.2020.3041404.

Tahir, A., Chen, Y., & Nilayam, P. (2022). FedSS: Federated learning with smart selection of clients. arXiv Prepr. arXiv2207.04569.

Tang, M., Ning, X., Wang, Y., Sun, J., Wang, Y., Li, H., & Chen, Y. FedCor: Correlation-based active client selection strategy for heterogeneous federated learning. In Proceedings of the 2022 IEEE/CVF Conference on Computer Vision and Pattern Recognition (CVPR); IEEE, June 2022; Vol. 2022-June, pp. 10092−10101.

Tang, Z., Zhang, Y., Shi, S., He, X., Han, B., & Chu, X. (2022). Virtual homogeneity learning: Defending against data heterogeneity in federated learning. In Proceedings of the International Conference on Machine Learning; PMLR; pp. 21111−21132.

Tripathy, S. S., Imoize, A. L., Rath, M., Tripathy, N., Bebortta, S., Lee, C.-C., Chen, T.-Y., Ojo, S., Isabona, J., & Pani, S. K. (2022). A novel edge-computing-based framework for an intelligent smart healthcare system in smart cities. *Sustainability*, 15, 735. Available from https://doi.org/10.3390/su15010735.

van der Velden, B. H. M., Kuijf, H. J., Gilhuijs, K. G. A., & Viergever, M. A. (2022). Explainable artificial intelligence (XAI) in deep learning-based medical image analysis. *Medical Image Analysis, 79*, 102470. Available from https://doi.org/10.1016/j.media.2022.102470.

Verma, P., Sood, S. K., & Kalra, S. (2018). Cloud-centric IoT based student healthcare monitoring framework. *Journal of Ambient Intelligence and Humanized Computing, 9*, 1293–1309. Available from https://doi.org/10.1007/s12652-017-0520-6.

Wang, S., Chen, M., Yin, C., Saad, W., Hong, C. S., Cui, S., & Poor, H. V. (2021). Federated learning for task and resource allocation in wireless high-altitude balloon networks. *IEEE Internet of Things Journal, 8*, 17460–17475. Available from https://doi.org/10.1109/JIOT.2021.3080078.

Wang, Y., Lin, L., & Chen, J. (2022). Communication-efficient adaptive federated learning. In Proceedings of the International Conference on Machine Learning; PMLR; pp. 22802–22838.

Wang, W., Li, X., Qiu, X., Zhang, X., Zhao, J., & Brusic, V. (2023). A privacy preserving framework for federated learning in smart healthcare systems. *Information Processing and Management, 60*. Available from https://doi.org/10.1016/j.ipm.2022.103167.

Wang, L., Wang, W., & Li, B. CMFL: Mitigating communication overhead for federated learning. In Proceedings of the 2019 IEEE 39th International Conference on Distributed Computing Systems (ICDCS); IEEE, July 2019; Vol. 2019-July, pp. 954–964.

Wen, J., Zhang, Z., Lan, Y., Cui, Z., Cai, J., & Zhang, W. (2023). A survey on federated learning: Challenges and applications. *International Journal of Machine Learning and Cybernetics, 14*, 513–535. Available from https://doi.org/10.1007/s13042-022-01647-y.

Wolfrath, J., Sreekumar, N., Kumar, D., Wang, Y., & Chandra, A. HACCS: Heterogeneity-aware clustered client selection for accelerated federated learning. In Proceedings of the 2022 IEEE International Parallel and Distributed Processing Symposium (IPDPS); IEEE, May 2022; pp. 985–995.

Xia, G., Chen, J., Yu, C., & Ma, J. (2023). Poisoning attacks in federated learning: A survey. *IEEE Access, 11*, 10708–10722. Available from https://doi.org/10.1109/ACCESS.2023.3238823.

Xia, Q., Ye, W., Tao, Z., Wu, J., & Li, Q. (2021). A survey of federated learning for edge computing: Research problems and solutions. *High-Confidence Computing, 1*, 100008. Available from https://doi.org/10.1016/j.hcc.2021.100008.

Yang, Q., Liu, Y., Chen, T., & Tong, Y. (2019). Federated machine learning: Concept and applications. *ACM Transactions on Intelligent Systems and Technology, 10*, 1–19. Available from https://doi.org/10.1145/3298981.

Yang, C., Xu, M., Wang, Q., Chen, Z., Huang, K., Ma, Y., Bian, K., Huang, G., Liu, Y., Jin, X., et al. (2022). FLASH: Heterogeneity-aware federated learning at scale. *IEEE Transactions on Mobile Computing*, 1–18. Available from https://doi.org/10.1109/TMC.2022.3214234.

Yoo, J. H., Jeong, H., Lee, J., & Chung, T. M. (2022). Open problems in medical federated learning. *International Journal of Web Information Systems, 18*, 77–99. Available from https://doi.org/10.1108/IJWIS-04-2022-0080.

Yuan, H., Hui, B., Yang, Y., Burlina, P., Gong, N. Z., & Cao, Y. (2022). *Addressing heterogeneity in federated learning via distributional transformation. Lecture notes in computer science (including subseries lecture notes in artificial intelligence and lecture notes in bioinformatics)* (13698, pp. 179–195). Springer, LNCS.

Yu, B., Mao, W., Lv, Y., Zhang, C., & Xie, Y. (2022). A survey on federated learning in data mining. *WIREs Data Mining and Knowledge Discovery*, *12*, 29−44. Available from https://doi.org/10.1002/widm.1443.

Zari, O., Xu, C., & Neglia, G. (2021). Efficient passive membership inference attack in federated learning. arXiv Prepr. arXiv2111.00430.

Zeng, Y., Mu, Y., Yuan, J., Teng, S., Zhang, J., Wan, J., Ren, Y., & Zhang, Y. (2022). Adaptive federated learning with non-IID data. *The Computer Journal*. Available from https://doi.org/10.1093/comjnl/bxac118.

Zeng, Y., Teng, S., Xiang, T., Zhang, J., Mu, Y., Ren, Y., & Wan, J. (2023). A client selection method based on loss function optimization for federated learning. *Computer Modeling in Engineering & Sciences*, *137*, 1047−1064. Available from https://doi.org/10.32604/cmes.2023.027226.

Zhang, J., Chen, J., Wu, D., Chen, B., & Yu, S. Poisoning attack in federated learning using generative adversarial nets. In Proceedings of the 2019 18th IEEE International Conference On Trust, Security And Privacy In Computing And Communications/13th IEEE International Conference On Big Data Science And Engineering (TrustCom/BigDataSE); IEEE, August 1 2019; pp. 374−380.

Zhang, D.Y., Kou, Z., & Wang, D. FairFL: A fair federated learning approach to reducing demographic bias in privacy-sensitive classification models. In Proceedings of the 2020 IEEE International Conference on Big Data (Big Data); IEEE, December 10 2020; pp. 1051−1060.

Zhang, K., Song, X., Zhang, C., & Yu, S. (2022). Challenges and future directions of secure federated learning: A survey. *Frontiers of Computer Science*, *16*, 165817. Available from https://doi.org/10.1007/s11704-021-0598-z.

Zhang, J., Zhu, H., Wang, F., Zhao, J., Xu, Q., & Li, H. (2022). Security and privacy threats to federated learning: Issues, methods, and challenges. *Security and Communication Networks*, *2022*, 1−24. Available from https://doi.org/10.1155/2022/2886795.

Zhan, Y., Li, P., & Guo, S. Experience-driven computational resource allocation of federated learning by deep reinforcement learning. In Proceedings of the 2020 IEEE International Parallel and Distributed Processing Symposium (IPDPS); IEEE, May 2020; pp. 234−243.

Zhao, Y., Li, M., Lai, L., Suda, N., Civin, D., & Chandra, V. Federated learning with non-IID data. arXiv Prepr. arXiv1806.00582 2018, https://doi.org/10.48550/arXiv.1806.00582.

Zhou, X., Lei, X., Yang, C., Shi, Y., Zhang, X., & Shi, J. (2022). Handling data heterogeneity in federated learning via knowledge fusion. arXiv Prepr. arXiv2207.11447, doi:10.48550/arXiv.2207.11447.

Zhu, Y., Markos, C., Zhao, R., Zheng, Y., & Yu, J.J. Q. FedOVA: One-vs-all training method for federated learning with non-IID data. In Proceedings of the 2021 International Joint Conference on Neural Networks (IJCNN); IEEE, July 18 2021; Vol. 2021-July, pp. 1−7.

Zou, S., Xiao, M., Xu, Y., An, B., & Zheng, J. FedDCS: Federated learning framework based on dynamic client selection. In Proceedings of the 2021 IEEE 18th International Conference on Mobile Ad Hoc and Smart Systems (MASS); IEEE, October 2021; pp. 627−632.

Federated learning challenges and risks in modern digital healthcare systems

11

Kassim Kalinaki[1,2], Owais Ahmed Malik[3], Umar Yahya[4] and Daphne Teck Ching Lai[5]

[1]Department of Computer Science, Islamic University in Uganda (IUIU), Mbale, Uganda
[2]Borderline Research Laboratory, Kampala, Uganda
[3]School of Digital Science, Universiti Brunei Darussalam, Gadong, Brunei
[4]Motion Analysis Research Laboratory (MARL), Islamic University in Uganda (IUIU), Kampala Campus, Kampala, Uganda
[5]School of Digital Science, Universiti Brunei Darussalam, Gadong, Brunei

11.1 Introduction

The intelligent healthcare sector has witnessed a significant revolution through recent advancements in communication technologies, accompanied by crucial support technologies such as the IoHT and artificial intelligence (AI) (Kalinaki & Fahadi, 2023; Kalinaki & Thilakarathne, 2023). However, in the face of the expanding size of healthcare networks and mounting privacy challenges, conventional AI techniques, such as machine learning (ML) and deep learning (DL), which rely on centralized data collection and processing, have proven unfeasible and unrealistic in healthcare (Chemisto et al., 2023; Nguyen, 2022). In light of these challenges, FL, an emerging distributed and collaborative technique, holds promise in mitigating the risks associated with traditional AI deployment. By enabling collaborative learning across distributed devices and networks, FL addresses the privacy and security challenges associated with conventional centralized data collection methods (Islam et al., 2022; Mothukuri, 2021; Rieke, 2020; Yin et al., 2021). Instead of sharing raw data, FL enables institutions to share only model updates while securely storing sensitive data. This approach minimizes the risk of data breaches and protects patient confidentiality, a critical challenge in the healthcare sector.

FL leverages a vast amount of data generated by various devices, wearables (e.g., fitness trackers and smartwatches), and implantable healthcare devices (e.g., pacemakers), as well as electronic health records (EHRs), to train ML models without compromising patient privacy (Dang, 2022; Fahim et al., 2023). For this reason, the healthcare industry is rapidly adopting FL as a powerful tool to improve patient care,

Federated Learning for Digital Healthcare Systems. DOI: https://doi.org/10.1016/B978-0-443-13897-3.00004-7

research outcomes, and treatment effectiveness. For instance, through collaborative training of ML models on diverse patient datasets from multiple healthcare institutions, FL enhances the accuracy of diagnostic models by incorporating insights from a wide range of data sources. As depicted in many studies in the field of radiology, FL has been employed to improve the accuracy of image classification and detection, leading to better diagnoses of diseases such as COVID-19, cancer, stroke, and heart conditions (Joshi et al., 2022; Kumar & Singla, 2021; Li, 2022). Additionally, FL allows healthcare providers to build personalized treatment models by analyzing patient data across different demographics, locations, and medical conditions (Nguyen, 2022; Pfitzner et al., 2021). This personalized approach helps deliver tailored treatment plans and interventions, improving patient outcomes.

Moreover, FL facilitates real-time monitoring of disease outbreaks by aggregating and analyzing data from various sources. For instance, in the case of infectious diseases, FL helps detect early warning signs, track transmission patterns, and facilitate timely public health interventions (Javed, 2023). Furthermore, FL enables collaborative research by securely pooling data from multiple healthcare institutions. This approach accelerates medical research and drug development, discovering new treatments and therapies. For instance, studies have demonstrated how FL has been utilized to study genomic data across multiple institutions, leading to advancements in personalized medicine and the identification of genetic risk factors for diseases (Aziz, 2022; Rieke, 2020).

With all the above benefits and other ways FL is transforming the healthcare sector, however, embracing this emerging and evolving technology is not without risks and challenges. This chapter aims to comprehensively review the various threats, risks, and challenges associated with FL in modern digital healthcare systems. The discussions within this chapter offer valuable insights for researchers, students, and key stakeholders interested in understanding and navigating the landscape of FL in the healthcare domain. Acknowledging these challenges and working toward robust solutions will enable researchers and key stakeholders to harness the full potential of FL in transforming healthcare and enhancing patient well-being.

11.1.1 Key contributions of the chapter

The contributions of this chapter are highlighted as follows.

1. Provision of an overview of FL in modern digital healthcare systems, highlighting its benefits in transforming the healthcare sector.
2. A comprehensive discussion of threats, challenges, and risks to FL in modern digital healthcare systems.
3. A summary of the solutions to mitigate challenges, threats, and risks to FL in modern digital healthcare systems.
4. A discussion on future trends aimed at minimizing the above threats and challenges.

11.1.2 **Chapter organization**

This chapter is ordered as follows: Section 11.1 consists of the introduction, which provides an overview of FL in modern digital healthcare systems, highlighting its benefits in revolutionizing the healthcare industry. Section 11.2 provides a comprehensive discussion of different challenges, threats, and risks of FL in modern digital healthcare systems, along with their corresponding solutions. Section 11.3 details the lessons learned in this chapter, highlighting the effectiveness of using real-life datasets in mitigating the challenges of FL. Section 11.4 highlights the future trends of FL in modern digital healthcare systems aimed at minimizing the challenges, threats, and risks. The conclusion is provided in Section 11.5.

11.2 **Federated learning risks and challenges in modern digital healthcare systems**

While FL offers several benefits in transforming the healthcare domain, as highlighted in the introduction, it poses certain risks and challenges in modern digital healthcare systems. Accordingly, this section presents the risks, threats, and challenges in the healthcare domain, categorized into privacy and security, data-related, and technical difficulties. Fig. 11.1 depicts FL's risks and challenges in modern digital healthcare systems.

11.2.1 **Privacy and security challenges of federated learning in modern digital healthcare systems**

FL in modern digital healthcare systems has several privacy and security challenges, such as adversarial, inference, poisoning, evasion, man-in-the-middle, gradient leakage, and free-riding attacks that have been reported in several FL implementations in digital healthcare systems and must be carefully addressed to protect patient data and ensure system integrity (Chen, 2018; Coelho, 2023; Joshi et al., 2022; Yang, 2023). For instance, hackers have targeted healthcare facilities in different countries, such as Boston Children's Hospital in 2014, the National Health Service of the UK in 2017, LifeLabs of Canada in 2019, and Düsseldorf University Hospital of Germany in 2020 (Coelho, 2023). Such attacks lead to unauthorized access to patient data, which can lead to loss of life and blackmail tarnishing the image of healthcare providers. Additionally, FL suffers from unintentional data leakage as a result of shared information, which may indirectly expose the private data of participating healthcare institutions (Alabdulatif et al., 2023; Alli, 2021; Kalinaki & Fahadi, 2023; Kalinaki & Thilakarathne, 2023; Mothukuri, 2021). Other studies have indicated the susceptibility of genomic data

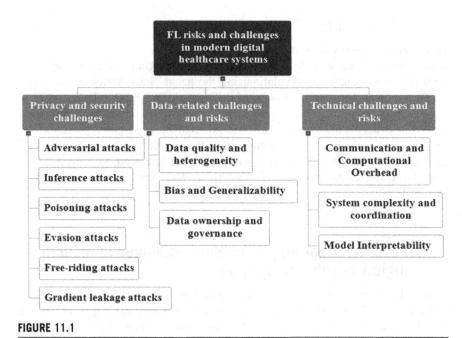

FIGURE 11.1

FL risks and challenges in modern digital healthcare systems. *FL,* Federated learning.

leakage due to using FL while training genome sequencing algorithms (Dhiman, 2022). To address the security and privacy challenges highlighted above, Table 11.1 summarizes FL attacks and the proposed solutions to mitigate them, especially in modern digital healthcare systems.

11.2.2 Data-related challenges of risks of federated learning in Internet of Health Things

In the world of the IoHT, where devices collect important patient-related data, there is a delicate balance between promise and danger. Healthcare organizations are embracing FL to uncover valuable insights from different datasets but must also face data risks and challenges. The following subsections explore the difficulties associated with data and their proposed solutions.

11.2.2.1 Data quality and heterogeneity

As healthcare devices generate torrents of data, ranging from vital signs to medical images, the quality and diversity of this data become paramount. Data quality refers to the accuracy, reliability, and completeness of the data collected from various healthcare devices. In IoHT, data quality varies due to sensor accuracy,

Table 11.1 Summary of attacks on federated learning (FL) and proposed solutions with references.

Category of FL challenges	Attacks	Summary of solution	References
Privacy and security	Adversarial attacks	Proposed a CycleGAN-based technique to detect adversarial attacks.	Shieh (2022)
	Inference attacks	Proposed an approach that integrates a level of DP, gracefully treading the fine line between preserving privacy and enhancing precision in the model, substantially diminishing the likelihood of this malicious onslaught.	Rahman (2020), Thapa et al. (2021)
	Poisoning attacks	A fusion of advanced aggregation models and DP was proposed to enhance resilience.	Grama (2020), Rahman (2020)
	Evasion attacks	Proposed region-based classification and ensemble learning approach to mitigate the attacks.	Cao and Gong (2017), Ahmed et al. (2022)
	Free-riding attack	Proposed federated optimization algorithm called Viceroy to mitigate the attacks.	Lewis (2023)
	Gradient leakage attacks	Unveiled an unconventional safeguard where they manipulate the essence of gradient perturbations to align with the looming threat of information leakage.	Wang (2022)

DP, *Differential privacy.*

noise interference, transmission errors, or inconsistencies in data collection protocols (Alhusein & Idrees, 2023; Antunes, 2022). Hence, erroneous or incomplete patient data can undermine the efficacy and reliability of the FL models trained on such data. Moreover, improper annotations of image data may lead to class imbalances in the trained models. Heterogeneity, however, alludes to the diversity and variability in the data collected from different sources within the IoHT ecosystem. Healthcare devices and sensors used in IoHT encompass a broad spectrum, each with its unique data format, structure, and protocols (Rieke, 2020). For example, wearable devices, EHRs, and imaging devices produce data in distinct formats and representations. Additionally, healthcare organizations may operate in different regions, leading to demographic variations, medical practices, and patient populations. Data heterogeneity can introduce challenges in harmonizing and integrating the data for effective FL. It necessitates addressing issues such as data interoperability, mapping disparate data formats, and accounting for variations in data distributions and characteristics across devices and locations.

Organizations can employ several strategies to address data quality and heterogeneity challenges in FL. First, establishing data quality standards and guidelines for data collection, including calibration and validation processes, can help ensure the accuracy and reliability of the collected data. Second, employing data preprocessing techniques, such as outlier detection and data cleaning algorithms, can improve the quality of the data used in FL. Third, developing data harmonization frameworks and standardized data exchange formats can aid in dealing with data heterogeneity, enabling seamless integration and interoperability. Additionally, utilizing transfer learning and domain adaptation techniques can assist in leveraging knowledge from different sources while accounting for the variations in the data. Several approaches have been proposed to address the data heterogeneity challenges of FL in modern digital healthcare systems. For instance, FedProx mitigates heterogeneity by granting every participant device the freedom to execute a distinctive workload, as proposed by Li (2018). Additionally, by implementing a novel approach, researchers successfully tackled the challenge of heterogeneity by ingeniously employing a part-data-sharing strategy, effectively boosting the training process for non-IID data (Zhao, 2018). This ingenious solution involved the creation of a compact subset of data shared globally among all edge devices. Remarkably, their experiments unveiled a staggering 30% surge in accuracy for the CIFAR-10 dataset, accomplished with a mere 5% of data shared across the network.

Furthermore, a new concept has emerged, introducing an inventive paradigm that revolutionized multisite fMRI analysis (Li, 2020a,b). The ingenious framework harnessed the power of privacy-preserving FL, eliminating the need for data-sharing. Remarkably, their breakthrough entailed pioneering the integration of domain adaptation techniques within the realm of FL for medical image analysis. This audacious approach sheds light on leveraging diverse medical data across multiple sites, enhancing performance not only in task execution but also in the identification of reliable and informative biomarkers. Their findings presented a promising remedy for training deep learning models using numerous small, heterogeneous, and privacy-sensitive medical datasets. More recently, a revolutionary primal-dual approach dubbed FedHybrid has emerged, allowing heterogeneous clients with varying capabilities to execute various updates (Niu & Wei, 2023). Theoretically, the authors presented a groundbreaking merit function that ingeniously merges the dual optimality gap with the primal tracking error. Remarkably, they established that FedHybrid achieves linear convergence toward the precise optimal point for strongly convex functions, irrespective of whether clients opt for gradient-type or Newton-type updates.

11.2.2.2 Bias and generalizability

Bias in FL arises from many sources, including demographic disparities, uneven device distribution, or differing healthcare practices across participating organizations. When training models using FL, the data from different devices or organizations may not fully represent the broader population or may inadvertently reflect existing biases present in the healthcare system (Rieke, 2020). Biased data can lead to models that disproportionately favor certain groups, resulting in disparities in diagnosis, treatment recommendations, or healthcare outcomes. For instance, a compelling case study revealed that when an AI system is trained on healthcare datasets, it acquires the ability to suggest reduced painkiller dosages specifically for individuals of African-American descent (Le Nguyen & Do, 2019). Remarkably, this recommendation reflects systemic bias rather than being grounded in biological characteristics. Consequently, detecting and mitigating biases is crucial in preventing unfair and discriminatory practices in healthcare. Generalizability is another critical challenge in FL. FL aims to create models that can generalize well and provide accurate predictions or recommendations in real-world scenarios beyond the training data (Li, 2023; Treleaven et al., 2022). However, the heterogeneity of data sources, patient populations, and healthcare practices within IoHT can challenge the generalizability of federated models. If the training data does not adequately capture the full range of variations and contexts, the resulting models may struggle to perform effectively in different settings or when encountering previously unseen data. Ensuring generalizability requires careful consideration of data diversity, including representative samples from different patient demographics, geographical regions, and healthcare contexts.

To address bias and generalizability issues, it is essential to carefully curate and balance the dataset used for training federated models, actively seek to include diverse patient populations, and account for demographic variations. Data preprocessing techniques, such as data augmentation or synthetic data generation, can help address data imbalances and improve generalizability. Fairness-aware algorithms and methods, such as bias detection and mitigation strategies, can be incorporated into the training process to reduce bias in model predictions. For example, a groundbreaking study harnesses the power of natural language processing (NLP) techniques to alleviate discrimination in patient treatment choices (Minot, 2022). Unveiling their revolutionary interpretable bias mitigation framework, the researchers skillfully detected and eliminated gendered language from not one but two comprehensive clinical-note datasets. Moreover, they revealed an innovative debiasing procedure that utilizes state-of-the-art BERT-based gender classifiers, blazing a trail toward a more equitable healthcare landscape.

11.2.2.3 Data ownership and governance

Data Ownership in FL refers to who holds the rights and control over the data generated by healthcare devices and systems. In IoHT, multiple entities, including healthcare organizations, patients, device manufacturers, and data aggregators, may contribute to the data used in FL (Ali, 2023; Patel, 2022). Determining ownership rights and responsibilities is paramount to establishing trust and avoiding data usage, access, and control disputes. Lack of clarity in data ownership can hinder collaboration, data sharing, and the effective implementation of FL initiatives. Data governance encompasses the policies, frameworks, and practices governing healthcare data collection, storage, sharing, and usage. In the context of FL, data governance plays a crucial role in ensuring compliance with relevant regulations, protecting patient privacy, and maintaining data integrity and security (Chalamala, 2022). It involves defining data sharing agreements, consent management, data anonymization or de-identification procedures, and mechanisms for auditing and enforcing data access controls. Failure to meet these requirements can result in legal and ethical implications. Robust data governance frameworks are essential to foster trust among participants, address ethical challenges, and safeguard the rights and interests of patients and other stakeholders involved.

To address data ownership and governance challenges, organizations must establish clear data ownership agreements and consent frameworks that outline the rights and responsibilities of all parties involved in FL initiatives. Collaboration agreements should specify data usage rights, retention policies, and dispute resolution mechanisms. Ensuring compliance with relevant regulations, such as HIPAA in the United States, GDPR in the European Union, and other national regulations, is crucial in protecting patient privacy and avoiding legal consequences.

Establishing comprehensive data governance frameworks is essential for guiding the ethical and responsible use of healthcare data. This includes implementing robust data anonymization or de-identification techniques, maintaining data security through encryption and access controls, and establishing data sharing and collaboration protocols. For instance, a judicious approach by life sciences and healthcare experts entails securely training algorithms within the confines of the fortified firewall of the hospitals, safeguarding the sanctity of the data, and limiting the exchange to sharing meticulously constructed models exclusively (Rieke, 2020). Regular audits and reviews of data governance practices can help ensure ongoing compliance, identify areas for improvement, and maintain data integrity. Furthermore, engaging patients and healthcare professionals in the data governance process through transparent communication and shared decision-making can enhance trust and enable patient-centric approaches to data management. Finally, empowering individuals to control their data through informed consent mechanisms and providing them access to their data can further strengthen data ownership and governance practices (Durga & Poovammal, 2022).

11.2.3 Technical challenges and risks of federated learning in modern digital healthcare systems

As healthcare organizations harness the power of distributed data and collaborative learning, they must navigate the intricate pathways of technical challenges that lie ahead. This section delves into the heart of these challenges, shedding light on the obstacles encountered and the steps taken to chart a course toward a future where FL thrives in the modern digital healthcare systems.

11.2.3.1 Communication and computational overhead

Communication overhead is the additional burden imposed on the network infrastructure while exchanging data and model updates between participating devices or organizations. In FL, data remains decentralized, residing on various devices or sources. To facilitate collaborative model training, communication is required to transmit data updates, model parameters, and aggregated results between these distributed entities (Coelho, 2023). However, transmitting large volumes of data over networks can strain bandwidth, consume substantial computational resources, and introduce latency issues (Joshi et al., 2022). Communication overhead hinders the efficiency and scalability of FL, prolonging the training process and potentially impacting the real-time nature of healthcare applications.

Conversely, computational overhead arises from the computational burden imposed on the participating devices during local model training and aggregation processes (Almanifi, 2023). Each device or organization performs local model updates in FL using its respective data before sharing aggregated information with other participants. The computational complexity of training sophisticated ML models on resource-constrained devices, such as wearables or edge devices, can be substantial. The computational overhead may result in increased power consumption, slower training times, and limited capacity to handle complex models or large datasets (Almanifi, 2023). These challenges can impede the scalability, feasibility, and practicality of FL in the IoHT ecosystem.

To mitigate communication and computation overhead challenges, several strategies can be employed. Optimization techniques, such as model compression and quantization, can reduce the size of data transmitted during communication, alleviating bandwidth constraints and reducing latency. Using efficient data transfer protocols and utilizing edge computing infrastructure can also help minimize communication overhead by reducing the distance data needs to travel and enhancing processing capabilities closer to the data sources (Almanifi, 2023; Li et al., 2021; Xiao et al., 2022). Furthermore, techniques such as FL with local model updates, adaptive aggregation algorithms, and selective participant involvement can help distribute computation more evenly, reducing the computation overhead on resource-constrained devices. Collaborative resource management approaches, such as task offloading and workload balancing, can optimize computation allocation across the participating devices or organizations (Dai, 2023; Rajagopal & Buyya, 2023). Hence, addressing communication and computation overhead in FL fosters efficient and scalable training processes, enabling real-time insights and applications within modern digital healthcare systems.

11.2.3.2 System complexity and coordination

System complexity arises from the intricate interplay of the various components involved in FL. The IoHT ecosystem encompasses multiple devices, data sources, and organizations with varying capabilities, protocols, and infrastructures (Rieke, 2020). Coordinating these heterogeneous entities and harmonizing their functionalities can be a daunting task. Challenges may include data preprocessing and harmonization, ensuring compatibility across different data formats, addressing disparities in computational resources, and handling variations in data distribution and characteristics. The complexity of integrating these diverse components and orchestrating their collective actions impacts the overall performance, scalability, and reliability of FL systems, especially in modern digital healthcare (Li, 2018). Coordination challenges, on the other hand, manifest as the need to synchronize and orchestrate the activities of participating devices or healthcare organizations in FL. Efficient coordination is necessary to ensure proper data sharing, model training, and aggregation processes. However, coordination becomes increasingly challenging due to the distributed nature of the data and the involvement of multiple entities with their respective local training schedules, network conditions, and computational capacities (Li, 2020a,b; Kumar & Singla, 2021). Achieving consensus on model updates, managing communication schedules, addressing potential conflicts or delays, and handling the dynamics of participant involvement pose significant coordination hurdles. Failure to effectively coordinate these activities leads to inefficiencies, suboptimal convergence, or even system instability.

Robust strategies and techniques can address system complexity and coordination challenges. Standardization efforts aid in establishing standard protocols, data formats, and interoperability frameworks that streamline system integration and reduce complexity (Cremonesi, 2023). Developing comprehensive data preprocessing pipelines, including data cleaning, normalization, and feature selection methods, can help harmonize heterogeneous data sources, ensuring compatibility and consistency (Li, 2018). Advanced coordination mechanisms, such as distributed optimization algorithms, decentralized decision-making frameworks, and consensus algorithms, can enable efficient coordination among participating entities (Imteaj, 2022; Nguyen, 2022; Zhu Juncen, 2023). Strategies such as federated averaging, DP, and adaptive communication schedules can help overcome coordination challenges by adapting to varying network conditions, resource availability, and participant involvement (Li, 2021). Additionally, leveraging advanced technologies such as edge computing and distributed computing frameworks can enhance coordination capabilities by enabling localized processing and reducing reliance on centralized infrastructure (Nair et al., 2023).

11.2.3.3 Model interpretability

As healthcare organizations embrace the power of distributed data and collaborative learning, the ability to interpret and understand the decisions made by FL models becomes crucial for trust, accountability, and effective healthcare delivery (Yoo, 2021). Interpretability challenges in FL have significant implications for modern digital healthcare systems. Clinicians, patients, and regulatory authorities need to trust the decisions made by the models and understand the reasoning behind them. Interpretability entails comprehending and explaining how an ML model arrives at its predictions or decisions after extensive training. In FL, where models are trained on data from various devices or organizations, achieving interpretability becomes more complex (Saraswat, 2022). The distributed nature of the data, coupled with privacy and security constraints, can hinder traditional model interpretation techniques that rely on accessing and analyzing the complete dataset. The black-box nature of FL models raises concerns about their

Table 11.2 Summary of data and technical challenges of federated learning in modern digital healthcare systems.

Category	Challenges	References
Data related	Data quality and heterogeneity	Rieke (2020), Antunes (2022), Alhusein and Idrees (2023)
	Bias and generalizability	Le Nguyen and Do (2019), Rieke (2020), Treleaven et al. (2022), Li (2023)
	Data ownership and governance	Chalamala (2022), Patel (2022), Ali (2023)
Technical related	Communication and computational overhead	Joshi et al. (2022), Almanifi (2023), Coelho (2023)
	System complexity and coordination	Li (2018, 2020a,b); Rieke (2020), Kumar and Singla (2021)
	Interpretability	Yoo (2021), Saraswat (2022)

transparency, making it challenging to understand the underlying factors or features that influence the predictions or recommendations made by the model (Saraswat, 2022). Table 11.2 summarizes data-related risks and technical challenges of FL in modern digital healthcare systems.

11.3 Lessons

Several valuable lessons have been learned in exploring challenges associated with FL in modern digital healthcare systems. Privacy and security challenges were identified as paramount, with the understanding that FL is susceptible to attacks such as membership inference, model inversion, and backdoor attacks. To safeguard privacy, different privacy-preserving techniques emerged as essential solutions while deploying real-life datasets. For instance, earlier endeavors aimed at protecting patient data privacy used real-life datasets such as the Medical Information Mart for Intensive Care (MIMIC III) data, Limited MarketScan Explorys Claims-EMR Data (LCED), and the Cancer Genome Atlas (TCGA). These efforts focused on preventing the sharing of raw data (Adnan & 2022, 2022; Choudhury, 2019). Recent research utilizing heart attack datasets proposed an asynchronous FL-DL-based approach for cardiovascular disease prediction, achieving improved prediction results compared with previous synchronous-communication-based FL approaches (Khan, 2023). Additionally, recent research utilizing logistic regression and support vector machines used the University of California, Irvin (UCI), benchmark dataset on heart diseases, demonstrating the applicability of the FL paradigm even where health data privacy is a huge concern (Kavitha Bharathi et al., 2022). Using real-life datasets is a testament to the compelling adaptability of FL in healthcare, especially when tested to ensure patient data privacy and security.

Moreover, as has been elaborated, data-related challenges also play a significant role in FL implementation. The lessons highlighted the importance of addressing data quality, heterogeneity challenges, and bias and generalizability issues. Effective data preprocessing, federated transfer learning, and bias correction methods were identified as critical solutions to ensure high-quality and representative datasets. Furthermore, technical challenges were thoroughly examined. The need to manage communication and computational overhead, system complexity, and interpretability challenges was also emphasized. Model compression, efficient communication protocols, coordination mechanisms, and interpretability techniques were found to be effective in overcoming these obstacles (Shah & Lau, 2021). In a nutshell, the lessons learned from this chapter provide crucial insights for the successful deployment of FL in modern digital healthcare systems. Stakeholders must prioritize privacy protection, address data-related risks, and tackle technical challenges to fully harness the benefits of FL while ensuring patient safety and data integrity. By implementing the identified solutions and exploring a possible other FL deployment configuration, healthcare practitioners, researchers, and policymakers can navigate these challenges and mitigate risks, paving the way for secure, efficient, and interpretable FL applications in the healthcare domain.

11.4 Future research directions

Future research should focus on several key areas to mitigate the challenges of adopting FL in modern digital healthcare systems. First, advanced privacy-preserving techniques, such as novel differential privacy mechanisms and secure multiparty computation, should be developed to protect patient data during training and aggregation. Second, robust security frameworks encompassing secure aggregation algorithms and threat modeling techniques must be established to detect and mitigate potential attacks on FL systems. Ethical considerations, including data ownership, consent, transparency, and fairness, should be thoroughly explored to ensure FL is deployed ethically. Additionally, the potential of federated transfer learning should be investigated to leverage knowledge transfer while preserving data privacy. Adaptive communication and computation strategies should be explored to minimize resource utilization, while research on model explainability and interpretability is crucial to foster trust and understanding. Regulatory and governance frameworks tailored for FL in healthcare, real-world deployment studies, and collaborative research networks are essential for practical implementation and scalability. Finally, data quality and heterogeneity management should be addressed through effective preprocessing techniques and standardized approaches. Pursuing these research directions will contribute to the responsible and secure implementation of FL in healthcare systems.

11.5 Conclusion

This chapter examined the challenges and risks of implementing FL in modern digital healthcare systems. Our exploration encompassed privacy and security challenges, data-related risks, and technical risks and challenges. Throughout the discussion, we have also presented potential solutions to mitigate these challenges, providing a comprehensive understanding of the complexities involved. Privacy and security challenges are paramount when dealing with sensitive healthcare data. We highlighted the various attacks that can compromise FL systems, including model inversion attacks, membership inference attacks, and poisoning attacks. Several techniques, such as differential privacy, secure aggregation, and encryption methods, were discussed to address these risks as adequate safeguards. Data-related risks and challenges are crucial aspects analyzed in this chapter. The challenges of ensuring data quality and managing heterogeneity have been explored, along with the implications of bias and the limited generalizability of FL models.

Furthermore, we delved into the intricate issues surrounding data ownership and governance, emphasizing the need for clear policies and frameworks to establish trust and accountability. The chapter also highlights the technical risks and challenges of FL in healthcare systems. Communication and computational overhead have been identified as potential obstacles, requiring optimization techniques and efficient algorithms to reduce resource consumption. System complexity and coordination obstacles have been addressed, emphasizing the importance of establishing proper coordination mechanisms and ensuring seamless integration across multiple stakeholders. Lastly, the issue of interpretability in FL models was examined, stressing the necessity of interpretability techniques to ensure transparency and trust in decision-making processes. By offering potential solutions to these challenges, we hope to guide researchers, practitioners, and policymakers in navigating the complexities of implementing FL while safeguarding privacy, ensuring data quality, and addressing technical constraints. As FL continues to evolve and be adopted in healthcare settings, it is imperative to remain vigilant and proactive in addressing these challenges. By leveraging the solutions discussed in this chapter and fostering collaboration among stakeholders, we can establish a robust foundation for the successful deployment of FL in modern digital healthcare systems. With careful consideration of privacy, data, and technical risks, we can unlock the immense potential of FL to revolutionize healthcare while ensuring the well-being and trust of patients, providers, and institutions alike.

References

Adnan, M., et al. (2022). Federated learning and differential privacy for medical image analysis. *Scientific Reports*, *12*(1), 1−10. Available from https://doi.org/10.1038/s41598-022-05539-7, *2022 12:1*.

Ahmed, U., Lin, J. C. W., & Srivastava, G. (2022). Mitigating adversarial evasion attacks of ransomware using ensemble learning. *Computers and Electrical Engineering*, *100*107903. Available from https://doi.org/10.1016/J.COMPELECENG.2022.107903.

Alabdulatif, A., Thilakarathne, N. N., & Kalinaki, K. (2023). A novel cloud enabled access control model for preserving the security and privacy of medical big data. *Electronics*, *12*(12), 2646. Available from https://doi.org/10.3390/electronics12122646.

Alhusein, D., & Idrees, A. K. (2023). *A comprehensive review of wireless medical biosensor networks in connected healthcare applications. Enabling technologies for effective planning and management in sustainable smart cities* (pp. 229—244). . Available from https://doi.org/10.1007/978-3-031-22922-0_9.

Ali, M., et al. (2023). Federated learning for privacy preservation in smart healthcare systems: A comprehensive survey. *IEEE Journal of Biomedical and Health Informatics*, *27*(2), 778—789. Available from https://doi.org/10.1109/JBHI.2022.3181823.

Alli, A. A., et al. (2021). Secure fog-cloud of things: Architectures, opportunities and challenges. In M. Ahmed, & P. Haskell-Dowland (Eds.), *Secure edge computing* (1st edn, pp. 3—20). CRC Press. Available from https://doi.org/10.1201/9781003028635-2.

Almanifi, O. R. A., et al. (2023). Communication and computation efficiency in Federated Learning: A survey. *Internet of Things*, *22*100742. Available from https://doi.org/10.1016/J.IOT.2023.100742.

Antunes, R. S., et al. (2022). Federated learning for healthcare: Systematic review and architecture proposal. *ACM Transactions on Intelligent Systems and Technology (TIST)*, *13*(4). Available from https://doi.org/10.1145/3501813.

Aziz, M. M. Al, et al. (2022). Generalized genomic data sharing for differentially private federated learning. *Journal of Biomedical Informatics*, *132*104113. Available from https://doi.org/10.1016/J.JBI.2022.104113.

Cao, X., & Gong, N. Z. (2017). *Mitigating evasion attacks to deep neural networks via region-based classification. ACM International Conference Proceeding Series, Part F132521* (pp. 278—287). . Available from https://doi.org/10.1145/3134600.3134606.

Chalamala, S. R., et al. (2022). Federated learning to comply with data protection regulations. *CSI Transactions on ICT*, *10*(1), 47—60. Available from https://doi.org/10.1007/S40012-022-00351-0, *2022 10:1*.

Chemisto, Musa, Gutu, Tar J. L., Kalinaki, Kassim, Mwebesa, Darlius Bosco, Egau, Percival, Kirya, Fred, Oloya, Ivan Tim, & Kisitu, Rashid (2023). Artificial intelligence for improved maternal healthcare: A systematic literature review. *IEEE Xplore*, 1—6. Available from https://doi.org/10.1109/AFRICON55910.2023.10293674.

Chen, S., et al. (2018). Automated poisoning attacks and defenses in malware detection systems: An adversarial machine learning approach. *Computers & Security*, *73*, 326—344. Available from https://doi.org/10.1016/J.COSE.2017.11.007.

Choudhury, O. *et al.* (2019). Differential privacy-enabled federated learning for sensitive health data. Available at: https://arxiv.org/abs/1910.02578v3 (Accessed: 17 May 2023).

Coelho, K. K., et al. (2023). A survey on federated learning for security and privacy in healthcare applications. *Computer Communications*, *207*, 113—127. Available from https://doi.org/10.1016/J.COMCOM.2023.05.012.

Cremonesi, F., et al. (2023). The need for multimodal health data modeling: A practical approach for a federated-learning healthcare platform. *Journal of Biomedical Informatics*, *141*104338. Available from https://doi.org/10.1016/J.JBI.2023.104338.

Dai, X., et al. (2023). Task co-offloading for D2D-assisted mobile edge computing in industrial internet of things. *IEEE Transactions on Industrial Informatics*, *19*(1), 480−490. Available from https://doi.org/10.1109/TII.2022.3158974.

Dang, T. K., et al. (2022). Federated learning for electronic health records. *ACM Transactions on Intelligent Systems and Technology*, *13*(5), 72. Available from https://doi.org/10.1145/3514500.

Dhiman, G., et al. (2022). Federated learning approach to protect healthcare data over big data scenario. *Sustainability*, *14*(5), 2500. Available from https://doi.org/10.3390/SU14052500, *2022, Vol. 14, Page 2500*.

Durga, R., & Poovammal, E. (2022). FLED-block: Federated learning ensembled deep learning blockchain model for COVID-19 prediction. *Frontiers in Public Health*, *10*892499. Available from https://doi.org/10.3389/FPUBH.2022.892499.

Fahim, K. E., Kalinaki, K., & Shafik, W. (2023). *Electronic devices in the artificial intelligence of the internet of medical things (AIoMT)*. *Handbook of Security and Privacy of AI-Enabled Healthcare Systems and Internet of Medical Things* (1st Edition). CRC Press. Available from https://doi.org/10.1201/9781003370321-3.

Grama, M. *et al.* (2020). Robust aggregation for adaptive privacy preserving federated learning in healthcare. Available at: https://arxiv.org/abs/2009.08294v1 (Accessed: 4 June 2023).

Imteaj, A., et al. (2022). A survey on federated learning for resource-constrained IoT devices. *IEEE Internet of Things Journal*, *9*(1), 1−24. Available from https://doi.org/10.1109/JIOT.2021.3095077.

Islam, T.U., Ghasemi, R., & Mohammed, N. (2022) Privacy-preserving federated learning model for healthcare data. 2022 *IEEE 12th Annual Computing and Communication Workshop and Conference*, CCWC 2022, pp. 281−287. Available at: https://doi.org/10.1109/CCWC54503.2022.9720752.

Javed, I., et al. (2023). Next generation infectious diseases monitoring gages via incremental federated learning: Current trends and future possibilities. *Computational Intelligence and Neuroscience*, *2023*, 1−12. Available from https://doi.org/10.1155/2023/1102715.

Joshi, M., Pal, A., & Sankarasubbu, M. (2022). Federated learning for healthcare domain − Pipeline, applications and challenges. *ACM Transactions on Computing for Healthcare*, *3*(4). Available from https://doi.org/10.1145/3533708.

Kalinaki, K., Fahadi, M., et al. (2023). '*Artificial intelligence of internet of medical things (AIoMT) in smart cities: A review of cybersecurity for smart healthcare. Handbook of security and privacy of AI-enabled healthcare systems and internet of medical things* (1st Edition). CRC Press. Available from https://doi.org/10.1201/9781003370321-11.

Kalinaki, K., Thilakarathne, N. N., et al. (2023). *Cybersafe capabilities and utilities for smart cities. Cybersecurity for smart cities* (pp. 71−86). Cham: Springer. Available from https://doi.org/10.1007/978-3-031-24946-4_6.

Kavitha Bharathi, S., Dhavamani, M. and Niranjan, K. (2022). A federated learning based approach for heart disease prediction. *Proceedings - 6th International Conference on Computing Methodologies and Communication*, ICCMC 2022, pp. 1117−1121. Available at: https://doi.org/10.1109/ICCMC53470.2022.9754119.

Khan, M. A., et al. (2023). Asynchronous federated learning for improved cardiovascular disease prediction using artificial intelligence. *Diagnostics*, *13*(14), 2340. Available from https://doi.org/10.3390/DIAGNOSTICS13142340, *2023, Vol. 13, Page 2340*.

Kumar, Y., & Singla, R. (2021). Federated learning systems for healthcare: Perspective and recent progress. *Studies in Computational Intelligence, 965*, 141−156. Available from https://doi.org/10.1007/978-3-030-70604-3_6.

Le Nguyen, T., & Do, T.T. H. (2019). Artificial intelligence in healthcare: A new technology benefit for both patients and doctors. *PICMET 2019 - Portland International Conference on Management of Engineering and Technology: Technology Management in the World of Intelligent Systems, Proceedings [Preprint]*. Available at: https://doi.org/10.23919/PICMET.2019.8893884.

Lewis, C., et al. (2023). Attacks against federated learning defense systems and their mitigation. *jmlr.org, 24*, 1−50. Available from https://www.jmlr.org/papers/volume24/22-0014/22-0014.pdf.

Li, C., Li, G., & Varshney, P. K. (2021). Communication-efficient federated learning based on compressed sensing. *IEEE Internet of Things Journal, 8*(20), 15531−15541. Available from https://doi.org/10.1109/JIOT.2021.3073112.

Li, H., et al. (2023). Review on security of federated learning and its application in healthcare. *Future Generation Computer Systems, 144*, 271−290. Available from https://doi.org/10.1016/J.FUTURE.2023.02.021.

Li, Q., et al. (2021). A survey on federated learning systems: Vision, hype and reality for data privacy and protection. *IEEE Transactions on Knowledge and Data Engineering [Preprint]*. Available from https://doi.org/10.1109/TKDE.2021.3124599.

Li, T. *et al.* (2018). Federated optimization in heterogeneous networks. Available at: https://arxiv.org/abs/1812.06127v5 (Accessed: 4 June 2023).

Li, T., et al. (2020a). Federated learning: Challenges, methods, and future directions. *IEEE Signal Processing Magazine, 37*(3), 50−60. Available from https://doi.org/10.1109/MSP.2020.2975749.

Li, X., et al. (2020b). Multi-site fMRI analysis using privacy-preserving federated learning and domain adaptation: ABIDE results. *Medical Image Analysis, 65*101765. Available from https://doi.org/10.1016/J.MEDIA.2020.101765.

Li, Z., et al. (2022). Integrated CNN and federated learning for COVID-19 detection on chest X-ray images. *IEEE/ACM Transactions on Computational Biology and Bioinformatics [Preprint]*. Available from https://doi.org/10.1109/TCBB.2022.3184319.

Minot, J. R., et al. (2022). Interpretable bias mitigation for textual data: Reducing genderization in patient notes while maintaining classification performance. *ACM Transactions on Computing for Healthcare, 3*(4). Available from https://doi.org/10.1145/3524887.

Mothukuri, V., et al. (2021). A survey on security and privacy of federated learning. *Future Generation Computer Systems, 115*, 619−640. Available from https://doi.org/10.1016/J.FUTURE.2020.10.007.

Nair, A. K., Sahoo, J., & Raj, E. D. (2023). Privacy preserving Federated Learning framework for IoMT based big data analysis using edge computing. *Computer Standards & Interfaces, 86*103720. Available from https://doi.org/10.1016/J.CSI.2023.103720.

Nguyen, D. C., et al. (2022). Federated learning for smart healthcare: A survey. *ACM Computing Surveys (CSUR), 55*(3). Available from https://doi.org/10.1145/3501296.

Niu, X., & Wei, E. (2023). FedHybrid: A hybrid federated optimization method for heterogeneous clients. *IEEE Transactions on Signal Processing, 71*, 150−163. Available from https://doi.org/10.1109/TSP.2023.3240083.

Patel, V. A., et al. (2022). Adoption of federated learning for healthcare informatics: Emerging applications and future directions. *IEEE Access [Preprint]*. Available from https://doi.org/10.1109/ACCESS.2022.3201876.

Pfitzner, B., Steckhan, N., & Arnrich, B. (2021). Federated learning in a medical context: A systematic literature review. *ACM Transactions on Internet Technology (TOIT)*, *21* (2). Available from https://doi.org/10.1145/3412357.

Rahman, M. A., et al. (2020). Secure and provenance enhanced internet of health things framework: A blockchain managed federated learning approach. *IEEE Access*, *8*, 205071−205087. Available from https://doi.org/10.1109/ACCESS.2020.3037474.

Rajagopal, S. M., Supriya, M., & Buyya, R. (2023). FedSDM: Federated learning based smart decision making module for ECG data in IoT integrated Edge−Fog−Cloud computing environments. *Internet of Things*, *22*100784. Available from https://doi.org/10.1016/J.IOT.2023.100784.

Rieke, N., et al. (2020). The future of digital health with federated learning. *npj Digital Medicine*, *3*(1), 1−7. Available from https://doi.org/10.1038/s41746-020-00323-1, *2020 3:1*.

Saraswat, D., et al. (2022). Explainable AI for Healthcare 5.0: Opportunities and challenges. *IEEE Access*, *10*, 84486−84517. Available from https://doi.org/10.1109/ACCESS.2022.3197671.

Shah, S. M., & Lau, V. K. N. (2021). Model compression for communication efficient federated learning. *IEEE Transactions on Neural Networks and Learning Systems [Preprint]*. Available from https://doi.org/10.1109/TNNLS.2021.3131614.

Shieh, C. S., et al. (2022). Detection of adversarial DDoS attacks using symmetric defense generative adversarial networks. *Electronics*, *11*(13), 1977. Available from https://doi.org/10.3390/ELECTRONICS11131977, *2022, Vol. 11, Page 1977*.

Thapa, C., Chamikara, M. A. P., & Camtepe, S. A. (2021). Advancements of federated learning towards privacy preservation: From federated learning to split learning. *Studies in Computational Intelligence*, *965*, 79−109. Available from https://doi.org/10.1007/978-3-030-70604-3_4.

Treleaven, P., Smietanka, M., & Pithadia, H. (2022). Federated learning: The pioneering distributed machine learning and privacy-preserving data technology. *Computer*, *55*(4), 20−29. Available from https://doi.org/10.1109/MC.2021.3052390.

Wang, J. *et al.* (2022) 'Protect privacy from gradient leakage attack in federated learning', *Proceedings - IEEE INFOCOM*, 2022-May, pp. 580−589. Available at: https://doi.org/10.1109/INFOCOM48880.2022.9796841.

Xiao, D., Tan, X. and Li, M. (2022) 'Communication-efficient and secure federated learning based on adaptive one-bit compressed sensing', *Lecture Notes in Computer Science (including subseries Lecture Notes in Artificial Intelligence and Lecture Notes in Bioinformatics)*, 13640 LNCS, pp. 491−508. Available at: https://doi.org/10.1007/978-3-031-22390-7_29.

Yang, H., et al. (2023). Gradient leakage attacks in federated learning: Research frontiers, taxonomy and future directions. *IEEE Network [Preprint]*. Available from https://doi.org/10.1109/MNET.001.2300140.

Yin, X., Zhu, Y., & Hu, J. (2021). A comprehensive survey of privacy-preserving federated learning. *ACM Computing Surveys (CSUR)*, *54*(6). Available from https://doi.org/10.1145/3460427.

Yoo, J.H. *et al.* (2021) 'Federated learning: Issues in medical application', *Lecture Notes in Computer Science (including subseries Lecture Notes in Artificial Intelligence and Lecture Notes in Bioinformatics)*, 13076 LNCS, pp. 3—22. Available at: https://doi.org/10.1007/978-3-030-91387-8_1.

Zhao, Y., et al. (2018). 'Federated learning with non-IID data'. *The Computer Journal [Preprint]*. Available from https://doi.org/10.48550/arXiv.1806.00582.

ZhuJuncen., et al. (2023). Blockchain-empowered federated learning: Challenges, solutions, and future directions. *ACM Computing Surveys*, *55*(11), 1—31. Available from https://doi.org/10.1145/3570953.

Case studies and recommendations for designing federated learning models for digital healthcare systems

Chun-Ying Wu[1,2,3,4,5], Pushpanjali Gupta[1,2,3] and Sulagna Mohapatra[3,6]

[1]*Institute of Biomedical Informatics, National Yang Ming Chiao Tung University, Taipei, Taiwan*
[2]*Institute of Public Health, National Yang Ming Chiao Tung University, Taipei, Taiwan*
[3]*Health Innovation Center, National Yang Ming Chiao Tung University, Taipei, Taiwan*
[4]*Division of Translational Research, Taipei Veterans General Hospital, Taipei, Taiwan*
[5]*Department of Public Health, China Medical University, Taichung, Taiwan*
[6]*Division of Gastroenterology, Taichung Veterans General Hospital, Taichung, Taiwan*

12.1 Introduction

In the era of artificial intelligence (AI), research on machine learning (ML) and deep learning (DL) has led to attention-seeking innovations in radiology, pathology, and other fields. Several deep learning models have been developed for better diagnosis and prediction. However, the models require a large number of curated datasets to achieve clinically acceptable results (Rieke et al., 2020). Several challenges are encountered during obtaining the optimum type of datasets. First, despite being a large healthcare system, the organization might not have different varieties of data that could be used to validate a model's robustness. For instance, when focusing on focal liver lesion (FLL), there are several types of FLLs, such as hepatocellular carcinoma (HCC), cholangiocarcinoma, liver metastasis, focal node hyperplasia, cirrhotic nodules, dysplastic nodules, cysts, hemangiomas, etc. However, it has been observed that most of the patients have HCC, cysts, or hemangioma. The rest of the FLLs occur rarely, resulting in fewer samples being obtained and consequently leading to poor model training (Nayantara et al., 2020). Second, several healthcare systems refrain from sharing data due to concerns about privacy and the sensitivity of healthcare information. Research has shown the potential ability to reconstruct a patient's face from computed tomography or magnetic resonance imaging (MRI) data. Third, and most importantly, healthcare data collection and curation are a time-consuming and labor-intensive job that also

Federated Learning for Digital Healthcare Systems. DOI: https://doi.org/10.1016/B978-0-443-13897-3.00007-2

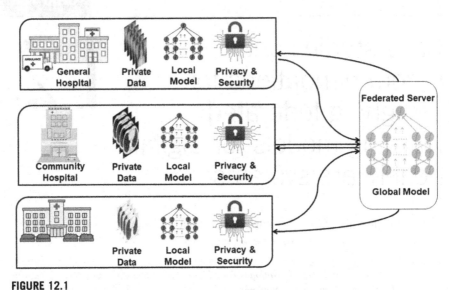

FIGURE 12.1

Federated learning in healthcare.

involves huge expenses. Consequently, such datasets might not be available for free, making them difficult to obtain for model training.

In recent years, the demand for training ML models using different sources of datasets while protecting patients' sensitive information has led to the emergence of federated learning (FL) as a paradigm. FL facilitates the development of global ML models in a fragmented fashion, as shown in Fig. 12.1, where the central federated server generates an initial training model and forwards it to the participating client healthcare institutions, such as hospitals and research institutes. Several individual models are trained within different participating institutions with their proprietary data (InfinitiesSoft, 2022). This process is followed by model updates through learning, and corresponding results are collected by the server and updated to all client models to create a better performance model. This type of learning offers significant advantages in the medical field, alleviating concerns about data privacy for researchers. It also distributes the tiresome task of collecting a large number of samples across participating institutions, enabling smooth and productive collaboration among healthcare institutions with shared research interests. Most importantly, the use of FL can minimize the concern related to the centralized processing of patient's healthcare information and the associated risk while transferring the data (Nagaraj et al., 2023).

12.1.1 Contributions

One of the major contributions of this chapter is the extensive and in-depth review of different challenges and solutions related to FL adaptation in digital healthcare. Furthermore, this review systematically explores different successful case studies of FL integrated with AI to provide faster and more accurate disease diagnosis in medical care. This chapter not only focuses on the published theoretical case studies but also on the real FL applications that have already been implemented across different countries of the world, such as the United States, Taiwan, etc. To the best of our knowledge, a complete study of implementation architecture with real use cases of FL with the collaboration of AI has not been included in a single chapter. Finally, this chapter highlights the existing and upcoming challenges as well as new future directions to enhance the working of FL in digital healthcare.

12.1.2 Chapter organization

Section 12.2 discusses the common challenges faced during the implementation and adoption of FL. Section 12.3 discusses case studies of how different frameworks of FL have addressed the issues. Section 12.4 discusses use cases of FL in the world and Taiwan. Section 12.5 discusses the open research direction. Section 12.6 summarizes the lessons learned in this chapter, and finally, Section 12.7 concludes the chapter with future directions.

12.2 Federated learning implementation: challenges and solutions

The applications of FL in digital healthcare can improve multimodal medical data integration, utilization, and diagnostic accuracy. However, ensuring privacy and security is a major concern in FL implementation. This section discusses the primary challenges and the respective solutions in terms of privacy protection, secure communication, integration of heterogeneous data, and selection of appropriate technology for FL integration in medical care.

12.2.1 Privacy and security concerns

Although FL is built with the primary aim of providing secure transmission of trained model parameters while protecting the sensitive data used for model derivation, there are still several potential ways to attack the FL system, threatening the data privacy of the participants (Cavusoglu & Kokcam, 2021). One way of protecting data involves encrypting the information exchanged between parties. In this method, the data owner encrypts the data, enabling the user to process it without direct access to its contents. It is important to note that the specific requirements may vary among different organizations. Therefore several modular privacy-enhancing technologies must be provided with the federated infrastructure to ensure maximum data insights with minimal privacy risks (Wen et al., 2023).

When the user's encrypted data is uploaded to the central server, a malicious attack can be made to infer the characteristics of the user group. To prevent such attacks on data privacy, a proper set of legal and contractual agreements must be set up for use in collaboration with different partners to secure the privacy and intellectual property of each party involved. To provide privacy and security of data, the most popular encryption mechanisms used are secure multiparty computing (SMC) (Xiong et al., 2022), homomorphic encryption (Ma et al., 2022), and differential privacy (Xu et al., 2022). Currently, a decentralized federated solution is also being developed to improve privacy-preserving communication. This novel distributed solution can handle several issues, such as single point failure, communication bottlenecks, harmful servers inferring gradients, and data leakage (Tian et al., 2023). The proxy-based FL (ProxyFL) can be used for privacy preservation where dual models known as the private model and the publicly shared proxy model are implemented. The designated models enable effective information transmission among the users without the requirement of a central server. This development allows model heterogeneity, where each participant can have their private model without depending on any architecture (Kalra et al., 2023).

12.2.2 Communication challenges

When multiple client servers want to send data to the central server for every update, the communication cost increases to a greater extent compared with the computation cost. This results in lower training efficiency of FL. Consequently, it is essential to implement model optimization threshold, where if the local model surpasses a previously set threshold for model optimization parameters, the resultant model will be transmitted to the central server for aggregation. Several researchers have studied the impact of communication between local and global servers. It is observed that increasing the local training and decreasing the global communication round boosted the convergence efficiency of the global model. Therefore increasing the time interval of model aggregation and reducing the frequency of communication between the local and global model, improves the performance of the FL model. Several algorithms are proposed based on this philosophy, such as federated averaging (FedAvg) (McMahan et al., 2017) and a federated adaptive weighting algorithm (Wu et al., 2022).

12.2.3 Choosing right federated learning platform

In recent years, there have been several FL solutions provided by different organizations for the research and development of customized FL environments. Nonetheless, when an organization wants to implement a professional and productive FL environment, there are a few concerns and complexities to be considered. First and foremost, the chosen frameworks must provide the owner organization full control of the data management, offering strict clarification of the level of access and reason for accessing the data (Neto et al., 2023). Second, deliveries of maximum and optimal outputs from the datasets must be guaranteed while prioritizing privacy, protecting the IP of involved partners, and adhering to the regulatory agreements as initially agreed upon by partners. Third, the selected FL framework must seamlessly integrate with the current established tools and ML working environment of the organization. Last, there should be tools provided for data cleansing and harmonization of data structures since real-world data involving tabular structure, voice, image, or video are not of homogenous structure.

12.2.4 Choosing right client

Along with the selection of the right FL platform, it is also necessary to select the right partner or client for the efficient working of the global FL model. In a synchronous FL environment, for global updates, the global FL model needs to wait for local updates to be uploaded by the participating clients. Due to a delay in the update from even one client, the efficiency and performance of the federated model are affected. Several approaches are proposed to focus on this issue. For instance, an orthogonal method is proposed by Liu et al. (2022) to identify whether local model updates are consistent with the trend of global model updates, allowing the avoidance of the transmission of irrelevant parameters from clients to servers. Reinforcement learning-based approaches have been proposed by Luo et al. (2021) and Liao and Li (2022) to automatically select high-quality clients for significantly improving the learning efficiency of the FL model within a limited budget. In practical scenarios, a client may have poor communication channels or limited computing resources, which affects their performance and ultimately affects the global model. Responding to such an issue (Nishio & Yonetani, 2019), FedCS has been proposed, which uses a predefined deadline-based approach for downloading, updating, and uploading local models. This greedy approach tries to aggregate as many client updates as possible within the deadline, making the training process efficient.

12.2.5 Data heterogeneity

When working in the FL scenario, the participating clients are geographically located, accumulating data of different volumes, varieties, structures, and formats, resulting in a heterogeneous distribution of data and volume. For instance, in the case of horizontal FL, all the clients have the same feature, whereas in the case of vertical FL, each client can have different data possessing variable features (Díaz & García, 2023). When multiple clients have heterogeneous data, the model drifting problem arises when trying to obtain an aggregation model. This can cause a delay in model convergence (Zhang et al., 2022). It is essential to provide a mechanism for an efficient FL environment while considering the specific distribution of data and the situation of the application environment. Wang et al. (2021) proposed the use of a polynomial time algorithm for scheduling workloads. A greedy approach is used to solve the problem of heterogeneous data, where the average cost minimization optimization problem is used to seek a balance between computation time and computation accuracy. Instead of using the greedy approach, priority-based scheduling is proposed (Taik et al., 2022) to prioritize clients with higher-quality data while minimizing federated edge learning with limited communication bandwidth. For non-IID data with noise, a framework based on federated Kalman filter (FKF) confidence is proposed by Hu et al. (2021), which uses a convolutional adversarial generative network with auxiliary classifiers used as a feature extractor to pretrain the data. FKF is generated by each client before model training, thus eliminating the issue of data heterogeneity and providing efficient FL aggregation. The influence of data heterogeneity on the FL environment still needs to be extensively analyzed.

12.2.6 Model heterogeneity

Like data heterogeneity, during the process of model parameter aggregation, uniformity in the local model structure is required. Nonetheless, such requirements are practically not realistic in an actual FL scenario since the client does not use the same configuration of hardware and software resources, and the data processed, volume of data used, computing power, and model structure used might be different. A client may have less volume of data resulting in a not so robust model; on the other hand, another client with weak computing resources can create a simpler shallow model. Such models, when aggregated to form a global federated model will, lower the performance of local models after model updates and during retraining. Although FL allows clients to develop different local models based on their data and resources, issues arise when direct integration of the generated models is not feasible. Model heterogeneity can occur due to different model types, such as neural networks or random forests, or different model structures, such as the depth of networks. To alleviate such issues, a multitasking FL algorithm is proposed (Mills et al., 2022), where a nonfederated batch normalization layer is used in a federated deep neural network. The clients can customize the training model, facilitating the speedy convergence of the model. Another work (Ni et al., 2022) proposed the use of a two-way codistilation algorithm to personalize the local model based on the output of the global model. Based on the suitability, a client can select the guide to build the local personalized model.

12.3 Case studies of federated learning frameworks

The emergent FL technology addresses several challenges in healthcare applications including collaborative integration of multiinstitutional diversified datasets at the same time maintaining the confidentiality and privacy of the accessed data. The efficient FL frameworks developed for intelligent and integrated healthcare data analysis are discussed as use cases in the following subsections.

12.3.1 Fed-BioMed

An open-source user-friendly, research and development initiative originally developed by Inria (*Institute National de Recherche en Informatique et Automatique*) and *Université Côte d'Azur* (UCA) is aimed at providing a trusted framework for the deployment of the state-of-the-art FL in sensitive healthcare systems (Silva et al., 2020). Fig. 12.2 demonstrates the strategies adopted by Fed-BioMed. Fed-BioMed provides a well-documented architecture and implementation addressing the challenges faced in using FL in the healthcare domain. On such issues, the handling of data is taken care of by providing the users with strict and tight control of data management and model training process. The straightforward design of Fed-BioMed is designed to simplify the development and deployment of FL in healthcare research, where the user can control the prototype of data and model through a graphical interface, which could also be easier for nontechnical users such as clinical data managers and physicians. Currently, Fed-BioMed is developed and coordinated by a small number of core developers. In the future, it has plans to add multiple developers, heterogeneous communities, and open-source contributors through a consortium. Fed-BioMed aims at satisfying four primary and four secondary requirements (Cremonesi et al., 2023), as summarized below.

12.3.1.1 Primary requirements
12.3.1.1.1 Data and model governance
In an environment where data privacy is of utmost importance, having the control to review, add, and revoke the availability of any given dataset for FL, the ability to monitor, audit, and approve the execution of specific workflows in FL, and the ability to customize, audit, and review the deployment of the software infrastructure has been provided, which can be implemented through a simple user interface with minimal learning efforts.

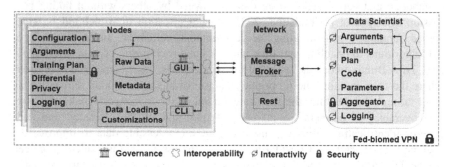

FIGURE 12.2

Strategies adopted by Fed-BioMed.

12.3.1.1.2 Integration with biomedical data sources

Although media data is highly unstructured and of substantial dimensions, the Fed-BioMed framework, which targets interoperability, offers direct and seamless integration for data management and analysis.

12.3.1.1.3 Researcher interactivity

Before the deployment of FL software, several research methods are experimented with during the ongoing software development. Fed-BioMed provides the ability to launch, stop, and manipulate the training process. It allows for on-the-fly modification of model parameters and resumption of training from checkpoints while preserving the FL paradigm of data privacy and node governance.

12.3.1.1.4 Security

Fed-BioMed uses network segmentation and secure network communication, providing a secure environment for minimizing surface attacks on the data provider's system through FL itself. Furthermore, it allows easy activation of gradient protection against model-targeted attacks, such as model poisoning, model inversion, and membership inference.

12.3.1.1.5 Secondary requirements

In addition to addressing several primary concerns in FL, Fed-BioMed addresses a few minor concerns. It facilitates pre/postprocessing of data or models in the FL approach, allows seamless integration with a variety of state-of-the-art ML libraries and Fl algorithms, provides flexible reproducibility of the development environment as obtained using containers or virtual environments, and finally provides resilience to node dropouts and unexpected failures.

12.3.2 TripleBlind

TripleBlind, a federated analytics solution, enables institutions or organizations to collaborate and gain insights from datasets owned by others without taking possession of any data (TripleBlind, 2023). The application of tripleblind is widespread, ranging from facilitating the development, validation, and deployment of ML algorithms to enabling real-time analytics and monitoring of data safety globally. TripleBlind divides the users into two categories: asset owner and asset user, where the asset is either the proprietary dataset or algorithm. Upon registering an asset through TripleBlind, the user categorized as the owner, can make the asset "discoverable" to partners (users) using TripleBlind. With automated and real-time de-identification, users can use previously inaccessible data to improve the accuracy of ML models. Considering privacy, the asset is not uploaded to or seen by the TripleBlind system; the data and algorithm interact in a peer-to-peer fashion, without involving the third-party TripleBlind. Like Fed-BioMed, TripleBlind also allows the owner to retain full control of dataset discoverability. The owner can generate different levels of agreement forms to provide other users private access to the data or algorithms where the assets are encrypted throughout the process. One-way encryption is adopted to prevent the leakage of raw data or algorithms if intercepted. TripleBlind provides cloud service; it also supports the integration of different ML frameworks and libraries where different types of data, such as tabular, image, voice, and video data, can be supported. TripleBlind provides regulatory boundaries within an organization and between organizations without the need for a hardware-dependent framework. TripleBlind never accesses the data or algorithm at any point in the data lifecycle, making it the fastest, most scalable privacy framework, which enables accurate analytics and a high degree of interoperability with a minimized risk of data privacy violations.

12.3.3 Federated Artificial Intelligence Technology Enabler

Federated AI Technology Enabler (FATE) initiated by WeBank's AI department, the world's first industrial-grade open-source FL framework, was open-sourced in 2019 (FATE, 2023). It provides a clear visual interface, scalable modeling pipeline with out-of-box usability, and excellent operational performance, supporting both standalone and cluster deployment setups. With the use of computation protocol based on homomorphic encryption and multiparty computation, it has attracted enterprises and institutions to have big data collaborations with data privacy and security maintained (Liu et al., 2021).

The core components of FATE are shown in Fig. 12.3, where the most significant component, FATEFlow, is responsible for end-to-end ML working and coordinated management. The jobs supported are data preprocessing, model training, testing, and

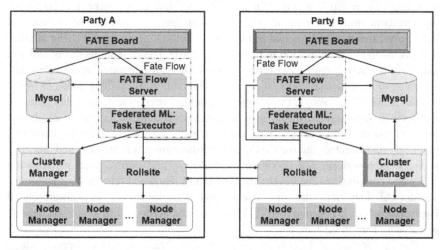

FIGURE 12.3

Core components of FATE. *FATE*, Federated Artificial Intelligence Technology Enabler.

deployment. Federated ML helps in the implementation of many privacy-preserving standards of ML algorithms, such as linear and logistic regression, XGBoost, and transfer learning, and tools such as intersect and OneHotEncoder. During model training, the computation efficiency is enhanced with the use of a distributed computation framework called ROLLSite. Another component called FATEBoard contains useful visualization/dashboarding tools for monitoring, exploring, and analyzing the pipelines. After the model is established, it needs to be deployed, FATE helps in serving FL through the dynamic loading of models, online inferencing, A/B testing, and caching. To provide enterprise-managed solutions over distributed infrastructure and across organizations, KubeFATE is developed, and to support cross-cloud deployment and management FATE-cloud is provided. FATE supports either manual or docker deployments in both Mac and Linux.

In FL, all parties involved could be honest-but-curious, therefore, to provide secure collaboration, in the case of horizontal federated learning, FATE assumes the server to be semihonest where only aggregated parameters are learned but not any individual's data. On the other hand, during vertical FL, intermediate results, and computations are exchanged in encrypted format allowing each party to only learn the final outputs such as local model parameters and local gradients. In this way, FATE guarantees lossless performance. In the current scenario, the applications of FATE are not vast. Accordingly, the future roadmap includes the integration of FATE and blockchain technology, building a lightweight version of FATE for edge-deployment, expanding the AI collaborative platform for computer vision enhancement, automatic speech recognition, etc.

12.3.4 Taiwan Artificial Intelligence Federated Learning Alliance

Taiwan AI Federated Learning Alliance (TAIFA), initiated by more than 50 founding members from famous enterprises, medical centers, research institutions, and local governments, was launched in January 2019 (NDC Taiwan, n.d.). The alliance is expected to boost innovation in healthcare, transportation, manufacturing, culture, and finance through the establishment of cross-domain integrated applications, talent cultivation, industrial cooperation, and collaboration. By joining hands to invest in joint learning technology, the industry, academia, government, and other research entities can focus on providing a high-quality environment for the research and development of Taiwan's precision health industry (Fig. 12.4). With the establishment of standardized international-level trustworthy data privacy and security services, high-performance operating mechanisms, and norms, each participating entity can contribute to the development of world-renowned federated AI solutions. Taiwan AI Lab built the first open-source FL framework in Taiwan called Harmonia, which is approved by TFDA and is used across multiple medical centers in Taiwan (Taiwan AI Lab, n.b.).

The key feature of TAIFA is comprehensive data governance support, which provides curation, preprocessing, labeling, and quality control over data through unified normalization approaches for standardized data usage, while ensuring

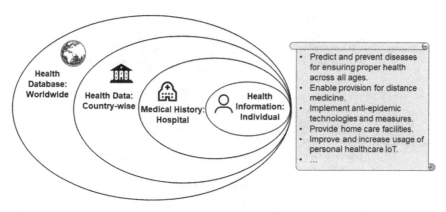

FIGURE 12.4

Toward precision health industry: TAIFA. *TAIFA*, Taiwan Artificial Intelligence Federated Learning Alliance.

data privacy. While dealing with sensitive data, TAIFA uses human-centered privacy compliance features for data deidentification, pseudonymization, or anonymization, and the right to object to data collection. To develop several ML applications through FL in an open-source framework, TAIFA has adopted Harmonia, which can replace data sharing with model sharing. Through the sharing of the model, FL has allowed the establishment of an AI platform that enables the sharing of the expertise of doctors from across the globe without the sharing of data itself. To enable global usage, cloud services are also launched for use in the United States, Japan, etc. TAIFA has currently several federated applications of AI such as Deepmets-Plus, which is an AI-assisted diagnosis system for brain metastases, Wuhan Pneumonia Chest X-ray Automatic Detection System, Intelligent Selection of the Optimal Phase of Coronary Artery in Heart Computer Tomography (TAIFA, 2020), etc.

12.3.5 SubstraFL

This FL framework, created by Owkin in 2016, is now hosted by Linux Foundation for AI and data. It focuses on data ownership and privacy in the medical field (Galtier & Marini, 2019). There is a wide variety of interfaces for different types of users; data scientists and data engineers, as shown in Fig. 12.5, can work with a different low-level Python library to create functions, datasets, and ML tasks; and admins can work with command-line interfaces and use SubstraFL to run complex experiments in a scalable manner. Project managers and high-level users can use web-based applications to monitor the training and explore the results. SubstraFL provides privacy using trusted environments (enclaves) that allow keeping private regions for code and data separate from the whole environment.

To prevent traceability, all operations on the platform are written to an immutable ledger. Finally, the encryption mechanism is used for datasets on the

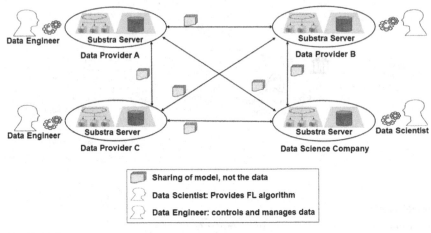

FIGURE 12.5

Federated learning using SubstraFL.

user side, during model updates, and during network communication. SubstraFL can be easily installed on the client side with the command *"pip install substrafl"* (Owkin, n.d.), while on the server side, there are two options: one can use *local deployment* for a quick test and development on a single machine; on the other hand, one can use *production deployment* for real deployments. The application of SubstraFL is more suitable for the production environment. Hence, it may not be an ideal option for basic workflows where there is a necessity for a large amount of interaction during model development. (FATE, 2023).

12.3.6 Federated learning application runtime environment

When different healthcare institutions collaborate to develop an FL model, it is a challenging task to manage the infrastructure needed for large-scale training and deployment. With an open-source platform and software development kit for FL, the NVIDIA federated learning application runtime environment (FLARE) provides its end users with the advantage of having distributed, multiparty collaboration for the development of robust AI platforms from simulation to production across different institutions (Roth et al., 2022). This domain-independent platform has several components, which include support for ML/DL libraries that allow researchers and data scientists to use existing ML/DL libraries, such as PyTorch, TensorFlow, scikit-learn, etc., in the FL paradigm. Several built-in FL algorithms, such as FedAvg, FedProx, FedOpt, Scaffold, and Ditto, are also provided. There is provision for horizontal and vertical FL, with support for both data analytics and ML lifecycle management. Basic requirements, such as privacy, are preserved with differential privacy and homomorphic encryption, and security is enforced through authorization and privacy policy (NVIDIA, n.d.).

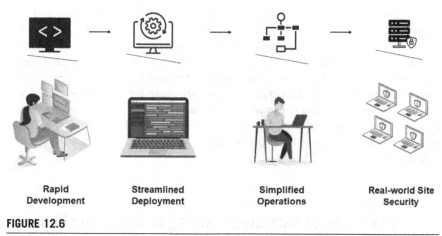

| Rapid Development | Streamlined Deployment | Simplified Operations | Real-world Site Security |

FIGURE 12.6

Federated learning research to production workflow in FLARE. *FLARE*, Federated learning application runtime environment.

In addition, it provides a simulator for rapid deployment and project management with a dashboard user interface and supports system fault tolerance and resiliency. The latest release of the NVIDIA FL platform, FLARE 2.3.0, is packed with exciting new features and enhancements, such as multicloud support using infrastructure-as-code, natural language processing examples, including BERT and GPT-2, and split learning for separating data and labels. With the new features incorporated into workflows, as shown in Fig. 12.6, the participating institutions can save time, reduce risk, improve accuracy, and boost AI workflows.

12.3.7 Classification of case studies based on scale of federation

Based on client participation and training scale, FL is divided into cross-device FL and cross-silo FL, respectively. In the case of cross-device FL, the respective clients are small and dispersed entities (e.g., smartphones, IoT devices, wearable gadgets, etc.), where each entity has a relatively small amount of local processing data. The clients belonging to cross-device FL are typically managed by the large number of edge devices participating in the training process of the system. In summary, in the case of cross-device, the number of participants is humongous with small computational power (Huang et al., 2022). In contrast, in the case of cross-silo, the participants or clients are mostly organizations (e.g., hospitals, financial institutions, etc.), where a global model is trained with an extensive amount of data and high computational power. Unlike cross-device FL, in cross-silo, the number of participants is small (from two to hundreds), and each participant has the right to participate in the training process. Further, in a cross-silo environment, the organization shares the global updated model with the respective local clients but not the data due to privacy issues (Hanser, 2023). Considering the discussed features of cross-device and cross-silo, the above-discussed use cases such as Fed-BioMed, TripleBlind, FATE, TAIFA, SubstraFL, and FLARE are classified into the cross-silo FL category as each use case comprises multiorganizational data, where the local organization is controlled by the global entity after aggregation. Even for use cases such as SubstraFL, there is a peer-to-peer connection among the developers; they can only share the model update but not the data for privacy issues. Similarly, for cases such as TripleBlind and Fed-BioMed, the owner has complete control over data visibility, where multiple agreements are generated for the private access of the data or algorithms for individual users.

12.4 Use cases of federated learning in digital healthcare

This section describes the significant role of FL in disease discovery, categorization, and outcome prediction. The integration of FL with AI has brought several benefits to healthcare management, including remote management of electronic health records, patient monitoring, disease detection, intelligent imaging diagnosis, etc.

12.4.1 Drug discovery

In the case of the pharmaceutical industry, it involves huge costs and time to introduce a new drug to the market. Even for large companies, R&D costs are becoming a great challenge to overcome. The use of AI and ML is not new in the drug design pipeline. ML models are trained to predict the outcomes of wet lab experiments. However, the predictive performance of ML models depends on the availability of enough qualitative training data. Due to intellectual property rights (IPR) and competitiveness, companies only use their data for model training. This led to tremendous efforts and investment toward the collection of organized data of high quality and quantity. Eventually, with the use of FL, ML ledger orchestration for drug discovery (MELLODDY), initiated in June 2019, aimed to increase efficiencies in drug discovery. It sought to achieve this by establishing a centralized FL-based predictive model developed using decentralized data from some of the world's largest pharmaceutical companies without leakage of proprietary information (Burki, 2019).

The goal of this model was to identify the most effective compounds for drug development by harnessing the collective knowledge of involved contributors while protecting their IPR and overall saving time and cost. The MELLODDY platform used billions of experimental data points documenting the behavior of more than 20 million small chemical molecules in over 40,000 biological assays to train the models. Local model training across proprietary data and partners boosted the predictive performance of the global model through the creation of a flexible, scalable, and secure framework for federated and privacy-preserving

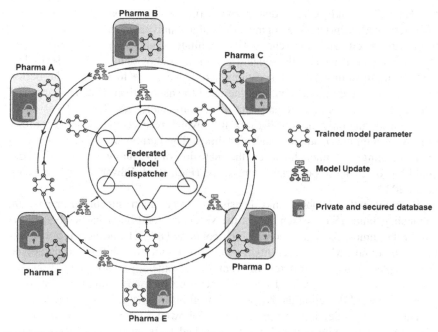

FIGURE 12.7

Federated learning in MELLODDY. *MELLODDY,* Machine learning ledger orchestration for drug discovery.

ML, as demonstrated in Fig. 12.7. The partners can build models that share the collective information of the chemical and biological spaces of the combination of other partners without disclosing the exact chemistry or assays.

MELLODDY being not disease-specific should apply to any area of pharmacology where small molecules are of relevance. In addition, current data is limited to preclinical details. In the future, patient-level data could be used to develop a precision medicine-based model (Heyndrickx et al., 2022).

12.4.2 Brain tumor classification system

In 2020, Intel Labs and Perelman School of Medicine at the University of Pennsylvania (Penn Medicine) joined hands to improve the tumor detection and treatment outcomes of the most common and fatal adult brain tumor, glioblastoma. Although the treatment options have been modified over the past 20 years, not many improvements have been observed in overall survival rates, where the current median survival is just 14 months after standard treatment. An AI software platform called federated tumor segmentation (FeTS) was developed using a federated global dataset from 71 institutions across six continents. The software improved brain tumor detection accuracy by 33%, demonstrating the ability of

FL. The FeTS toolkit serves a dual purpose: first, it aims to provide access to several pretrained deep learning segmentation algorithms, along with their fusion, to healthcare experts and researchers. This facilitates comparative evaluation of the algorithms and enhances the quantification of radiographic scans. Second, it enables multiinstitutional collaborations across the globe to improve the performance of pretrained models without the need to share patient data. This approach effectively addresses privacy and data-ownership challenges. The FeTS is developed considering only the mpMRI modality, which restricts the model generalization using multiple anatomies and modalities. The current model can only work by segmenting the tumor; however, the inclusion of the tumor type classification could be considered a potential future scope (Pati, Baid, Edwards, Sheller, Foley, et al., 2022).

Radiologically, glioblastomas consist of three main subcompartments, the enhancing tumor (ET), which shows the vascular blood-brain barrier breakdown within the tumor; the tumor core, which contains both ET and necrotic part; and the whole tumor, which represents the union of tumor core, infiltrated tissue, and the complete tumor considered relevant to radiotherapy (Pati, Baid, Edwards, Sheller, Wang, et al., 2022). This study therefore focused on multiclass learning problems toward detecting the boundaries of glioblastoma subcompartments.

During data collection, expert clinicians followed a reference standard involving manual annotation protocols and established harmonized preprocessing pipelines to deal with the varying characteristics of mpMRI data due to scanner hardware and acquisition protocol differences. The model training consisted of different phases: initially, a *public model* was trained using data from 231 cases across 16 sites, followed by a *preliminary model* involving data from 2471 cases across 35 sites. The final model was developed using data from 6314 cases across 71 sites. To evaluate the performance of the trained model, a *local validation dataset* was created separating 20% of the total cases contributed by each participating site (Intel, 2022). In addition, external validation was performed using data from six sites (332 cases) not involved in the training stage. The performance of the model was evaluated using the dice similarity coefficient, assessing the spatial agreement between the reference standard and the model's prediction for each of the subcompartments. The radiologists of participating institutions annotated the data and trained the ML algorithm through FL. FL has opened the door to future studies and collaboration while protecting sensitive patient information, preserving data integrity, and ensuring security through confidential computing. The team accomplished a remarkable result by processing a high volume of data in a decentralized system where only models derived from the raw data were transmitted to the central server, eliminating the need to transmit the data itself.

This use case is a remarkable proof to demonstrate how FL can be used for worldwide collaboration to securely access multiple institutional data, enabling the creation of the most diverse dataset of the largest size of GPM patients (N = 6,314) ever used in the literature, with all the data remaining in the database of its proprietary owner.

12.4.3 **Prediction of clinical outcomes in COVID-19 patients**

During the COVID-19 pandemic crisis, several medical, scientific, academic, and data science communities united to find a solution for rapid and secure assessment of models for validation across multiple healthcare institutions. To address this challenge, a global FL-based model called the electronic medical records chest X-ray AI model (EXAM) was developed using data from 20 institutes across the globe. The generated model takes as input the vital signs, laboratory data, demographic data (age), and chest X-rays to predict the future oxygen requirements of symptomatic patients with COVID-19, thus assisting patient triaging (Dayan et al., 2021). Although healthcare providers are known to prefer models validated on their datasets, such models often lack diversity, potentially resulting in lower generalizability and overfitting. A previously developed AI-based SARS-COV-2 clinical decision support model, generated at Mass General Brigham (MGB), was retrained using the FL approach with diverse data, resulting in the creation of the EXAM model. The model was trained with a very large multicontinental dataset of 16,148 cases, where the average AUC achieved was >0.92 when predicting outcomes at 24 H and 72 H from the time of visiting the emergency room. It was found that local models trained with unbalanced cohorts greatly benefited from the FL approach, resulting in a 16% improvement in average AUC across all participating institutions. Furthermore, when predicting the mechanical ventilation treatment or death at 24 H at the site with the largest independent dataset, the model achieved a sensitivity of 0.95 and a specificity of 0.88.

A collaborative method integrating FL with the convolutional neural network is developed to classify COVID-19, bacterial infections, and other viral infections using chest X-ray images. The developed system has five primary steps, including data collection, augmentation, intermittent clients, client-side model training, and model aggregation on the server side. Accordingly, the X-ray image is collected from remote hospitals, where the number of training images varies. To handle this data imbalance issue, a data augmentation method is employed locally. After the completion of the training at each local machine, the generated local model update (LMU) in the form of local training weights is transmitted to the central server. The central server is responsible for aggregating all LMUs from the remote hospital to generate a global model update (GMU). Finally, the GMU is transmitted to the local hospitals for global weight updating to achieve an accurate classification outcome (Ullah et al., 2023).

12.5 **Open Research Directions**

The revolutionary integration of FL in digital healthcare has brought tremendous improvement in the transfer, storage, and analysis of medical data with the utmost perseverance of patient privacy. However, there are several open challenges and research issues related to the application of FL in digital healthcare that need to be carefully addressed.

12.5.1 Real-world evidence

FL, considered the next generation of learning for generating ML models, can be useful for different use cases, which include gathering feedback related to any healthcare products such as drugs or appliances. The consumers could provide anonymous feedback regarding the usage, benefits, and issues related to the products, allowing the developing companies to gather real-world data outside of experimental settings. Nonetheless, such data collection methods should comply with the privacy and security requirements of FL, allowing consumers to have full control over the data to be shared.

12.5.2 Clinical trial recruitments

Another important use of FL could be in the recruitment of patients for clinical trials. It is often observed that the recruitment of patients during clinical trials of phase III studies worldwide typically takes 1–3 years. In most cases, the deadline is missed for enrollment due to medical institutions not sharing information related to the medical conditions of patients as per HIPAA and GDPR. With the use of FL, deidentification of patient information and participating institutions, secure collaborations can help recruit patients qualifying for clinical trials globally.

12.5.3 Supervised/unsupervised learning

The current trend of FL is focused on data security and privacy, where clients or participating institutions are assumed to have labeled data. Nevertheless, in a practical scenario, it could be possible that not all the labels are correct or that not all the data is labeled. In addition, there might be an interesting collection of large databases without the data being labeled. In such a scenario, if the training can be performed using a semisupervised or unsupervised method (Herabad, 2023), it might effectively increase the size of the dataset and the number of participating institutions. Few works have suggested the use of unsupervised learning, where the model is trained solely on unsupervised data (van Berlo et al., 2020). Recently, self-learning, also referred to as *contrastive learning* has been gaining popularity where the model tries to learn similar and distinct patterns from an unlabeled data distribution (Henaff et al., 2020). However, very few studiea have focused on this issue, making it an open direction for FL-related research.

12.5.4 Fault tolerance in federated learning

The standard working of FL considers the synchronous FL environment, where the central aggregation server waits for updated local parameters from all clients before performing any global updates on the parameters. However, as discussed earlier, communication delays from one client result in an overall delay in model updates, leading to poor FL. There is a high chance that one client might drop out of FL due to poor communication bandwidth (Wu et al., 2021). In addition, when multiple clients face data heterogeneity and model heterogeneity, the FL might crash. Therefore it is critical to address such issues. New algorithms and research must be carried out to tackle the problem of synchronous environments, and FL environments crashing due to sudden dropouts of clients or participating institutions.

12.5.5 Participant's mobility behavior in federated learning

In the case of distributed FL, the dynamic behavior of the participants poses higher threats to maintaining the authenticity of the global system as they may leave or join the system at their convenience. Therefore it is necessary to develop robust management strategies to handle the dynamic flow of users while ensuring the stability of the entire platform (Beltrán et al., 2023).

12.6 Summary

This section focuses on the summarization of different key lessons that we have learned from the challenges, existing solutions, and future research insights related to the applications of FL in digital healthcare. For instance, the challenges discussed related to communication, privacy, and security are efficiently handled by the FL frameworks such as Fed-BioMed, TripleBlind, FLARE, and SubstraFL, where effective measures are taken to preserve the privacy of individuals, groups, and organizations. Most importantly, the data and model heterogeneity challenges are successfully managed by different use cases, such as TAIFA, which includes diversified users and a variety of data generated from enterprises, medical centers, research institutions, and local governments. Similarly, a coordinated data analysis framework can be seen in SubstraFL and Fed-BioMed where multiple developers can work simultaneously and independently on private medical data to develop intelligent NLP models for effective healthcare. Furthermore, the successful use cases discussed, such as MELLODDY, FeTS, and EXAM, are aimed at providing faster and more accurate disease diagnosis in medical care. Especially in the current pandemic situation, the development of EXAM can enable rapid COVID screening and treatment responses for individuals. Finally, we believe the discussed open research directions for the advancement of FL will enhance its application in digital healthcare to a new height.

12.7 Conclusions

The main advantage of using FL in healthcare is that it provides effective collaboration among different health institutions across the globe, where a global federated model can be trained using the data of all institutions without having to see or access the data, providing data privacy and security of the utmost importance. With FL, faster and more robust data analysis can be performed for the development of complex models that could be used for accurate diagnosis and treatment. Furthermore, when multiple institutions work together in an FL environment, the cost of infrastructure, training, and maintenance is reduced. This could ultimately reduce the cost of medical data analytics. The research and implementation for enhancing the capability and capacity of FL in digital healthcare and traditional medicine will continue for many decades, as many fundamental issues still need to be addressed. Currently, one of the primary challenges is the secure and collaborative integration of FL with AI for analyzing multiinstitutional wide-ranging demographic, clinical, and imaging data. The integration of FL with blockchain technology could be considered an excellent solution for establishing a highly secure and appropriate collaborative domain for medical data analysis. Also, the introduction of FL in precision medicine will have a high impact on future health. Overall, the research on utilizing the potentiality of FL in digital healthcare is in the growing stage, where novel applications, motivations, scopes, and technologies are in the stage of exploration.

References

Beltrán, E.T. M., Pérez, M.Q., Sánchez, P.M. S., Bernal, S.L., Bovet, G., Pérez, M.G., Pérez, G.M., & Celdrán, A.H. (2023). Decentralized federated learning: Fundamentals, state of the art, frameworks, trends, and challenges. *IEEE Communications Surveys & Tutorials*.

Burki, T. (2019). Pharma blockchains AI for drug development. *The Lancet*, *393*(10189), 2382.

Cavusoglu, U., & Kokcam, A. H. (2021). A new approach to design S-box generation algorithm based on genetic algorithm. *International Journal of Bio-Inspired Computation*, *17*(1), 52−62.

Cremonesi, F., Vesin, M., Cansiz, S., Bouillard, Y., Balelli, I., Innocenti, L., Silva, S., Ayed, S.-S., Taiello, R., & Kameni, L. (2023). Fed-BioMed: Open, transparent and trusted federated learning for real-world healthcare applications. arXiv preprint arXiv:2304.12012.

Dayan, I., Roth, H. R., Zhong, A., Harouni, A., Gentili, A., Abidin, A. Z., Liu, A., Costa, A. B., Wood, B. J., & Tsai, C.-S. (2021). Federated learning for predicting clinical outcomes in patients with COVID-19. *Nature Medicine*, *27*(10), 1735−1743.

Díaz, J. S.-P., & García, Á. L. (2023). Study of the performance and scalability of federated learning for medical imaging with intermittent clients. *Neurocomputing*, *518*, 142−154.

FATE (2023). https://fate.fedai.org/.

Galtier, M.N., & Marini, C. (2019). Substra: A framework for privacy-preserving, traceable and collaborative machine learning. arXiv preprint arXiv:1910.11567.

Hanser, T. (2023). Federated learning for molecular discovery. *Current Opinion in Structural Biology*, *79*, 102545.

Henaff, O.J., Srinivas, A., De Fauw, J., Razavi, A., Doersch, C., Eslami, S.M. A., & van den Oord, A. (2020). Data-efficient image recognition with contrastive predictive coding. International Conference on Machine Learning, Vol 119 119.

Herabad, M. G. (2023). Communication-efficient semi-synchronous hierarchical federated learning with balanced training in heterogeneous IoT edge environments. *Internet of Things, 21.*

Heyndrickx, W., Mervin, L., Morawietz, T., Sturm, N., Friedrich, L., Zalewski, A., Pentina, A., Humbeck, L., Oldenhof, M., & Niwayama, R. (2022). MELLODDY: Cross pharma federated learning at unprecedented scale unlocks benefits in QSAR without compromising proprietary information.

Huang, C., Huang, J., & Liu, X. (2022). Cross-silo federated learning: Challenges and opportunities. arXiv preprint arXiv:2206.12949.

Hu, K., Wu, J. S., Weng, L. G., Zhang, Y. W., Zheng, F., Pang, Z. C., & Xia, M. (2021). A novel federated learning approach based on the confidence of federated Kalman filters. *International Journal of Machine Learning and Cybernetics, 12*(12), 3607−3627.

InfinitiesSoft (2022). https://blog.infinix.co/tw/2022/03/14/nvidia-flare%E8%88%87%E8%81%AF%E5%90%88%E5%AD%B8%E7%BF%92-federated-learning.

Intel (2022). https://www.intel.com/content/www/us/en/newsroom/news/intel-penn-medicine-federated-learning-brain-tumor-detection.html.

Kalra, S., Wen, J., Cresswell, J. C., Volkovs, M., & Tizhoosh, H. (2023). Decentralized federated learning through proxy model sharing. *Nature Communications, 14*(1), 2899.

Liao, Z., & Li, S. (2022). Solving nonlinear equations systems with an enhanced reinforcement learning based differential evolution. *Complex System Modeling and Simulation, 2*(1), 78−95.

Liu, Y., Fan, T., Chen, T., Xu, Q., & Yang, Q. (2021). Fate: An industrial grade platform for collaborative learning with data protection. *The Journal of Machine Learning Research, 22*(1), 10320−10325.

Liu, S. L., Yu, G. D., Yin, R., Yuan, J. T., Shen, L., & Liu, C. H. (2022). Joint model pruning and device selection for communication-efficient federated edge learning. *IEEE Transactions on Communications, 70*(1), 231−244.

Luo, L., Zhao, N., & Lodewijks, G. (2021). Scheduling storage process of shuttle-based storage and retrieval systems based on reinforcement learning. *Complex System Modeling and Simulation, 1*(2), 131−144.

Ma, J., Naas, S. A., Sigg, S., & Lyu, X. X. (2022). Privacy-preserving federated learning based on multi-key homomorphic encryption. *International Journal of Intelligent Systems, 37*(9), 5880−5901.

McMahan, H. B., Moore, E., Ramage, D., Hampson, S., & Arcas, B. A. Y. (2017). Communication-efficient learning of deep networks from decentralized data. *Artificial Intelligence and Statistics, Vol 54*(54), 1273−1282.

Mills, J., Hu, J., & Min, G. Y. (2022). Multi-task federated learning for personalised deep neural networks in edge computing. *IEEE Transactions on Parallel and Distributed Systems, 33*(3), 630−641.

Nagaraj, D., Khandelwal, P., Steyaert, S., & Gevaert, O. (2023). Augmenting digital twins with federated learning in medicine. *The Lancet Digital Health, 5*(5), e251−e253.

Nayantara, P. V., Kamath, S., Manjunath, K. N., & Rajagopal, K. V. (2020). Computer-aided diagnosis of liver lesions using CT images: A systematic review. *Computers in Biology and Medicine, 127.*

NDC Taiwan "TAIFA." https://www.taifa.org/ (n.d.).

Neto, H. N. C., Hribar, J., Dusparic, I., Mattos, D. M. F., & Fernandes, N. C. (2023). A survey on securing federated learning: Analysis of applications, attacks, challenges, and trends. *IEEE Access, 11*, 41928−41953.

Nishio, T., & Yonetani, R. (2019). Client selection for federated learning with heterogeneous resources in mobile edge. ICC 2019 - 2019 Ieee International Conference on Communications (Icc).

Ni, X. M., Shen, X. Y., & Zhao, H. M. (2022). Federated optimization via knowledge codistillation. *Expert Systems with Applications, 191*.

NVIDIA https://developer.nvidia.com/flare (n.d.).

Owkin https://owkin.com/substra (n.d.).

Pati, S., Baid, U., Edwards, B., Sheller, M. J., Foley, P., Reina, G. A., Thakur, S., Sako, C., Bilello, M., & Davatzikos, C. (2022). The federated tumor segmentation (FeTS) tool: an open-source solution to further solid tumor research. *Physics in Medicine & Biology, 67*(20), 204002.

Pati, S., Baid, U., Edwards, B., Sheller, M., Wang, S.-H., Reina, G. A., Foley, P., Gruzdev, A., Karkada, D., & Davatzikos, C. (2022). Federated learning enables big data for rare cancer boundary detection. *Nature Communications, 13*(1), 7346.

Rieke, N., Hancox, J., Li, W. Q., Milletari, F., Roth, H. R., Albarqouni, S., Bakas, S., Galtier, M. N., Landman, B. A., Maier-Hein, K., Ourselin, S., Sheller, M., Summers, R. M., Trask, A., Xu, D. G., Baust, M., & Cardoso, M. J. (2020). The future of digital health with federated learning. *Npj Digital Medicine, 3*(1).

Roth, H.R., Cheng, Y., Wen, Y., Yang, I., Xu, Z., Hsieh, Y.-T., Kersten, K., Harouni, A., Zhao, C., & Lu, K. (2022). NVIDIA FLARE: Federated learning from simulation to real-world. arXiv preprint arXiv:2210.13291.

Silva, S., Altmann, A., Gutman, B., & Lorenzi, M. (2020). Fed-biomed: A general open-source frontend framework for federated learning in healthcare. Domain Adaptation and Representation Transfer, and Distributed and Collaborative Learning: Second MICCAI Workshop, DART 2020, and First MICCAI Workshop, DCL 2020, Held in Conjunction with MICCAI 2020, Lima, Peru, October 4−8, 2020, Proceedings 2, Springer.

TAIFA (2020). https://www.taifa.org/press-conference-launched-by-the-taiwan-united-learning-medical-alliance/.

Taik, A., Mlika, Z., & Cherkaoui, S. (2022). Data-aware device scheduling for federated edge learning. *IEEE Transactions on Cognitive Communications and Networking, 8*(1), 408−421.

Taiwan AI Lab "Harmonia." https://ailabs.tw/healthcare/harmonia-an-open-source-federated-learning-framework/ (n.b.).

Tian, Y., Wang, S., Xiong, J., Bi, R., Zhou, Z., & Bhuiyan, M. Z. A. (2023). Robust and privacy-preserving decentralized deep federated learning training: Focusing on digital healthcare applications. *IEEE/ACM Transactions on Computational Biology and Bioinformatics*.

TripleBlind (2023). https://tripleblind.com/.

Ullah, F., Srivastava, G., Xiao, H., Ullah, S., Lin, J. C.-W., & Zhao, Y. (2023). A scalable federated learning approach for collaborative smart healthcare systems with intermittent clients using medical imaging. *IEEE Journal of Biomedical and Health Informatics*.

van Berlo, B., Saeed, A., & Ozcelebi, T. (2020). Towards federated unsupervised representation learning. Proceedings of the Third ACM International Workshop on Edge Systems, Analytics and Networking (Edgesys'20): 31−36.

Wang, C., Yang, Y. Y., & Zhou, P. Z. (2021). Towards efficient scheduling of federated mobile devices under computational and statistical heterogeneity. *IEEE Transactions on Parallel and Distributed Systems, 32*(2), 394−410.

Wen, J., Zhang, Z. X., Lan, Y., Cui, Z. H., Cai, J. H., & Zhang, W. S. (2023). A survey on federated learning: challenges and applications. *International Journal of Machine Learning and Cybernetics, 14*(2), 513−535.

Wu, W. T., He, L. G., Lin, W. W., Mao, R., Maple, C., & Jarvis, S. (2021). SAFA: A semi-asynchronous protocol for fast federated learning with low overhead. *IEEE Transactions on Computers, 70*(5), 655−668.

Wu, X., Zhang, Y. T., Shi, M. Y., Li, P., Li, R. R., & Xiong, N. N. (2022). An adaptive federated learning scheme with differential privacy preserving. *Future Generation Computer Systems-the International Journal of Escience, 127*, 362−372.

Xiong, L. Z., Han, X., Yang, C. N., & Shi, Y. Q. (2022). Robust reversible watermarking in encrypted image with secure multi-party based on lightweight cryptography. *IEEE Transactions on Circuits and Systems for Video Technology, 32*(1), 75−91.

Xu, Y., Peng, C. G., Tan, W. J., Tian, Y. L., Ma, M. Y., & Niu, K. (2022). Non-interactive verifiable privacy-preserving federated learning. *Future Generation Computer Systems-the International Journal of Escience, 128*, 365−380.

Zhang, J. H., Cheng, X. Y., Wang, C., Wang, Y. C., Shi, Z., Jin, J. H., Song, A. B., Zhao, W., Wen, L. S., & Zhang, T. T. (2022). FedAda: Fast-convergent adaptive federated learning in heterogeneous mobile edge computing environment. *World Wide Web-Internet and Web Information Systems, 25*(5), 1971−1998.

Government and economic regulations on federated learning in emerging digital healthcare systems

13

Abdulwaheed Musa[1,2,3], Abdulhakeem Oladele Abdulfatai[1],
Segun Ezekiel Jacob[1] and Daniel Favour Oluyemi[1]

[1]*Department of Electrical and Computer Engineering, Kwara State University, Malete, Nigeria*
[2]*Centre for Artificial Intelligence and Machine Learning Systems, Kwara State University, Malete, Nigeria*
[3]*Institute for Intelligent Systems, University of Johannesburg, Johannesburg, South Africa*

13.1 Introduction

The emerging digital healthcare systems have become an area of significant interest due to the potential benefits they offer in terms of improved patient care and outcomes. The digital healthcare system has revolutionized the way healthcare services are delivered. With the increase in the adoption of digital technologies in healthcare, the volume of healthcare data generated has also increased significantly. These systems utilize advanced technologies such as artificial intelligence (AI) and machine learning (ML) to analyze large volumes of medical data and give insights into patient care. However, using such technology in the healthcare industry comes with its challenges, such as security and privacy concerns, ethical issues, and regulatory policies. In healthcare, data privacy and security are paramount, and patients may not feel comfortable sharing their sensitive information with centralized systems. Additionally, traditional ML models can be costly to store and process and may not be scalable.

Federated learning (FL), a decentralized machine learning (ML) approach, has emerged as a promising solution for analyzing healthcare data while preserving patient privacy. It is gaining traction in digital healthcare systems as it is capable of overcoming some of the challenges of using AI and ML in healthcare, such as privacy concerns, data ownership issues, and the high cost of developing and maintaining ML models. FL allows for decentralized data processing, data privacy, scalability, and generalizability. Recent applications of FL have yielded positive results in various healthcare domains, including disease diagnosis, drug discovery, medical imaging analysis, and personalized medicine. For example, FL has been used to automate breast cancer diagnosis using mammography images,

Federated Learning for Digital Healthcare Systems. DOI: https://doi.org/10.1016/B978-0-443-13897-3.00012-6

detect and monitor cardiac arrhythmia using electrocardiogram (ECG) signals, and predict drug-target interaction using gene expression data.

FL is a type of ML technique that lets multiple servers or devices collaboratively train a model while keeping their training data locally. This technique allows healthcare organizations to train ML models on sensitive patient data without compromising patient privacy. In FL, the central server begins the process of model training by sending an initial model to the servers or devices. Each server or device then trains the model locally using its data and sends only the updated model parameters back to the central server. The central server combines these updates and uses them to update the global model. It is particularly well-suited to situations where data is sensitive, such as medical or financial transactions.

However, as the model training process in FL involves several parties, it can be challenging to determine accountability for any errors or issues that may occur. Thus, the use of FL in healthcare has also raised some concerns about regulations, data privacy, security, and ownership. To address this concern, policymakers could develop regulations on various security measures, such as data handling, encryption, model testing, validation, and monitoring. The regulations on FL in digital healthcare systems are important to ensure patient privacy and security. Thus, this study explores the regulations (government and economic) on FL in emerging digital healthcare systems.

For FL to be widely adopted in healthcare, government and economic regulators must create policies and regulations that address ethical concerns. These should cover data privacy, security, and fairness, and offer incentives for organizations to adopt FL technologies. The regulations also ensure fair competition among healthcare organizations and prevent monopolistic behavior. Resources should also be made available to make it easy for healthcare organizations to adopt FL and ensure that AI models are secure and compliant. Also, investment opportunities are needed to incentivize research and development in the field of FL in healthcare applications.

Furthermore, this study gives an overview of emerging digital healthcare systems and their challenges. It also explores the details of FL in the context of emerging digital healthcare systems. The security and privacy issues surrounding the use of FL in healthcare are discussed. To achieve the goals of this study, a review of academic articles, reports, and policy documents on FL and relevant regulatory policies was conducted. This includes studies that evaluate the potential benefits and challenges associated with FL in the healthcare industry.

13.1.1 Key contributions of the chapter

Some of the significant contributions of this chapter include the following:

1. Surveys past related works on FL in digital emerging healthcare systems.
2. Examines FL in emerging digital healthcare systems, highlighting its benefits and applications.
3. Investigates the market potential and investment opportunities for FL in emerging digital healthcare systems.

4. Presents government and economic regulations on FL in emerging digital healthcare systems.
5. Proposes a commercialization and cost–benefit analysis.

13.1.2 Chapter organization

After the related works presented in Section 13.2, the other parts of the chapter are organized such that Section 13.3 provides an overview of the emerging digital healthcare systems; Section 13.4 focuses on FL in healthcare, including its characteristics and benefits; Section 13.5 discusses security and privacy issues surrounding the use of FL in healthcare; Section 13.6 presents the existing regulatory policies related to FL in healthcare; Section 13.7 examines the market potential and investment opportunities for FL in emerging digital healthcare systems; and Section 13.8 provides a cost–benefit analysis of commercializing FL in healthcare. The future role of FL in emerging digital healthcare systems is discussed in Section 13.9. In Section 13.10, the limitations of the study are highlighted, before a conclusion is drawn in Section 13.11.

13.2 Related works

FL has emerged as a viable method for implementing innovative and cost-effective healthcare systems while safeguarding privacy, as demonstrated by various studies. In addition, Zheng et al. (2022) presented a comprehensive taxonomy for FL in smart cities, covering various aspects such as architecture, communication, and privacy. The study also sheds light on the challenges and outstanding issues in privacy, security, scalability, and heterogeneity that must be addressed for the full realization of FL's potential in smart cities.

Mothukuri et al. (2021) also explored the challenges related to security and privacy in FL systems, along with potential methods to assess malicious threats within the networks. Lim et al. (2022) examined the implementation of FL-based AI, touching upon factors such as communication costs, resource allocation, and security concerns.

O'Dell and Jahankhani (2021) provided an overview of the evolution of AI and its role as a manager in the fourth industrial revolution, emphasizing the significance of FL in developing intelligent systems. Gadekallu et al. (2021) conducted a survey on FL for vast data, identifying opportunities, applications, and future directions.

Ali et al. (2021) proposed the combination of blockchain and FL for the Internet of Things (IoT), discussing advances and challenges in this integration. The connection between FL-AI and IoT, considering issues such as scarification, security, and extensibility was explored, and an analysis of FL-based AI technology in the context of IoT was conducted (Islam et al., 2021). Furthermore, Pham et al. (2021) provided an overview of FL applications in Industrial IoT, focusing on the characteristics and fundamentals of FL.

Meanwhile, emerging digital technologies, including FL, are being developed to address the challenges facing the healthcare sector (Hesse et al., 2021). Consequently, studies on FL in emerging digital healthcare systems have been conducted. One such study is by Lalmuanawma et al. (2020), who reviewed the applications of ML and AI for the COVID-19 pandemic, finding that FL could be employed to develop predictive models for COVID-19. Pfitzner et al. (2021) also conducted a review of FL in the medical context, identifying the opportunities and challenges it presents. Dayan et al. (2021) proposed FL for predicting medical outcomes in COVID-19 patients, underscoring the importance of FL in developing predictive models for this health crisis.

Singh et al. (2022) put forth a framework for preserving the privacy of IoT healthcare data using FL and blockchain, emphasizing the critical role of privacy preservation in deploying FL in IoT systems.

FL has indeed emerged as a promising technique for harnessing distributed data while ensuring patient privacy. Gaobotse et al. (2022) illustrated how non-invasive smart implants and sensors have revolutionized healthcare services, opening the door to emerging digital health technologies such as wearables. Smuck et al. (2021) identified key factors, such as patient education, data security, interoperability, and clinical workflows, that healthcare providers must consider when implementing wearables.

Dinh-Le et al. (2019) conducted a scoping review on the integration of wearable health technology and electronic health records (EHRs), highlighting the potential for improved clinical decision-making and patient outcomes. Mbunge et al. (2021) emphasized the importance of virtual care in delivering healthcare services remotely.

In addition to these advancements, other researchers have explored the technical challenges of implementing FL in healthcare. For instance, Choi et al. (2021) proposed a secure FL framework for healthcare that leverages homomorphic encryption and differential privacy (DP). Rieke et al. (2020) focused on the challenges and issues in adopting FL-based AI in digital healthcare.

Several studies have explored the regulatory and economic implications of FL in healthcare. Zhao et al. (2020) investigated the impact of data privacy regulations, including the General Data Protection Regulation (GDPR), on FL adoption in healthcare, revealing potential hindrances due to high compliance costs and legal uncertainties.

In essence, FL stands as a transformative technology that finds applications in healthcare, IoT, and a variety of other sectors. While it boasts numerous advantages, it also presents a range of challenges, including concerns related to privacy, security, scalability, and adherence to regulatory standards. Addressing these challenges diligently is imperative for the successful integration of FL. The significance of FL within the healthcare domain and its role in enhancing patient outcomes and service provision cannot be emphasized enough, establishing it as a pivotal focal point for research and development in the ever-evolving digital landscape.

Recent examinations and reviews in the field highlight FL's potential as a promising approach for implementing healthcare and IoT systems, as it enables the utilization of dispersed data resources while preserving data privacy. This decentralized method is viewed as a viable solution to address issues stemming from centralization and holds considerable promise for a range of applications, encompassing healthcare and smart city initiatives. Table 13.1 presents further previous-related studies on FL for healthcare systems.

Table 13.1 Surveys on federated learning for healthcare systems.

References	Summary	Advantages	Limitations
Sheller et al. (2020)	Federated learning results in models reaching 99% of model quality achieved with centralized data.	Model quality, generalizability, and learning patterns	Synchronization used in this study may be insufficient for data such as electronic health records and clinical notes, as well as genomics, where more variances might be present across international institutions.
Dang et al. (2022)	Multi-institutional collaboration is more feasible with the wide adoption of electronic health records.	Performance of current state-of-the-art FL algorithms	Data homogeneity. All local datasets come from hospitals located in the United States and thus share certain characteristics.
Mehrjou et al. (2022)	A large number of local training epochs improve performance while reducing communication costs in many settings.	Predicting survival of each ICU stay	Representativeness of dataset.
Mondrejevski et al. (2022)	Federated learning can be seen as a valid and privacy-preserving alternative to central machine learning for classifying intensive care unit mortality	Precision-Recall Curve and F1-Score	Data used is from a single medical center, and selection bias is unavoidable.
Linardos et al. (2022)	Federatively trained models exhibit increased robustness and are more sensitive to domain shift effects.	Accuracy of disease diagnosis, as well as robustness and sensitivity of federatively trained models	Small size data.

FL, *Federated learning.*

Overall, the literature suggests that FL can be a valuable tool in emerging digital healthcare systems. However, the implementation of FL faces various challenges, including regulatory, legal, and ethical considerations. Governments and economic regulations need to be put in place to ensure the ethical and responsible use of FL in healthcare, as discussed in this study.

13.3 Emerging digital healthcare systems

This section provides an overview of emerging digital healthcare systems, highlighting their major components, benefits, and challenges to healthcare providers and patients. Digital healthcare technologies are designed to enhance the quality of healthcare, accessibility, and efficiency. Technology is catching up with the healthcare industry at a fast pace, revolutionizing the way healthcare services are delivered.

Modern technologies such as AI, ML, and IoT have paved the way for emerging digital healthcare systems. They are a marked departure from the traditional healthcare system, as they are more nimble, cost—efficient, and automated. They are increasingly gaining traction and are fast becoming the norm across multiple healthcare settings. The emerging digital healthcare system typically consists of several components, as discussed below.

13.3.1 Artificial intelligence and machine learning

AI and ML are revolutionizing the healthcare industry, because of their ability to analyze large volumes of patient data and identify trends and patterns. AI and ML have been applied in various healthcare settings, including diagnosis, treatment, and prognosis. One of the areas where AI and ML have been extensively used is in fighting the COVID-19 pandemic (Lalmuanawma et al., 2020). Researchers have pointed out the challenges and limitations of using AI and ML in healthcare, including data privacy and ethical considerations. O'Dell and Jahankhani (2021) have discussed the potential of AI in enhancing decision-making processes, reducing costs, and improving efficiency in healthcare delivery.

13.3.2 Blockchain technology

Blockchain technology can be referred to as an emerging digital healthcare system that involves the use of a decentralized database to store patient data securely. It has been found to enhance the privacy and security of patient data, reduce errors, and enhance data interoperability.

13.3.3 Telemedicine

Telemedicine is the use of telecommunication and information technology for remote healthcare services using digital technologies such as video conferencing and remote monitoring. It is a rapidly growing field, accelerated by the COVID-19 pandemic. It is used for diagnosis, treatment, and consultation. It has been found to improve access to healthcare services, reduce costs, and improve patient outcomes. Thus, healthcare providers, policymakers, and technology developers must continue to invest in and explore the development of telemedicine technologies and systems.

13.3.4 Electronic health records

EHRs refer to digital versions of patients' health records that are saved in a centralized database. In the realm of modern healthcare systems, EHRs play a pivotal role, providing healthcare professionals with quick access to patients' medical histories. EHRs have been found to reduce medical errors and improve the satisfaction of patients (Liu et al., 2021). Nonetheless, EHRs face challenges, including privacy and security, interoperability, and standardization. Emerging digital healthcare systems, such as noninvasive smart implants, wearables, and blockchain technology, can help overcome these challenges.

13.3.5 Wearable devices

Wearable devices are now widely used in healthcare. They can track different health-related data, such as sleep patterns, physical activity, and heart rate, while the data is integrated with EHRs to give a more complete picture of a patient's health situation (Smuck et al., 2021). However, there are issues and challenges with the adoption of wearable devices in healthcare. These include data security and privacy and the need for interoperability standards. Mbunge et al. (2021) have stressed the importance of interoperability and the need for standardized data formats to ensure that data from wearable devices can be easily integrated into EHRs. However, it is necessary to address ethical, legal, and privacy concerns associated with using wearable devices in healthcare.

13.4 Federated learning in emerging digital healthcare systems

According to McMahan et al. (2017) and Yang et al. (2019), FL is a learning paradigm that aims at addressing data governance and privacy problems by collaboratively training algorithms without sharing the data itself. Originally developed for various areas, such as use cases for mobile and edge devices, it is gaining prominence for healthcare applications. It is a distributed interactive AI that shows promise for digital healthcare, as it lets customers (e.g., hospitals) participate in AI training while retaining privacy.

FL as an emerging ML architecture holds significant implications in the context of government and economic regulations. This decentralized approach to model training allows data to remain localized on users' devices or within organizations, reducing the need for centralized data repositories. This, in turn, has a direct impact on data privacy and regulatory compliance. FL enables compliance by keeping sensitive user data on-device, preventing large-scale data transfers and minimizing the risks of privacy breaches.

From an economic regulation perspective, FL plays a role in preserving competitive markets. In some industries, data monopolies have emerged, where a few major players have access to vast amounts of valuable data. FL distributes the model training process and encourages a more level playing field, allowing smaller companies to participate in ML without the need for massive centralized data resources.

In Fig. 13.1, client devices are at the forefront, representing individual users' smartphones, IoT devices, or computers. Each of these devices hosts a local ML

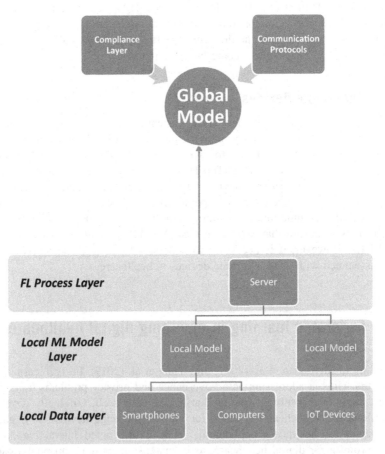

FIGURE 13.1

Federated learning architecture.

model, serving as the initial point for data analysis. These local models are trained on the data residing on the respective devices, allowing for personalized insights without the need for data to leave the user's control. The server acts as the central coordinator, collecting periodic model updates from all client devices. It then aggregates these insights to update the global model, which benefits from the collective knowledge without directly accessing or storing user data. In the context of government and economic regulations, an additional layer for compliance and privacy safeguards is essential to ensure that the architecture aligns with relevant legal frameworks. Secure communication protocols ensure data privacy during the update process, making FL an innovative approach to ML that combines individual insights while preserving data privacy and regulatory requirements.

Fig. 13.2 illustrates how FL works in healthcare. Consider a research institution collaborating with several hospitals to train a model for predicting the progression of a particular disease. Instead of transferring the EHRs from the hospitals to the research institution, the parties involved would collaborate to train the model in a distributed way. Each hospital would use its data to train the local

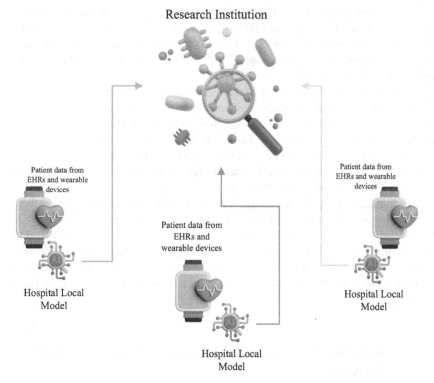

FIGURE 13.2

Federated learning framework in digital healthcare.

model and then share the updated model variables with the research institution. The research institution would then combine the updated model parameters from each party to create a more accurate and robust model. This innovative approach keeps data locally stored at each entity, sharing only the model updates with a central server.

FL has the potential to improve healthcare outcomes by preserving data privacy and security, reducing costs of computation and storage, and improving the efficiency of machine learning models. However, the implementation of FL in digital healthcare systems also poses certain threats, concerns, and risks, such as data heterogeneity, bias, and privacy breaches, as well as the need for standardized frameworks and regulations to ensure transparency, fairness, and accountability (Pfitzner et al., 2021; Dayan et al., 2021).

Governments and regulatory bodies must play a crucial role in addressing these concerns and risks. Clear guidelines and regulations around the use of FL in digital healthcare systems should be implemented. These guidelines should address issues around data privacy, ethical considerations, regulatory compliance, and model bias, ensuring that healthcare organizations are using FL safely and ethically (Abdulrahman et al., 2020; Elayan et al., 2021). Additionally, economic incentives such as funding and tax breaks can be provided to healthcare organizations that adopt FL in their digital healthcare systems, encouraging widespread adoption of this technology (Rieke et al., 2020). The integration of other emerging technologies, such as blockchain and IoT, can also improve FL in digital healthcare systems (Nguyen et al., 2021).

A successful implementation of FL can therefore have remarkable potential for enabling large-scale accuracy in medicine, which would result in models that provide accurate decisions and are sensitive to scarce diseases while preserving privacy and governance concerns. However, FL still needs more serious engineering considerations for optimal algorithm performance without compromising patient security or privacy. Yet, FL can overcome the limitations of methods that require a single pool of centralized data.

As highlighted in the literature, FL has various applications in healthcare, such as predicting clinical outcomes in patients, preserving the privacy of IoT healthcare data, and improving personalized medical care (Gadekallu et al., 2021; Ali et al., 2021; Singh et al., 2022; Zheng et al., 2022). The diverse applications and benefits of FL, as well as the potential threats, concerns, and risks associated with its implementation in modern healthcare systems, can be further explored.

13.4.1 Applications of federated learning in digital healthcare systems

The healthcare industry, in particular, has seen a growing interest in FL due to its potential to enable more accurate diagnoses, personalized treatments, and drug discovery while protecting patient data. FL has enormous potential for application in the digital healthcare system. By using FL, healthcare providers can train machine learning models on distributed patient data without needing to transfer sensitive data to a centralized location. This allows for the development of more accurate diagnostic tools and personalized treatment plans while maintaining patient privacy. FL can enable real-time monitoring of vital signs and symptoms, such as heart rate or blood glucose levels, in patients with chronic conditions such as diabetes or heart disease, allowing medical providers to handle them before serious health issues arise. FL can be applied in digital healthcare systems in a variety of ways, including the following.

13.4.1.1 Disease diagnosis

FL can be used to develop ML models for disease diagnosis by aggregating data from multiple healthcare organizations. The model can then be used to diagnose patients at the point of care, allowing for faster and more accurate diagnoses.

13.4.1.2 Disease prediction

Hospitals could use FL models to predict how likely it is for a patient to develop a disease or get an infection after being admitted into their care. This could help doctors make better decisions about how best to treat each patient based on their individual needs. Additionally, these predictions could be used by other organizations working with hospitalized patients (such as insurance companies) to provide better coverage options for those who need it most.

13.4.1.3 Understanding disease

Another area where FL can benefit healthcare is by helping to understand better diseases such as cancer or Alzheimer's at the genetic level. By using FL models on DNA samples collected from thousands of people with these diseases, researchers can gain insights into which factors are most contributory to the development of these diseases from patient datasets that they can then use in their research efforts without requiring direct access to protected health information (PHI).

13.4.1.4 Personalized treatment plans

FL can be used for personalized treatment models for patients based on their health data. By training models on data from multiple healthcare organizations, the model can learn from a diverse range of patient data and develop personalized treatment plans for each patient. This can result in resource efficiency and reduced costs.

13.4.1.5 Remote patient monitoring

Remote patient monitoring is a notable application of FL in digital healthcare. This method allows healthcare providers to monitor patients from afar without requiring them to physically visit a healthcare facility.

13.4.1.6 Drug discovery

FL can be used to develop ML models for drug discovery by training on data from multiple sources. This can enhance new drug identification more quickly and accurately, leading to faster drug development and approval.

13.4.2 Benefits of federated learning in digital healthcare systems

One study demonstrated the feasibility and accuracy of using FL for medical image analysis. The study used FL to train a convolutional neural network for identifying skin lesions and showed that FL can achieve performance comparable to centralized learning methods while retaining data privacy (Majeed et al., 2020). Another study showed that FL can be used to train ML models for predicting hospital readmissions based on EHR data from multiple institutions. The results indicated that FL could improve model accuracy compared with a centralized approach while preserving data privacy (Sheller et al., 2020). FL also promotes inclusivity and equity in healthcare by allowing healthcare providers to leverage data from diverse patient populations, as discussed below.

13.4.2.1 Improved data privacy

FL enables healthcare organizations to collaborate and train ML models without having to centralize the data. This means that patient data remains local to each entity, reducing the risk of data breaches, thereby ensuring patient privacy.

13.4.2.2 Increased accuracy

By training ML models on data from numerous sources, FL can improve the accuracy of diagnoses and treatment plans. This is because the model can learn from a diverse range of patient data and develop a more comprehensive understanding of the health status of each patient.

13.4.2.3 Cost-effective

FL can reduce the cost of developing and deploying ML models in digital healthcare systems. By enabling collaboration between healthcare organizations, FL can reduce the need for each organization to develop its models, reducing the time and cost associated with model development.

13.4.2.4 Enhanced patient care

FL can help improve patient care by enabling more accurate diagnoses, personalized treatment plans, and faster drug development. This can bring about better healthcare outcomes for patients and a more efficient healthcare system.

Overall, FL has a variety of benefits in the field of digital healthcare systems. With its ability to facilitate secure data sharing and analysis, FL can improve the accuracy and efficiency of data analysis, reduce data exposure risk, and improve patient outcomes.

13.4.3 Threats, concerns, and risks of federated learning in digital healthcare systems

While FL has many potential benefits for digital healthcare systems, some threats, concerns, and risks need to be considered. They highlight the need for careful planning, design, and implementation of FL in digital healthcare systems, with a focus on ensuring data privacy, security, and ethical considerations.

13.4.3.1 Data security

FL raises concerns and issues over data security as patient data is distributed across multiple healthcare organizations. This means that healthcare organizations must put in place robust security measures to protect patient data from cyber threats and data breaches.

13.4.3.2 Ethical considerations

FL raises ethical considerations around data ownership and patient consent. Healthcare organizations must ensure that patients are aware of how their data is being put to use and obtain their consent before using their data for ML.

13.4.3.3 Regulatory compliance

FL raises concerns around regulatory compliance, as healthcare organizations must comply with privacy and data protection regulations, such as Health Insurance Portability and Accountability Act (HIPAA) in the United States and GDPR in the European Union. Healthcare organizations must ensure that they are complying with these regulations when using FL in their digital healthcare systems.

13.4.3.4 Model bias

FL raises concerns about model bias, as models may learn from biased data and produce biased results. Healthcare organizations must ensure that their models are trained on a variety datasets to reduce bias.

13.5 Security and privacy issues in healthcare ecosystem

The healthcare ecosystem is facing numerous security and privacy issues that affect patients, healthcare providers, and medical facilities (Rasool et al., 2022). Typically, privacy issues can be divided into three categories: input data privacy, trained model privacy, and model output privacy. Security issues involve data poisoning attacks, model poisoning attacks, or evasion attacks. In data poisoning, clients deliberately use malicious data samples to mislead the global model by deploying their local models that are trained on malicious data. For model poisoning, the goal of the attackers is to mislead the ML models to give malicious results. In the case of an evasion attack, adversaries confuse the ML model used by giving misguided and modified patterns that appear to look like original test patterns.

Security and privacy issues are the main motivation behind FL, since it can provide practical solutions to various issues in the healthcare ecosystem (Markakis et al., 2019). FL ensures the confidentiality of the input data, since it only works with learned variables from the private data and not with the original data. However, a governance framework that integrates technical, organizational, and legal aspects is necessary to ensure ethical and secure healthcare (Hatzivasilis et al., 2019).

One of the most significant issues is unauthorized access to sensitive patient information (Ruotsalainen & Blobel, 2020). FL allows healthcare organizations to share data without sharing patient information, and each organization trains its local model on their data, sharing only the model weights with the central server (Hathaliya & Tanwar, 2020).

FL also provides a practical solution to data breaches by allowing healthcare organizations to collaborate on model development without sharing patient data (Ruotsalainen & Blobel, 2020). By training local models on their data and sharing only model weights, healthcare organizations can collaborate securely and reduce the risk of data breaches (Hathaliya & Tanwar, 2020). FL can help address model bias by allowing healthcare organizations to collaborate on model development using diverse datasets (Hathaliya & Tanwar, 2020). By training local models on diverse datasets and sharing model weights, healthcare organizations can ensure that their models are more accurate and unbiased, leading to better patient outcomes for all groups (Ruotsalainen & Blobel, 2020).

FL can address privacy concerns when collecting data from wearable devices and other sources (Markakis et al., 2019). By training local models on patient data that is stored locally on their devices and sharing only model weights, FL allows patients to retain control over their data while still allowing healthcare organizations to use that data for model development (Hathaliya & Tanwar, 2020). A multilayered approach combining technical, organizational, and legal measures is necessary to ensure ethical, secure, and trustwoerthy healthcare system (Hatzivasilis et al., 2019).

Despite the use of FL to secure customers' private data by sharing only the trained variables, researchers have shown that useful information on the client's training data can reasonably be derived from updating the trained model with a high accuracy of about 90%. It shows that although private data is not used in FL, customer private information is still vulnerable. Thus solutions have been proposed to fix different types of vulnerabilities.

To protect information from other clients, DP is one of the most commonly deployed privacy protection techniques because of its simple algorithm and relatively low communication overhead (Abadi et al., 2016). Another effective approach to maintaining the privacy of distributed datasets is secure multiparty computation, in which many parties collectively calculate a function using their inputs without exposing their private inputs to other parties. This needs a minimal amount of user data to be shared.

Similarly, fully homomorphic encryption (FHE) and its variants are also used in improving the trained model's security. In this approach, clients participating can only see the encrypted data, and they are required to perform calculations on the encrypted data. Typically, homomorphic encryption is divided into partially homomorphic encryption, somewhat homomorphic encryption, and fully homomorphic encryption, depending on the number of operations on the encrypted data.

13.6 Regulatory policies on federated learning in emerging digital healthcare systems

This section discusses regulations and regulatory policies governing the utilization of FL in emerging digital healthcare systems. The use of FL in healthcare raises concerns about data privacy, ethical considerations, regulatory compliance, and model bias. To address these concerns, a comprehensive regulatory framework that considers the legal, ethical, and social implications of the technology and provides guidance to stakeholders in the healthcare ecosystem to ensure the safe and responsible use of FL is needed. Table 13.2 demonstrates why regulations are needed in FL.

Table 13.2 Need for regulations in federated learning.

Reasons for regulation	Explanation
Data protection and privacy	FL involves the sharing of sensitive patient data across multiple institutions, making it vulnerable to privacy breaches and misuse. Regulations, such as HIPAA and GDPR, help protect patient data by enforcing strict privacy policies.
Fairness and bias	FL relies on diverse data from different sources. If the data is biased or incomplete, it can lead to unfair outcomes, such as inadequate diagnosis or treatment for certain groups. Regulations can ensure that data used in FL is diverse, unbiased, and representative of entire population.
Transparency and accountability	FL involves complex algorithms and models that can be difficult to interpret or understand. Regulations can require that algorithms used in FL are transparent and explainable, enabling regulators to hold developers and organizations accountable for any potential harm caused by the technology.
Ethical concerns	FL raises ethical concerns around data ownership, consent, and autonomy. Regulations can establish guidelines and ethical frameworks for use of FL in healthcare, ensuring that technology is used in a way that preserves patient rights and interests.

FL, *Federated learning*; GDPR, *General Data Protection Regulation*; HIPAA, *Health Insurance Portability and Accountability Act.*

13.6.1 **Government and economic regulations**

Government regulations play a crucial role in ensuring that FL is used safely and ethically in digital healthcare systems (Ahmad et al., 2020; Fernandes & Chaltikyan, 2020; Myles & Leslie, 2023; Pesapane et al., 2021). Government regulations aim to protect patient privacy and autonomy and oversee the development and deployment of digital healthcare systems. Data privacy laws such as HIPAA and GDPR ensure patient data privacy and security, while the FDA's Digital Health Software Precertification Program provides regulatory oversight for the development and deployment of digital healthcare systems.

Economic regulations, such as pricing policies, reimbursement models, and other incentives, ensure that healthcare organizations can effectively utilize FL technology and access the necessary data for its deployment. Economic regulations, such as funding and tax breaks, can facilitate the adoption of FL in healthcare (Ahmad et al., 2020; Fernandes & Chaltikyan, 2020; Myles & Leslie, 2023; Pesapane et al., 2021). Additionally, tax credits and grants provide economic incentives to foster the adoption and development of digital healthcare systems. Therefore regulatory policies and guidelines are necessary to promote the safe and responsible use of FL in emerging digital healthcare systems.

In this context, there have been several government and economic initiatives (that can impact the cost, funding, and sustainability of the technology) to regulate the use of FL in emerging digital healthcare systems, as discussed below.

13.6.1.1 *Health Insurance Portability and Accountability Act*

HIPAA refers to a US federal law that provides regulations around the use and disclosure of PHI. It regulates the use and disclosure of PHI by covered entities, including providers of healthcare using FL, health plans, and healthcare clearinghouses, to ensure compliance with HIPAA regulations, protecting patient privacy and confidentiality. HIPAA requires covered entities to get written authorization from individuals for any disclosure or use of PHI for research purposes.

13.6.1.2 *General Data Protection Regulation*

GDPR refers to a comprehensive data protection regulation that provides strict regulations around the collection, storage, and utilization of personal data, including patient data. Healthcare organizations that use FL in the European Union must comply with these regulations to protect the privacy of patients (Ahmad et al., 2020; Fernandes & Chaltikyan, 2020; Myles & Leslie, 2023).

13.6.1.3 *Chinese government regulations*

The Chinese government has issued guidelines for the use of AI in healthcare, which include recommendations for the development and validation of AI algorithms. The guidelines also address the use of FL, stating that "the development of FL in healthcare should be guided by the principles of data security, privacy protection, and compliance with laws and regulations".

13.6.1.4 Food and Drug Administration

The Food and Drug Administration (FDA) regulates food, drugs, and medical devices, including those that use FL algorithms, to ensure their safety and effectiveness (Ahmad et al., 2020). The FDA-published guidelines address the use of FL, including recommendations for the development and validation of the algorithms. For instance, the regulations state that: "FL can be used where the underlying data cannot be moved or shared, but still must be used to develop a predictive algorithm" (FDA, 2019).

13.6.1.5 World Health Organization

The World Health Organization (WHO) has published guidelines on ethics in digital health technology, including federated learning. The guidelines recommend that the development and deployment of digital health technologies, including FL models, be guided by autonomy, privacy, and confidentiality.

13.6.1.6 Precision Medicine Initiative

The Precision Medicine Initiative (PMI) is a US government initiative that provides funding for research on personalized medicine, including the use of FL to improve patient care.

13.6.1.7 Small Business Innovation Research (SBIR) and Small Business Technology Transfer (STTR) programs

The US government programs provide funding for small businesses that are developing innovative healthcare technologies, including those that use FL (Pesapane et al., 2021).

13.6.2 Other regulations

Other important regulations on FL in emerging digital healthcare systems include the following.

13.6.2.1 International Organization for Standardization (ISO)

The International Organization for Standardization (ISO) has developed a standard for the use of AI in healthcare, which includes guidelines for the development and validation of AI algorithms. The standard also includes provisions on the use of federated learning, stating that: "federated learning should be guided by principles of data privacy and protection, informed consent, transparency, and accountability".

13.6.2.2 Healthcare Information and Management Systems Society (HIMSS)

The Healthcare Information and Management Systems Society (HIMSS) focuses on improving healthcare using information technology globally. HIMSS has a policy paper on the use of AI in healthcare, including FL. The policy paper outlines key principles for the ethical use of AI in healthcare, including transparency, accountability, and privacy.

13.6.2.3 National Institute of Standards and Technology (NIST)

The National Institute of Standards and Technology (NIST) provides guidelines and standards for the development and deployment of secure and interoperable healthcare technologies, including FL. The NIST framework is a tool that organizations can use to manage privacy risks arising from their activities. It provides a set of privacy protection outcomes and a risk management structure that can be applied to different data processing activities, including FL.

In addition to these specific regulations, there are also broader ethical and legal considerations that must be noted when using FL in digital healthcare systems. These considerations include issues such as data privacy, informed consent, transparency, and accountability. Healthcare organizations must work with regulatory bodies to ensure that their FL algorithms are developed and deployed responsibly and ethically.

Regulatory policies are important in the development and deployment of digital healthcare systems, regarding the use of FL technology. These policies provide a framework to ensure the safety and ethical use of technology, including protecting patient data, ensuring patient autonomy, and providing oversight for the development and deployment of digital healthcare systems. Ultimately, regulatory policies play an important role in preventing the misuse or exploitation of technology, holding organizations accountable, and promoting their safe and responsible use. By providing clear guidelines and incentives, governments can encourage the adoption of this technology while also protecting patient privacy and confidentiality.

13.7 Market potential and investment opportunities for federated learning in emerging digital healthcare systems

The global digital healthcare market is on a remarkable growth trajectory, with forecasts predicting substantial expansion in the coming years. According to a report, the digital healthcare market is expected to reach an impressive valuation of approximately $2.6 billion by 2026. This substantial growth can be attributed to several key factors. First, there is a notable increase in the prevalence of chronic diseases worldwide, which has created an urgent need for more efficient and technologically advanced healthcare systems. Additionally, advancements in digital health technologies, telemedicine, and wearable devices have played a pivotal role in shaping the healthcare landscape.

Furthermore, the digital health market is expected to continue its growth well into 2027. This persistent growth underscores the sustained demand for innovative healthcare solutions and the immense market potential within the digital health sector.

One promising and emerging technology within this landscape is FL in healthcare. Researchers in the field, as documented by Alberola-López et al. (2021), anticipate a substantial surge in this area, with a projected value of $2.1 billion by 2025. This growth is driven by the unique capabilities of FL, which allows healthcare institutions to collaboratively train ML models on decentralized data without compromising patient privacy. This presents a significant and lucrative opportunity for investors looking to explore and invest in this groundbreaking technology.

The potential for investment in FL in healthcare is further bolstered by the growing adoption of digital health technologies. A report (Intelligence, 2020) highlights the upward trend in healthcare institutions embracing digital solutions to enhance patient care and streamline operations. These technologies synergize effectively with FL, creating a compelling ecosystem for investors.

Investing in FL means investing in a brighter healthcare future. For healthcare providers, it can lead to more accurate diagnoses, personalized treatments, and even drug discovery breakthroughs. And for investors, the return on investment (ROI) is promising. As this technology gains traction, it opens doors to lucrative opportunities in the development of secure data sharing platforms, compliance tools, and advanced security systems, all tailored to protect patient data. As these solutions become more essential in healthcare, the potential for long-term profitability is substantial.

So, not only does FL promise to elevate the quality of healthcare, but it also offers smart investors a chance to be part of a growing market with strong ROI potential. This technology is not just about numbers; it is about health, security, and a promise of the future.

Investors have a unique opportunity to shape the future of healthcare through a groundbreaking market segment. By investing in the creation of a marketplace or platform that connects healthcare institutions with ML experts and data scientists specializing in federated learning, transformative changes in the industry can be catalyzed.

This platform has immense potential to streamline collaborations between healthcare providers and data experts. It bridges the gap between those who possess valuable healthcare data and those who have the expertise to unlock its potential without compromising privacy. As the demand for FL solutions in healthcare continues to rise, this platform becomes the bridge, accelerating the adoption of this technology and making it more accessible for a broader range of healthcare institutions.

The ROI here is not only financial but also extends to the realm of improving healthcare services and patient outcomes. As federated learning becomes more mainstream, your investment can pave the way for innovative partnerships and data-driven healthcare solutions that can have a far-reaching impact on society. This is not just an investment in technology; it is an investment for better, more efficient healthcare and a safer, more private future for all.

To provide a comprehensive overview, Table 13.3 illustrates the estimated market size of FL in healthcare, expressed in USD billions, for the years spanning from 2020 to 2025.

Table 13.3 Projected market potential for federated learning (FL) in healthcare.

Year	Market value	Source	Factors driving growth
2021	$820 million	Tiwari et al. (2021)	Increasing awareness of FL's benefits
2022	$1.2 billion	Lim et al. (2022)	Expanding applications of FL in medical research
2023	$1.8 billion	Li et al. (2021)	Proven success in preserving patient privacy
2024	$2.0 billion	Pournajaf et al. (2022)	Regulatory support for FL in healthcare
2025	$2.1 billion	Alberola-López et al. (2021)	Integration with digital health systems

Investment in FL technology can also benefit healthcare providers by reducing the costs associated with data collection and analysis. This is particularly relevant in emerging economies, where access to healthcare is limited and resources are scarce.

13.7.1 Market potentials

With the increasing availability of medical data, the market potential for FL in emerging digital healthcare systems is significant, as it can help healthcare providers improve medical research, diagnostics, and treatment while reducing healthcare costs and improving patient outcomes. In this context, FL is expected to play a critical role in the development of the next generation of healthcare services and products, with significant implications for the global healthcare market.

13.7.1.1 Improved patient outcomes

In digital healthcare systems FL can lead to improved patient outcomes through personalized medicine and disease prevention. This can result in reduced healthcare costs and better patient satisfaction.

13.7.1.2 Reduced data privacy risks

In digital healthcare systems FL allows for data to be analyzed without being shared or exposed to third parties, thus reducing the risk of data breaches and privacy violations.

13.7.1.3 Enhanced efficiency

In digital healthcare systems FL can lead to enhanced efficiency by reducing the need for data storage and transfer and enabling real-time monitoring and decision-making.

13.7.1.4 Increased accessibility

In digital healthcare systems FL can increase accessibility to healthcare services and reduce healthcare disparities by giving access to quality healthcare services in remote areas.

13.7.2 **Investment opportunities**

With the potential to transform the healthcare industry, FL presents a unique investment opportunity for investors who are looking to capitalize on the growing demand for innovative healthcare solutions.

13.7.2.1 *Federated learning-based healthcare solutions*

There is a significant market opportunity for healthcare solutions that leverage FL technology to provide personalized healthcare services, disease prevention, and real-time monitoring. Such solutions include predictive analytics, remote patient monitoring, and patient-centric care models.

13.7.2.2 *Federated learning-based healthcare devices*

FL-based healthcare devices, such as wearables, implantable devices, and IoT sensors, can provide real-time monitoring and data collection, enabling personalized medicine and disease prevention.

13.7.2.3 *Data analytics and management*

With the increasing adoption of FL in digital healthcare systems, there will be a growing need for data analytics and management solutions to analyze and manage the massive amounts of data generated by these systems.

13.7.2.4 *Cybersecurity and privacy solutions*

As FL in digital healthcare systems becomes more prevalent, there will be a growing need for cybersecurity and privacy solutions to protect patient data and ensure regulatory compliance.

13.7.2.5 *Investment in healthcare start-ups*

Several healthcare start-ups, are using FL to develop innovative healthcare solutions. Investing in healthcare start-ups can provide opportunities for early-stage investment in the growing healthcare industry.

13.7.2.6 *Investment in research and development*

There is a need for research and development in FL for healthcare applications. Investing in research and development can lead to the development of novel FL algorithms and applications in the healthcare industry.

Overall, the market potential and investment opportunities for FL in emerging digital healthcare systems are vast. FL can enable the development of novel healthcare solutions while maintaining data privacy and security. Investors and entrepreneurs who can develop and deploy innovative FL-based healthcare solutions and services stand to benefit from this rapidly growing market.

13.8 Commercialization and cost—benefit analysis of fl in emerging digital healthcare systems

In this section, the commercialization and cost—benefit analysis of FL in emerging digital healthcare systems are discussed. Commercialization requires an analysis of cost—benefits. According to Liu et al. (2021), FL can significantly lower the costs of constructing and maintaining a centralized data center while guaranteeing data security and privacy.

Yang et al. (2019) analyzed the cost benefits of using FL for clinical decision support (CDS) in a healthcare system setting. Yang et al. (2019) found that FL can reduce the cost of developing CDS by enabling the use of data from multiple sources without the need for data sharing or centralization. They found that FL could save the healthcare system $10 billion per year, while also improving the accuracy of CDS by 10%.

Rieke et al. (2020) analyzed the potential cost—benefits of FL in healthcare across various use cases. The review found that FL has the potential to reduce costs associated with data storage and transmission while enabling the use of data from multiple sources to improve the performance of the model.

Du et al. (2020) also analyzed the cost—benefits of FL for electroencephalography (EEG) analysis in healthcare, revealing a substantial reduction in the required number of training samples and, consequently, a decrease in the overall cost of EEG analysis. In addition, Choi et al. (2021) studied the cost-effectiveness of using FL for diabetic retinopathy diagnosis and found it to be more cost-effective than a centralized approach. Furthermore, Guo et al. (2021) examined the cost—benefits of FL for medical image analysis in healthcare and found that it can enhance the accuracy of medical image analysis while reducing the associated cost and time of data transfer.

The commercialization of FL in healthcare requires addressing several challenges. One of such challenges is the need for standardization and interoperability of healthcare data. Another challenge is the need for regulatory compliance with healthcare data privacy regulations, such as GDPR.

A cost—benefit analysis can help healthcare providers and organizations decide whether the technology is worth investing in. One such method is net present value (NPV) or ROI analysis. NPV analysis considers the project's discounted cash flows, while ROI analysis focuses on expected returns on investment. Either method can be used to weigh the costs and benefits of implementing FL in healthcare.

Fig. 13.3 shows that the initial investment costs for implementing FL in healthcare systems are high due to the need for specialized hardware and software. However, over time, the operational costs of FL are significantly lower than that of traditional ML approaches since the data remains local to each healthcare system. The direct benefits of FL include improved patient outcomes and

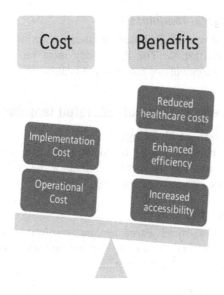

FIGURE 13.3

Cost—benefit analysis of federated learning.

reduced costs, such as the reduction of hospital stays and medication errors. The indirect benefits can include increased efficiency and productivity due to the use of a collaborative approach that enables healthcare professionals to work together effectively.

13.8.1 Commercialization of federated learning

The commercialization of FL in digital healthcare systems involves the development and deployment of FL-based healthcare solutions and services that address the needs of providers, patients, and other stakeholders. Key factors that influence the commercialization of FL in digital healthcare systems are discussed below.

13.8.1.1 Regulatory compliance

Regulatory compliance is critical to the commercialization of FL-based healthcare solutions and services. FL solutions must comply with healthcare data privacy regulations to ensure the protection of patient's data.

13.8.1.2 Market demand

The market demand for FL-based healthcare solutions and services is a key factor that influences the commercialization of technology. Healthcare providers and patients must see the value of FL in improving patient outcomes and reducing healthcare costs.

13.8.1.3 Technological readiness

The technological readiness of FL-based healthcare solutions and services is critical to their commercialization. The technology must be scalable, reliable, and secure to meet the requirements of patients and healthcare providers.

13.8.2 Cost—benefit analysis of federated learning

The cost—benefit analysis of FL in digital healthcare systems involves comparing the costs of implementing FL-based healthcare solutions and services with the benefits that they provide. Key factors that influence the cost—benefit analysis of FL in digital healthcare systems are discussed below.

13.8.2.1 Implementation costs

The implementation costs of FL-based healthcare solutions and services include the costs of developing and deploying the technology. These costs must be weighed against the benefits that FL provides.

13.8.2.2 Operational costs

The operational costs of FL-based healthcare solutions and services include the costs of maintaining the technology and providing ongoing support. These costs must be weighed against the benefits that FL provides.

13.8.2.3 Benefits

The benefits of FL-based healthcare solutions and services include improved patient outcomes, reduced healthcare costs, enhanced efficiency, and increased accessibility to healthcare services. These benefits must be weighed against the costs of implementing and operating FL-based healthcare solutions and services.

The commercialization of FL in healthcare has major implications for the industry and requires addressing challenges related to data standardization, interoperability, and regulatory compliance. The cost—benefit analysis of FL in healthcare suggests that it has the potential to reduce costs associated with data storage and transmission while enabling the use of data from multiple sources to improve model performance. The next section explores the role of FL in future digital healthcare systems.

13.9 Role of federated learning in future digital healthcare systems

As the healthcare industry continues to evolve and embrace digital technologies, FL offers a unique approach to ML that can enable the development of more personalized, efficient, and effective healthcare solutions. In future digital healthcare systems, FL could play a significant role in advancing patient care and medical research. Here are some of the potential roles in future digital healthcare systems:

13.9.1 Personalized medicine

FL can enable the development of more personalized healthcare solutions by allowing ML models to be trained on distributed data sources. This can lead to better diagnosis and treatment of individual patients based on their unique health data.

13.9.2 Improved data privacy

FL can enhance data privacy in healthcare systems by allowing ML models to be trained on data distributed across many sources. This eliminates the need for data to be centrally stored, reducing the risk of data and access breaches.

13.9.3 Real-time monitoring

FL can enable real-time monitoring of patient data by allowing ML models to be trained on data that is continuously generated by wearables and other connected devices. This can lead to more proactive and preventive healthcare interventions.

13.9.4 Efficient resource allocation

FL can enable more efficient resource allocation in healthcare systems by allowing ML models to be trained on data generated by multiple healthcare providers. This can lead to better utilization of healthcare resources and improved healthcare outcomes.

13.9.5 Enhanced research capabilities

FL can enable more efficient and effective healthcare research by allowing ML models to be trained on distributed data sources. This can lead to a faster discovery of new treatments and interventions.

13.9.6 Improved patient engagement

FL can enable better patient engagement by allowing ML models to be trained on patient-generated data, such as health behaviors and preferences. This can lead to more patient-centered healthcare solutions.

13.9.7 Improving disease prediction and diagnosis

FL can enable healthcare providers to collaboratively train predictive models on large datasets without having to share patient data. This can lead to more accurate disease prediction and diagnosis. (Li et al., 2021)

13.9.8 Enhancing clinical decision-making

FL can help healthcare providers make better clinical decisions by providing access to collective knowledge from multiple datasets.

13.9.9 Advancing drug discovery

FL can help increase the discovery of drugs by enabling pharmaceutical companies to collaborate on training predictive models on large datasets while protecting the privacy of their data.

13.9.10 Improving public health

FL can help improve public health by enabling the collaborative analysis of data from multiple sources, such as EHRs, social media, and wearables. This can provide insights into disease outbreaks, treatment effectiveness, and population health trends.

13.10 Limitations of the study

This study provides valuable insights into the potential of FL in emerging digital healthcare systems. It is, however, important to acknowledge the limitations of the study.

First, the scope of this study is limited to government and economic regulations of FL in healthcare. While this is an important and rapidly growing area, the research fails to consider other forms of regulations, such as ethical or legal considerations, including data ownership, consent, and transparency. While these issues are acknowledged in the chapter, they are not fully explored or integrated into the analysis.

Second, the study relies heavily on theoretical or hypothetical scenarios without providing empirical data or real-world case studies to support its arguments. While these scenarios are based on existing research and industry trends, they may not fully capture the complexity and nuances of real-world healthcare systems. The study did not gather the perspectives of healthcare providers or patients, leaving a gap in the understanding of how they view this technology.

Third, the study presents some viewpoints or biases that are not balanced or representative of all stakeholders in the healthcare ecosystem. For example, the study discusses the benefits of FL for healthcare providers and investors, without much consideration of the potential risks or concerns for patients or regulatory bodies.

These limitations suggest that further studies of the topic are still required, especially in the areas of ethical and legal implications, potential impact on healthcare access and affordability, and practical challenges that may arise.

13.11 Conclusion and future works

In this chapter, we have explored some of the key regulatory issues in FL in healthcare and provided recommendations on how to address them. FL is a new approach to ML that enables the training of models on decentralized data sources, such as those found in emerging digital healthcare systems. However, this approach also raises important issues related to government and economic regulations.

The chapter discusses FL in emerging digital healthcare systems as well as security and privacy issues in the healthcare ecosystem. It also discusses market potentials and investment opportunities for FL in emerging digital healthcare systems, as well as commercialization and cost−benefit analysis.

Finally, this study recommends the need for regulatory policies and investment opportunities to encourage the adoption of FL technologies in healthcare systems. The study recommends the use of advanced encryption techniques, such as HE, to protect data privacy while still allowing for collaborative ML. There is also a need for continued research and development in FL, particularly in the context of emerging digital healthcare systems. This includes the development of new FL algorithms and techniques, as well as the integration of FL with other emerging technologies, such as blockchain and edge computing. Also, the importance of interdisciplinary collaboration between healthcare professionals, data scientists, and policymakers cannot be overemphasized for the responsible and ethical implementation of FL.

References

Abadi, M., McMahan, H. B., Chu, A., Mironov, I., Zhang, L., Goodfellow, I., & Talwar, K. (2016). Deep learning with differential privacy. *Proceedings of the ACM Conference on Computer and Communications Security, 2016*, 308−318.

Abdulrahman, S., Tout, H., Ould-Slimane, H., Mourad, A., Talhi, C., & Guizani, M. (2020). A survey on federated learning: The journey from centralized to distributed on-site learning and beyond. *IEEE Internet of Things Journal, 8*(7), 5476−5497.

Ahmad, O. F., Stoyanov, D., & Lovat, L. B. (2020). Barriers and pitfalls for artificial intelligence in gastroenterology: ethical and regulatory issues. *Techniques and Innovations in Gastrointestinal Endoscopy, 22*(2), 80−84.

Alberola-López, C., Sánchez-Monedero, J., & Serrano-Aguilera, J. F. (2021). Federated learning in healthcare: opportunities and challenges. *Journal of Medical Systems, 45*(6), 1−7.

Ali, M., Karimipour, H., & Tariq, M. (2021). Integration of blockchain and federated learning for Internet of Things: Recent advances and future challenges. *Computers & Security, 108*, 102355.

Choi, Y., Jung, H. W., Lee, J. H., & Park, H. J. (2021). Cost-effectiveness analysis of a diabetic retinopathy diagnosis system using a federated learning model. *International Journal of Medical Informatics, 104393*, 148.

Dang, T. K., Lan, X., Weng, J., & Feng, M. (2022). Federated learning for electronic health records. *IEEE Transactions on Network Science and Engineering, 2023*. Available from https://ieeexplore.ieee.org/abstract/document/10132014/.

Dayan, I., Zhong, A., Kaggie, J., & Quraini, A. (2021). Federated learning for predicting clinical outcomes in patients with COVID-19. *Nature Medicine, 27*(10), 1735—1743.

Dinh-Le, C., Chuang, R., Chokshi, S., & Mann, D. (2019). Wearable health technology and electronic health record integration: Scoping review and future directions. *JMIR mHealth and uHealth, 7*(9), e12861.

Du, Y., Song, Y., Xu, Q., & Guo, L. (2020). Federated learning for electroencephalography (EEG) analysis: A feasibility study. *Journal of Biomedical Informatics, 107*, 103479.

Elayan, H., Aloqaily, M., & Guizani, M. (2021). Deep federated learning for IoT-based decentralized healthcare systems. In *2021 international wireless communications and mobile computing (IWCMC)*. Harbin City, China.

Fernandes, F. A., & Chaltikyan, G. V. (2020). Analysis of legal and regulatory frameworks in digital health: a comparison of guidelines and approaches in the European Union and United States. *Journal of the International Society for Telemedicine and eHealth, 8*, e11.

Gadekallu, T. R., Pham, Q.-V., Huynh-The, T., Bhattacharya, S., Maddikunta, P. K., & Liyanage, M. (2021). *Federated learning for big data: A survey on opportunities, applications, and future directions*. arXiv preprint arXiv:2110.04160.

Gaobotse, G., Mbunge, E., Batani, J., & Muchemwa, B. (2022). Non-invasive smart implants in healthcare: Redefining healthcare services delivery through sensors and emerging digital health technologies. *Sensors International, 3*, 100156.

Guo, Y., Li, W., Li, Q., Li, H., Li, Y., Li, Y., & Huang, Y. (2021). A federated learning framework for medical image analysis. *IEEE Journal of Biomedical and Health Informatics, 25*(3), 937—946.

Hathaliya, J. J., & Tanwar, S. (2020). An exhaustive survey on security and privacy issues in Healthcare 4.0. *Computer Communications, 153*, 311—335.

Hatzivasilis, G., Soultatos, O., Ioannidis, S., Verikoukis, C., Demetriou, G., & Tsatsoulis, C. (2019). Review of security and privacy for the Internet of Medical Things (IoMT). *2019 15th international conference on distributed computing in sensor systems (DCOSS)*. Santorini, Greece.

Hesse, B. W., Kwasnicka, D., & Ahern, D. K. (2021). Emerging digital technologies in cancer treatment, prevention, and control. *Translational Behavioral Medicine, 11*(11), 2009—2017.

Intelligence, M. (Hrsg.). (2020). *Digital health market - growth, trends, and forecasts (2020 - 2025)*. https://www.mordorintelligence.com/industry-reports/digital-health-market.

Islam, S., Papastergiou, S., & Mouratidis, H. (2021). Dynamic cyber security situational awareness framework for healthcare ICT infrastructures. *25th Pan-Hellenic Conference on Informatics*.

Lalmuanawma, S., Hussain, J., & Chhakchhuak, L. (2020). Applications of machine learning and artificial intelligence for Covid-19 (SARS-CoV-2) pandemic: A review. *Chaos, Solitons & Fractals, 139*, 110059.

Li, Y., Jiang, Q., Wang, Y., Wu, Z., & Huang, J. (2021). Federated learning in healthcare: A review and research agenda. *IEEE Access, 9*, 121690—121709.

Lim, E. G., Wang, S. J., & Jeong, J. (2022). Federated learning for healthcare informatics. *Journal of Medical Systems, 46*(4), 1—10.

Linardos, A., Kushibar, K., Walsh, S., & Gkontra, P. (2022). Federated learning for multi-center imaging diagnostics: a simulation study in cardiovascular disease. *Scientific Reports, 12*, 673. Available from https://www.nature.com/articles/s41598-022-07186-4.

Liu, H., Hu, Q., Xie, B., & Zhang, X. (2021). Commercialization potential of federated learning in healthcare. *Journal of Medical Systems*, *45*(1), 1−6.

Majeed, T., Rashid, R., Ali, D., & Asaad, A. (2020). Problems of deploying CNN transfer learning to detect COVID-19 from chest X-rays. *medRxiv: The Preprint Server for Health Sciences*, 1−11. Available from https://doi.org/10.1101/2020.05.12.20098954.

Markakis, E., Nikoloudakis, Y., Pallis, E., & Manso, M. (2019). Security assessment as a service cross-layered system for the adoption of digital, personalised and trusted healthcare. *2019 IEEE 5th World Forum on Internet of Things (WF-IoT)*.

Mbunge, E., Muchemwa, B., Jiyane, S., & Batani, J. (2021). Sensors and healthcare 5.0: Transformative shift in virtual care through emerging digital health technologies. *Global Health Journal*, *5*(4), 169−177.

McMahan, H. B., Moore, E., Ramage, D., Hampson, S., & Agüera y Arcas, B. (2017). Communication-Efficient Learning of Deep Networks from Decentralized Data. In: Proceedings of the 20th International Conference on Artificial Intelligence and Statistics (AISTATS), 54. https://arxiv.org/abs/1602.05629.

Mehrjou, A., Soleymani, A., Buchholz, A., Hetzel, J., Kukreja, S. L., & Schölkopf, B. (2022). Federated learning in multi-center critical care research: A systematic case study using the eicu database. arXiv preprint arXiv:2204.09328. https://arxiv.org/abs/2204.09328.

Mondrejevski, L., Armengol, N. L., & others. (2022). Early prediction of the risk of ICU mortality with Deep Federated Learning. In: 2023 IEEE 36th International Symposium on Computer-Based Medical Systems (CBMS). https://ieeexplore.ieee.org/abstract/document/10178863/.

Mothukuri, V., Parizi, R. M., Pouriyeh, S., Huang, Y., Dehghantanha, A., & Srivastava, G. (2021). A survey on security and privacy of federated learning. *Future Generation Computer Systems*, *115*, 619−640. Available from https://www.sciencedirect.com/science/article/pii/S0167739X20329848.

Myles, S., & Leslie, K. (2023). Regulating in the public interest: Lessons learned during the COVID-19 pandemic. *Healthcare Management Forum*, *36*(1), 36−41.

Nguyen, D. C., Ding, M., Pathirana, P. N., Seneviratne, A., Li, J., Niyato, D., & Poor, H. V. (2021). Federated learning for industrial internet of things in future industries. *IEEE Wireless Communications*, *28*(6), 192−199.

O'Dell, L. M., & Jahankhani, H. (2021). *The evolution of AI and the human-machine interface as a manager in Industry 4.0. Strategy, leadership, and AI in the cyber ecosystem* (pp. 3−22). *Academic Press*.

Pesapane, F., Bracchi, D. A., Mulligan, J. F., Linnikov, A., Maslennikov, O., Lanzavecchia, M. B., & Cassano, E. (2021). Legal and regulatory framework for AI solutions in healthcare in EU, US, China, and Russia: new scenarios after a pandemic. *Radiation*, *1*(4), 261−276.

Pfitzner, B., Steckhan, N., & Arnrich, B. (2021). Federated learning in a medical context: A systematic literature review. *ACM Transactions on Internet Technology (TOIT)*, *21*(2), 1−31.

Pham, Q.-V., Fang, F., Ha, V. N., Piran, M. J., Le, M., Le, L. B., Ding, Z., & Le-Ngoc, T. (2021). A survey of multi-access edge computing in 5G and beyond: Fundamentals, technology integration, and state-of-the-art. *IEEE Access*, *8*, 116974−117017. Available from https://ieeexplore.ieee.org/document/9112708.

Pournajaf, S., Kanhere, S. S., Sherratt, J., & Nepal, S. K. (2022). Privacy-preserving federated learning for healthcare applications: Opportunities and challenges. *IEEE Journal of Biomedical and Health Informatics*, *26*(2), 703−712.

Rasool, R. U., Ahmad, H. F., Rafique, W., Qayyum, A., & Qadir, J. (2022). Security and privacy of internet of medical things: A contemporary review in the age of surveillance, botnets, and adversarial ML. *Journal of Network and Computer Applications, 201*, 103332.

Rieke, N., Hancox, J., Li, W., Milletarì, F., Roth, H. R., Albarqouni, S., & Maximilian. (2020). The future of digital health with federated learning. *NPJ Digital Medicine, 3*(1), 119.

Ruotsalainen, P., & Blobel, B. (2020). Health information systems in the digital health ecosystem—problems and solutions for ethics, trust and privacy. *International Journal of Environmental Research and Public Health, 17*(9), 3006.

Sheller, M. J., Edwards, B., Reina, G., Martin, J., & Pati, S. (2020). Federated learning in medicine: facilitating multi-institutional collaborations without sharing patient data. *Scientific Reports, 10*, 1−12.

Singh, S., Rathore, S., Alfarraj, O., Tolba, A., & Yoon, B. (2022). A framework for privacy-preservation of IoT healthcare data using Federated Learning and blockchain technology. *Future Generation Computer Systems, 129*, 380−388.

Smuck, M., Odonkor, C. A., Jonathan, K. W., Schmidt, N., & Michael, A. , S. (2021). The emerging clinical role of wearables: Factors for successful implementation in healthcare. *NPJ Digital Medicine, 4*(1).

Tiwari, A., Madria, S., Kumar, A., Mishra, A. R., Shafiq, M. Z., Gupta, N., & Kim, J. H. (2021). Federated learning in healthcare: Recent advances and future directions. *IEEE Journal of Biomedical and Health Informatics, 25*(3), 838−858.

Yang, Q., Liu, Y., Chen, T., & Tong, Y. (2019). Federated machine learning: Concept and applications. *ACM Transactions on Intelligent Systems and Technology (TIST), 10*(2), 1−19. Available from https://dl.acm.org/doi/abs/10.1145/3298981.

Zheng, Z., Zhou, Y., Sun, Y., Wang, Z., Liu, B., & Li, K. (2022). Applications of federated learning in smart cities: recent advances, taxonomy, and open challenges. *Connection Science, 34*(1), 1−28.

Legal implications of federated learning integration in digital healthcare systems

14

Agbotiname Lucky Imoize[1], Mohammad S. Obaidat[2,3,4,5] and Houbing Herbert Song[6]

[1]*Department of Electrical and Electronics Engineering, Faculty of Engineering, University of Lagos, Lagos, Nigeria*

[2]*The King Abdullah II School of Information Technology, The University of Jordan, Amman, Jordan*

[3]*School of Computer and Communication Engineering, University of Science and Technology Beijing, Beijing, P.R. China*

[4]*Department of Computational Intelligence, School of Computing, SRM University, SRM Nagar, Kattankulathur, Tamil Nadu, India*

[5]*School of Engineering, The Amity University, Noida, Uttar Pradesh, India*

[6]*Department of Information Systems, University of Maryland Baltimore County (UMBC), Baltimore, MD, United States*

14.1 Introduction

The digital healthcare system is fast proliferating in the global space, and research on federated learning (FL) integration in healthcare systems has recently garnered considerable attention from academia and industry (Ali et al., 2022; Gu et al., 2023; Rahman et al., 2023). Digital healthcare systems facilitate patient convenience, reduce the cost of medical care services, and improve ease of access (Ayoade et al., 2022). However, as the healthcare system becomes increasingly complex, accommodating several participants, including patients and medical professionals, a noticeable surge in security and privacy concerns is observed. These concerns could be data theft, breach, compromise, unauthorized access, or unlawful interception.

Globally, billions of medical devices are deployed in various healthcare systems to improve the medical conditions of patients suffering from acute to critical health challenges (Li et al., 2021; Rieke et al., 2020). The interconnection of these devices, aiding the transmission of valuable patient data over open communication channels, poses security and privacy concerns (Imoize et al., 2021; Wang et al., 2023). As more devices are connected to meet the requirements of the growing number of patients who need urgent medical attention, the likelihood of

Federated Learning for Digital Healthcare Systems. DOI: https://doi.org/10.1016/B978-0-443-13897-3.00014-X

security breaches during device-to-device communication via insecure channels increases, and the need to secure sensitive user data over these channels becomes increasingly important.

Data breaches, unlawful interception, privacy violations, and security risks in digital healthcare systems are growing at an alarming rate, and the need to address these security risks and vulnerabilities in digital healthcare systems cannot be overemphasized. Addressing these security concerns requires stringent regulatory frameworks such as the Health Insurance Portability and Accountability Act (HIPAA) adopted in the United States (Atchinson & Fox, 1997; Ostin, 2001; Sonosky & Giordano, 2006) and the General Data Protection Regulation (GDPR) (Voigt & Von dem Bushehr, 2017) projected by the European Union. These regulatory frameworks are further discussed in this chapter.

Healthcare service providers and other stakeholders in the healthcare ecosystem need to abide by the governing regulations to achieve success in the fight against cyber threats and emerging security vulnerabilities in healthcare systems (Imoize et al., 2023). In addition, the idea of striking a balance between data accessibility and digital innovation while adhering to the rules on privacy preservation needs to be discussed thoroughly by all stakeholders in the healthcare ecosystem. Stakeholders in the healthcare sector should prioritize healthcare systems that protect users' privacy, rights, confidentiality, and trust. This involves preventing data breaches and unauthorized access to sensitive user information, leveraging the potential of advanced machine learning (ML) and artificial intelligence (AI) tools (Awotunde et al., 2022; Rani et al., 2023).

In recent times, FL is fast changing the dynamics of modern healthcare systems. The integration of FL in healthcare systems has witnessed the rapid development of novel techniques for the treatment and management of critical health challenges. Interestingly, FL models facilitate unprecedented collaborations among organizations, leading to rapid contributions and training with their local datasets (Antunes et al., 2022). In a typical FL setting, several collaborators participate in resource sharing and management (Rahman et al., 2023). In this case, the raw data from each participant does not necessarily leave the devices of the hosts. Instead, useful model parameters of local ML models are shared with a coordinating server, which aggregates these parameters efficiently and sends the output to the various participants to update their own models.

FL technology finds useful applications in complex data ecosystems, enhancing efficient data management, and addressing critical data sovereignty issues (Boscarino et al., 2022). In recent times, the rapid advancement in wireless communication systems has orchestrated the proliferation of dense massive devices that transmit critical information via open channels. However, the security of these channels is critical to data sovereignty and privacy, which is a significant political and social issue for all stakeholders, including government agencies, institutions, and citizens in the healthcare ecosystem (Farahani & Monsefi, 2023).

Data sovereignty requires that data collected within a country and its usage follow strict regulations and government policies (Boscarino et al., 2022;

McCartney et al., 2022). More importantly, sensitive user data must be adequately protected from malicious attacks and unlawful exploitation using cutting-edge FL tools, blockchain technology, quantum computing, and other emerging technologies. Thus, the government needs to be more proactive and explicit about the governing laws regulating the use of FL in sensitive data processing. There is no denying the fact that standard regulatory laws for healthcare data handling and protection are imperative for promoting trust, preserving patients' privacy rights, ensuring seamless healthcare data processing, and facilitating the delivery of effective healthcare services. The core requirements call for examining the integration of FL in modern healthcare systems.

Currently, FL provides robust platforms to help strengthen the governing regulations for improved security, privacy, transparency, and accountability (Treleaven, Smietanka, & Pithadia, 2022). Therefore leveraging the enormous potential of FL would facilitate the safety, security, and privacy of sensitive user information and encourage compliance with all legal requirements by all parties in the healthcare ecosystem. However, most existing works overlook examining the legal implications of integrating FL into digital healthcare systems. Additionally, addressing the escalating security and privacy concerns in modern healthcare systems comprehensively has not been given adequate treatment in the literature. To address this gap, this chapter presents the legal implications of FL with the aim of examining its integration into emerging digital healthcare systems.

14.1.1 Key contributions of the chapter

The chapter includes the following noteworthy contributions:

- Presents FL integration in healthcare systems and highlights the legal guidelines for guaranteeing healthcare data security and privacy in an FL setting.
- Explores the legal considerations for healthcare data protection and preservation in an FL setting.
- Examines the challenges and risks associated with healthcare data protection and preservation in an FL setting.
- Examines the legal and ethical perspectives on integrating FL in healthcare systems.
- Presents the prevailing issues in existing FL models and their application in healthcare systems.
- Explores the legal and regulatory procedures for FL integration in healthcare across selected jurisdictions, with a specific focus on the European Union, China, Nigeria, Russia, Singapore, the United Kingdom, and the United States.
- Highlights the need for legal frameworks for integrating FL into emerging healthcare systems.

- Presents key requirements for creating effective legal frameworks for integrating FL into digital healthcare systems.
- Discusses the legal implications of noncompliance with the regulatory requirements by participants and stakeholders in the FL healthcare ecosystem.
- Provides key recommendations for developing effective legal frameworks for integrating FL into digital healthcare systems.

14.1.2 Chapter organization

The remaining part of this chapter is organized as follows. Section 14.2 presents FL integration in healthcare systems. Section 14.3 examines the legal guidelines for guaranteeing healthcare data security and privacy in an FL setting. Section 14.4 focuses on the legal considerations for healthcare data protection in FL. It examines the challenges and risks associated with healthcare data protection within FL frameworks, while also addressing the legal and ethical perspectives of integrating FL into healthcare systems. The section also explores the legal issues inherent in FL model application in healthcare systems. Section 14.5 explores the legal and regulatory procedures for FL integration in healthcare in some selected jurisdictions, with a specific focus on the European Union, China, Nigeria, Russia, Singapore, the United Kingdom, and the United States. Section 14.6 examines the need for legal frameworks for FL integration in modern healthcare systems. Also, the requirements for creating effective legal frameworks for integrating FL into healthcare systems are highlighted. Additionally, the section discusses the legal implications of noncompliance with the regulatory requirements by participants and stakeholders in the FL ecosystem. The section provides recommendations for developing effective legal frameworks for FL integration in emerging healthcare systems. Finally, Section 14.7 concludes the chapter and provides the future scope.

14.2 Federated learning integration in healthcare systems

FL offers a viable solution to the inherent issues such as security, privacy, ownership, and strict regulation challenges in deep learning (DL) (Li et al., 2020; Zhang et al., 2021). DL models trained with single institutional data are highly susceptible to institutional data bias, showing very high accuracy. However, the models show limited performance when applied to data from another institution or from different departments in the same institution. As shown in the first part of Fig. 14.1, the model was trained using healthcare data from a single institution. However, it is not feasible to train DL models in a centralized data lake while also complying with strict government regulations and ensuring that security and privacy are maintained. The latter part of Fig. 14.1 provides a shared global DL model, leveraging a central aggregator server. In this case, the local parties can

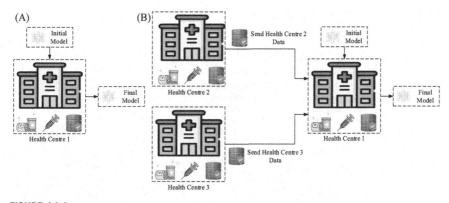

FIGURE 14.1

A typical description of institutional participants and collaborative learning. Part (A) describes institutional learning with one participant, and part (B) shows several participants in a collaborative learning scenario.

retain patient data, guaranteeing data anonymity and security (Prayitno et al., 2021).

In recent times, FL has been integrated into modern healthcare systems. The basic architecture for FL with healthcare data is described in Fig. 14.2. The architecture enables access to patient medical datasets for healthcare institutions for use in distributed data analysis while maintaining the privacy and confidentiality of patients. The institutions categorized as data owners keep medical records of various forms and features (Antunes et al., 2022). The data owners communicate via a secure communication infrastructure. The architecture also connects all data holders with the FL manager. These datasets are highly confidential, comprising sensitive patient information protected by law and the associated regulations. Permission from the participating patients' needs to be approved before the datasets are used lawfully. However, acquiring such consent from the parties involved poses an orthogonal issue to the design of FL architecture in modern healthcare systems. Thus the need to examine the orthogonal issues affecting the design of FL architecture for integration in modern healthcare systems cannot be overemphasized.

The information contained in the local datasets is sensitive in nature and cannot be shared with third parties. In this case, each data owner uses their private datasets to train a private model, comprising parameters obtained from other data owners provided by the FL manager. Since the security of information sharing among the data owners is critical, the data confidentiality module is employed to protect outgoing information. The dataset processing module employs special techniques to allow external learning algorithms to gain access to the local datasets and interpret the results. In addition, the model aggregation combines the

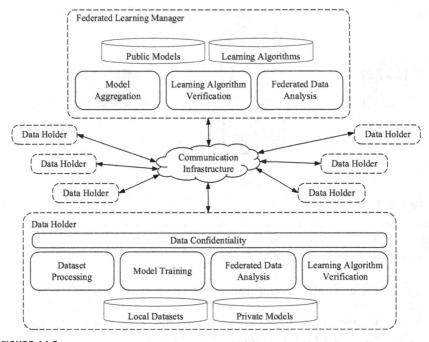

FIGURE 14.2

A basic architecture for federated learning with patient data.

private entities into a global learning model for all participants, thereby unleashing the benefits of FL integration in healthcare systems (Antunes et al., 2022).

14.3 Legal guidelines for healthcare data security and privacy in federated learning setting

In general, there are variations in legal definitions and corresponding regulatory frameworks for healthcare systems, and this sometimes depends on the jurisdiction examined. In the United States, for example, clinically focused applications such as clinical decision support diagnostic tools (Beeler, Bates, & Hug, 2014; Pawloski et al., 2019; Carayon et al., 2020) entail additional requirements provided by the Food and Drug Administration (FDA) (Bosch & Lee, 1994).

Currently, the United States is still in the process of evaluating the necessary federal legislation to govern the use of personally identifiable information in relation to FL technology (Ni, 2017; Schwartz & Solove, 2011). However, some private organizations are taking the initiative toward effective and efficient guidelines for the application of ML and FL in digital healthcare systems. ML and FL technologies are growing rapidly and will keep evolving. However, rapid changes in FL development add to the difficulty of establishing effective legislation and regulations governing its applications in emerging digital healthcare systems.

The HIPAA enacted in 1996 and the Health Information Technology for Economic and Clinical Health Act introduced in 2009 by the US government have facilitated rapid progress in healthcare service delivery (Atchinson & Fox, 1997; Gostin, 2001; Nosowsky & Giordano, 2006). Similar reports governing healthcare infrastructure management have been provided by the Center for Open Data Enterprise (Boulton et al., 2011) and the Department of Health and Human Services in the United States (McGuire, 2011). The aforementioned reports contain recommendations for handling healthcare data, leveraging emerging ML technologies. However, the HIPAA and other reports require frequent updating to provide real-time guidelines, especially as FL models are advancing alarmingly.

The application of FL in emerging digital healthcare systems has proven to be advantageous globally. The legal frameworks for adopting FL in emerging digital healthcare systems maintain different jurisdictions, following distinct regulatory policies that vary from one country to another. However, it is interesting to note that there exist a few broad concepts and guidelines for adopting FL and guaranteeing data security in emerging digital healthcare systems. These guidelines are briefly described in the following sections.

14.3.1 General data protection regulations

Several laws governing data protection exist in different countries. For example, the California Consumer Privacy Act (CCPA) (Bukaty, 2019; Mulgund et al., 2021) is in place in the state of California in the United States, while the GDPR is enforced in the European Union (Voigt & Von dem Bussche, 2017). These laws provide relevant information and standards regarding the collection, storage, and handling of personal data related to healthcare information (Schwartz & Solove, 2014). In practice, these rules and regulations should be strictly adhered to while managing patient data in emerging digital healthcare systems.

The GDPR legislates that personal data can only be collected legally, following stringent regulatory requirements, for lawful usage. Another aspect of the GDPR that is worth mentioning is the protection of data ownership (Truong et al., 2021). In this case, the GDPR obligates data controllers to grant fundamental rights to data subjects, including control over their data. The legislation provides useful information governing the processing of personal data of residents in the European Union/United Kingdom. Interestingly, the regulation clarifies several questions on the best practices for handling personal data in the European Union. Specifically, the legislation aims to safeguard personal information from misuse and ensures data privacy. The legislation has global relevance and all companies or institutions processing the personal data of EU residents. The legislation grants rights to data owners to control their data. It also spells out the possible penalties for major breaches. The legislations are briefly summarized in Fig. 14.3.

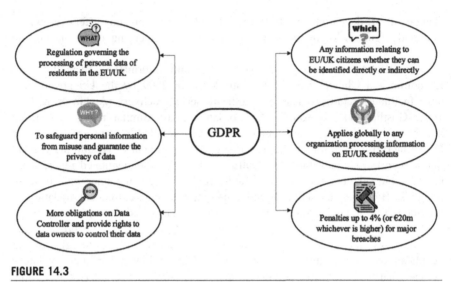

FIGURE 14.3

A brief description of the GDPR legislation. *GDPR*, General Data Protection Regulation.

14.3.2 Informed consent

A key requirement in healthcare data handling and management is informed consent. It is imperative to seek the consent of patients before obtaining their medical information. It is unethical to gather, retain, or share sensitive patient data without the consent of the participants. An acceptable consent requires that the patients are fully aware of the scope, goal, and potential hazards associated with the processing of their health data for use in digital healthcare systems. Another point worth mentioning is that the patients should be given the liberty or freedom to revoke any signed consent when they decide to change their agreement. Last, the procedure for providing consent should be free, fair, and transparent.

14.3.3 De-identification and anonymization

In modern digital healthcare systems, the potential of FL, cryptography, and blockchain technologies can be productively harnessed to guarantee anonymity or to de-identify critical patient information to guarantee privacy preservation (Imoize et al., 2022). On the one hand, anonymization involves eliminating all identifying patient information from the data before processing. On the other hand, de-identification involves replacing all identifying patient information in the data using pseudonyms. Interestingly, these techniques find useful applications in minimizing the risk of re-identifying patient information and ensuring confidentiality and trust.

14.3.4 **Ethical considerations**

The application of FL models in digital healthcare systems poses an ethical dilemma. What if the FL models are compromised and used unlawfully? What happens when the stipulated standards are not followed appropriately? To avoid biases and ensure fairness, accountability, and openness in ethical procedures for handling critical user data, FL models integrated into healthcare systems must adhere to standard design criteria and comply with relevant laws and regulations (Imoize et al., 2020).

14.3.5 **Data retention and destruction**

FL models used in healthcare systems should be designed to follow explicit guidelines and regulations for data retention and deletion. There should be a time lapse for data retention and destruction, and the timeline should be followed strictly when processing or handling patient data. Medical data should be deleted from the healthcare system when the allotted time has lapsed. In addition, it is important that patients are well informed about their data storage and removal procedures at all times throughout the data handling cycle.

14.3.6 **Proactive security measures**

Proactive and provably secure measures are critical to safeguarding user data in digital healthcare systems. Appropriate encryption, audit trails, authentication, and privacy preservation mechanisms need to be put in place during healthcare data transmission, storage, and access. It is worth mentioning that only authorized users of the platform should access the data using multiple authentication systems, and unauthorized users should be denied access at all times. Last, it is imperative to conduct regular security audits and vulnerability assessments of the system to identify intruders or eavesdroppers and prevent any unlawful interception or exploitation of healthcare information.

14.3.7 **Compliance with healthcare regulations**

The design and implementation of digital healthcare systems should comply with the available national and international healthcare laws. A few examples of the existing healthcare laws are HIPAA in the United States (Atchinson & Fox, 1997; Gostin, 2001; Nosowsky & Giordano, 2006), and the Personal Data Protection Act (PDPA) in Singapore (Chik & Pang, 2014; Chik, 2013). These laws regulate healthcare activities in these countries. In addition, specifics and standards for the handling of patient data are highlighted clearly in these regulations (Dwyer et al., 2004). The specifics comprise security precautions, breach reporting procedures, and patient rights preservation mechanisms.

14.3.8 Data ownership and control

Data ownership and control are still an issue in conventional healthcare systems. The design of modern healthcare systems should specify data ownership and control procedures with the utmost clarity. Patients should be able to know whether they are allowed to view, revise, or even request to remove their personal health information in accordance with the prevailing laws and regulations. The FL-based healthcare systems should state the terms and conditions under which, with whom, and how data is shared and transferred to the participants in the healthcare system.

14.3.9 Data minimization and purpose limitation

The purpose for which healthcare data is harvested needs to be followed and maintained. Only the essential information should be extracted for use at any time. To enhance efficiency and accuracy, it is essential to discard nonessential information, which helps in data minimization. In other words, only the required data should be retained, and others should be discarded appropriately. The implementation of data minimization and purpose limitation principles reduces the risk of unauthorized access, data compromise, and unlawful interception.

14.3.10 Transparent governance and accountability

FL-empowered healthcare systems should be designed and operated in open governance platforms for transparency and accountability. All stakeholders in the healthcare ecosystem should ensure responsible conduct in handling and processing patient information. The handling procedures should be well outlined without any ambiguity. Transparent protocols and processes should be put in place to handle potential breaches or unlawful access to sensitive patient data. In the same vein, healthcare system providers should act in a responsible manner at all times to enforce regulatory compliance and openness in their data protection procedures.

14.3.11 International data transfer

In healthcare data handling and management, there is a need for regulatory bodies to enforce strict compliance with international data transfer laws, especially when transferring sensitive healthcare data across international boundaries. The applicable data protection laws, such as the Standard Contractual Clauses (SCCs) or Privacy Shield (Kaur et al., 2022; Popowicz-Pazdej, 2021), need to be put in place to guarantee the security and privacy of the transferred data.

14.3.12 Continuous monitoring and auditing

Regular monitoring and auditing of healthcare systems are key to identifying any potential vulnerabilities, breaches, or noncompliance with regulatory data protection schemes. FL provides a seamless platform for real-time monitoring and auditing of healthcare systems, leading to prompt rectification of any abnormalities and enabling continuing adherence to the standard privacy and security requirements.

14.4 Legal considerations for healthcare data protection in federated learning setting

In modern healthcare systems, data protection is a critical requirement. To promote effective data protection mechanisms, several legal considerations need to be explored. The following legal considerations describe healthcare data protection briefly. These comprise legal and ethical obligations, patient confidentiality, trust and adoption, and data integrity and accuracy.

14.4.1 Legal and ethical obligations

According to the prevailing laws and regulations, all stakeholders in the healthcare system are obligated to protect sensitive patient information in a robust manner. When there is a security breach in healthcare systems, patients could be in critical danger, and medical records could be exploited unlawfully, leading to data compromise, and loss of public trust. The existing laws stipulate that adequate data protection procedures and measures should be put in place to avoid jeopardizing the health conditions of patients, medical equipment, and other valuable resources.

14.4.2 Patient confidentiality

In digital healthcare systems, medical examination records are often shared among various participants. Most often, physicians share medical test results with patients. Medical personnel need to maintain confidentiality, which is a critical requirement for upholding ethical standards in emerging digital healthcare systems.

14.4.3 Trust and adoption

Patient trust in a digital healthcare setting is not negotiable. Effective and efficient data protection measures comprising stringent security regulations and privacy protection should be put in place to boost patient trust in the systems. Once the patients have trust and their privacy is guaranteed, they will be encouraged to embrace the use of digital healthcare systems, and even persuade others to do the same.

14.4.4 Data integrity and accuracy

Data integrity and accuracy are critical to making informed clinical decisions based on patients' medical results. Inaccurate medical examination data increases the of doctors prescribing the wrong medications, which can pose severe risks to the health of the patient, including the possibility of untimely death. The circumstances and manner in which medical tests are conducted are also very important. Laboratory technicians need to report their observations as accurately as possible without relying on guess figures or estimated outcomes.

14.4.5 Challenges and risks associated with data protection

Healthcare data protection involves putting appropriate processes and policies in place to ensure the confidentiality, security, trust, and integrity of patient data in healthcare service delivery. Primarily, healthcare data is sensitive data that needs to be protected at all costs. The security of healthcare data is not negotiable, and securing healthcare data is critical to ensuring patient confidence, enforcing strict adherence to legal regulations, and preventing data breaches and unlawful access. Several reasons abound why the protection of patient data is key in healthcare systems. Healthcare data protection enhances patients' privacy, maintains confidentiality, and helps to build trust.

Patients are often concerned about the safety of their health records. When patients' sample are taken for diagnostics and analysis, they need to be assured that their sensitive health records are safe and maintained strictly confidential. This trust is critical to boosting the doctor—patient relationship and promoting an environment for patients to relate freely with their doctors. In addition, healthcare service providers must adhere strictly to the governing legal regulations. For example, in the United States, HIPAA (Atchinson & Fox, 1997; Gostin, 2001; Nosowsky & Giordano, 2006) outlines the requirements for security and privacy in healthcare systems and highlights how patient health records should be handled with the highest confidentiality.

The benefits of safeguarding patient records in healthcare systems cannot be overemphasized (Iguoba et al., 2022; Nwaneri et al., 2022). Apart from enabling trust and protecting the rights of patients, the possibility of unauthorized access to confidential patient information is reduced significantly. Thus, the likelihood of data breaches and compromises is minimized. A brief summary of the challenges and risks associated with data protection in digital healthcare systems is described below.

14.4.5.1 Data interoperability

When medical records are shared on various platforms for use by participating medical personnel, the risk of data interoperability can occur. Depending on the data exchange platform, the healthcare information may need conversion from one format to another. In practice, data format inconsistencies could potentially increase the risk of data loss, interception, or corruption, making it quite difficult to share healthcare data securely.

14.4.5.2 Data breaches

A data breach is one of the most common security threats to healthcare systems. Data breaches can potentially expose healthcare systems to unauthorized access or the unauthorized disclosure of patient records. As a result, this may involve healthcare data being manipulated, putting the lives of patients in critical danger.

14.4.5.3 Cybersecurity risks

Cybersecurity threats are fast proliferating in the healthcare system in unprecedented dimensions. Every minute, ransomware, phishing, malware, or distributed denial-of-service attacks (Dong et al., 2019; Osanaiye et al., 2016) occur in digital healthcare systems, posing an unimaginable threat to patient security, confidentiality, and privacy.

14.4.6 Lack of robust authentication and authorization mechanisms

The authentication and authorization schemes deployed in digital healthcare systems should be robust and active in real-time (Meshram et al., 2022; Meshram et al., 2023). Adversaries are most likely to penetrate systems with limited authentication and authorization. As a consequence, healthcare systems should be empowered with multifactor authentication and authorization mechanisms to avoid unauthorized access and data breaches.

14.4.6.1 Legal and regulatory compliance

Data privacy laws regulating healthcare service delivery, such as the GDPR (Goddard, 2017; Truong et al., 2021; Voigt & Von dem Bussche, 2017) and HIPAA, should be strictly adhered to by all parties in the healthcare platform. Medical equipment design, deployment, implementation, and use must comply with the existing regulatory framework. It is interesting to note that these regulations should be carefully drafted, considering patient consent, data storage, and patient privacy rights.

14.4.6.2 Security of digital healthcare devices

Due to the recent advancement in wireless communication systems, billions of mobile devices are rapidly proliferating in the medical space. These devices comprise smartphones, tablets, and wearable technology-based devices (Arikumar et al., 2022). Standard and legal medical practice entails that all devices used in digital healthcare are provably secure against any potential adversary. As these devices are more susceptible to theft, loss, or unauthorized access, exposing sensitive medical records, the need to secure them becomes increasingly important.

14.4.6.3 Human errors and training

In medical practice, the training of all medical personnel is a key requirement. The training of healthcare workers and their associated support staff helps to facilitate the efficient management of medical records and other healthcare information. However, no matter the level of training acquired, human errors do occur in medical results handling and management. These errors may be unintentional healthcare data exposure to an adversary or incorrect data handling, which could potentially jeopardize the health condition of the patient (Rufai et al., 2022). However, it should be emphasized clearly that medical personnel with limited medical training or inadequate awareness of data protection procedures are most likely to present healthcare data with errors.

14.4.7 Legal and ethical perspectives on the integration of federated learning

Ethical conduct is a key requirement and an integral part of healthcare systems. The concept of ethics in healthcare systems has been emphasized by Bujalkova (2001), citing the pioneering work of Hippocrates on medical ethics. To fully understand the workings of FL in healthcare systems and the associated challenges, ethics should be prioritized to ensure that patient's rights are respected according to the governing laws. The work by Markose et al. (2016) focuses on having a consistent and proactive discussion on the ethical foundation of healthcare systems that provides a culture of respect, prioritizing the patient and their fundamental human rights.

To promote ethics in medical practice, the American College of Radiology and other international radiology organizations, presented a unified working document, reflecting the importance of developing ethical standards for ML systems, including FL (Bautista et al., 2009; Geis et al., 2019). Recently, thousands of toolkits have been developed by the American College of Radiology's AI-LAB aimed at promoting a unified framework to develop ML algorithms that focus on patient populations. Additionally, these algorithms could be extended to accommodate other emerging requirements (Allen et al., 2019). The United Nations recently discussed the intersection of AI, ethics, and health, pushing a "Global Dialogue" for various concerns regarding the application of ML models in healthcare systems and possible solutions (Azoulay, 2019). According to a recent guideline released by the World Health Organization, several industry experts, academics, and public sector officials are keenly advocating the careful design of ML and FL models to enhance the protection of human rights, autonomy, equity, transparency, and greener environments to meet the UN sustainable development goals (Chotchoungchatchai et al., 2020; Lu et al., 2015).

14.4.8 Legal issues inherent in federated learning model applications

There are several issues observed in current FL models, and more will be identified as the technology advances in its applications. Some of the prominent issues include a lack of transparency regarding algorithms, instances of algorithmic discrimination, absence of contestability for nonoptimal results, legal uncertainties surrounding FL in healthcare systems, and insufficient mechanisms for accountability in case of damage. These points are elabortaed as follows.

14.4.8.1 Lack of transparency for algorithms

The problem of transparency still exists in FL algorithms. The models created to solve a particular healthcare issue may not work efficiently when applied to address another ailment other than the one for which they were developed. Most times, the participants are quite aware of the implications but could prioritize their interests over the well-being of patients and other parties involved. This occurrence is prevalent and fast gaining entrance into the healthcare system. For optimal performance, the design and development of modern FL-enabled systems should prioritize transparency, ensuring clarity and ease of comprehension without ambiguity.

14.4.8.2 Algorithmic discrimination

The likelihood of potential bias in FL models owing to the disproportionate representation of minority groups is an endemic problem that needs to be addressed promptly (Köchling & Wehner, 2020). For instance, in 2015, there was a sharp criticism of Google's facial recognition algorithm that identified black people as apes. To address this issue, Google modified the algorithm and prevented it from classifying gorillas (Mulshine, 2015). Even with the recent technological advancement, there is no denying the fact that FL models are still prone to errors, giving the wrong output in various areas of application, including healthcare systems. For example, ML-based melanoma detection algorithms are primarily trained on light-skinned people. However, black people, who are less likely to develop melanoma, are more likely to be negatively impacted by the deadly disease (Noor, 2020; Nwaneri et al., 2022).

14.4.8.3 Lack of contestability for nonoptimal results

Results from medical tests should undergo contestability tests to ensure their accuracy and optimize trust among participants. When patients' results are not tested adequately, especially to know whether they are optimal or not, trust issues become inevitable.

14.4.8.4 Questions on legality of federated learning in healthcare systems

In recent times, several questions have been raised regarding the legality of FL in healthcare systems. Some of the most frequent questions revolve around personhood, whether FL could be attributed with human-like qualities. The laws regulating the application of FL in healthcare systems still need to be communicated clearly to all parties involved. The stakeholders should be able to share vital information with the users regarding the application and legality of FL in healthcare systems.

14.4.8.5 Lack of accountability for damage

The key questions are: What are the repercussions when patients' medical records are compromised due to faults in the FL algorithm? Who should be held responsible? What compensations are due to the victim? Issues of liability for damages are an integral component of the existing laws and regulations. However, these issues are still being continuously contested. In many jurisdictions, particularly for ML-based algorithms, it is often challenging to directly attribute damage to the algorithm itself, since it is not a human being. Consequently, claimants may find themselves uncertain about their course of action and who should be held liable: The person or parties behind the mask? The issue becomes even more complicated when the claimant wants to sue for damages without knowing the liable party. It becomes even more complex in FL systems due to several participating institutions. If the FL model shows some form of protected health information (PHI), leveraging reverse engineering, establishing the culpable party from the pool of multiple institutions becomes very difficult (Moore & Frye, 2019). To address this problem, Hallevy (2015) recommends holding AI liable just like corporations and proposes different punishments for AI-based models. However, many questions are still begging for answers when it comes to the issue of liability in FL-supported healthcare systems.

14.5 Legal and regulatory procedures in selected jurisdictions

In general, the legal and regulatory frameworks around FL differ from one country to another, and this fact is worth noting when designing FL-based models for healthcare systems. Reliable information about the precise legal and regulatory framework in a given jurisdiction can be obtained from official government sources and legal professionals.

The design, development, and deployment of digital healthcare systems within a specific jurisdiction are governed by prevailing laws and policies, collectively known as the legal and regulatory framework of that jurisdiction. The legal and regulatory framework adopted in healthcare systems outlines the rights, duties, and responsibilities of all stakeholders in the healthcare ecosystem. Though different jurisdictions may have distinct regulatory policies, some generalized rules and guidelines are put in place to guarantee data privacy in FL-empowered healthcare systems. The legal and regulatory procedures adopted in some selected jurisdictions are outlined as follows.

14.5.1 People's Republic of China

The Chinese government is at the forefront of AI innovation. China has presented several laws and regulations to ensure the ethical and transparent use of AI and

FL, especially in healthcare systems. Notable among these laws is the People's Republic of China's cybersecurity law for AI (Pyo, 2021). The law addresses technical issues emanating from the unethical use of cybersecurity and the protection of user information. Another law has recently been put in place for the responsible use of blockchain technology. The law provides a legal framework for the development, deployment, and regulation of blockchain technology in China. In particular, the law on promoting the development of the blockchain industry emphasizes the protection of the rights and interests of all users of the technology. The law also spells out procedures for avoiding fraud and unlawful exploitation of sensitive user information. In addition, the Chinese government has drafted China's National New Generation Artificial Intelligence Development Plan (Wu et al., 2020). The plan is aimed at promoting the strategic development objectives for AI and associated technologies. The plan covers topics comprising technical research, industrial applications, talent harvesting, responsible use of AI, and ethical considerations (Kumar et al., 2021).

14.5.2 European Union

The European Union has introduced the GDPR and emerging AI-specific regulations. These are described briefly as follows.

14.5.2.1 General data protection regulation

The European Union has projected a detailed data protection law referred to as the GDPR (Goddard, 2017; Truong et al., 2021; Voigt & Von dem Bussche, 2017). This regulation specifies the handling and safeguarding of personal data. The regulation provides standards, legal protections, and responsibilities for all stakeholders using AI and FL algorithms.

The GDPR establishes six core principles as rational guidelines for service providers to manage personal data (Truong et al., 2021). The principles provide critical information on how to process personal data in a responsible manner. Notably, the security, privacy, legitimacy, retention, updating, relevancy, transparency, and accuracy of user data are emphasized. These core principles are highlighted in Fig. 14.4.

14.5.2.2 Emerging artificial intelligence-specific regulations

Another important law provided by the European Union is the Artificial Intelligence Act to control the design, implementation, deployment, and use of AI systems in the European Union, with a specific focus on high-risk AI applications.

14.5.3 Federal Republic of Nigeria

The legal system in Nigeria is evolving rapidly from a textbook-based system to AI practice (Ilegieuno, Chukwuani, & Adaralegbe, 2021). Recently, in Lagos and

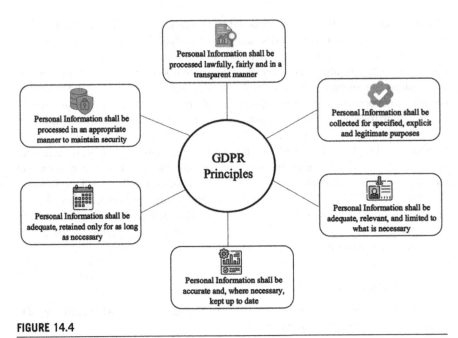

FIGURE 14.4

Core principles of GDPR. *GDPR*, General Data Protection Regulation.

Rivers States, electronic filing of court cases has been adopted. In particular, corporate law practice is currently embracing digitization, leaning toward standards obtainable in developed jurisdictions. In Nigeria, several factors, such as inadequate data protection laws, have slowed the usage of AI and blockchain technology in legal practice (Ilegieuno, Chukwuani, & Eigbobo, 2021). Nigeria currently does not have a detailed AI regulatory framework. However, in recent times, several initiatives have been put in place to promote AI adoption and address the potential risks associated with the use of AI and other related emerging technologies, such as FL.

In accordance with the National Information Technology Development Agency (NITDA) Act 2007 (Lateef, Taiwo, & Adeyoju, 2022), the contributions of the relevant stakeholders are being considered to facilitate the development of the National Artificial Intelligence Policy (NAIP). The act compels the agency to create a framework necessary for the planning, research, development, standardization, application, coordination, monitoring, evaluation, and regulation of information technology practices, activities, and systems in Nigeria. The act also stipulates that in all matters related to the use and application of AI technologies, NITDA should come up with an NAIP, following the mandate from the Federal Ministry of Communications and Digital Economy. The goal is to maximize the benefits, mitigate possible risks, and address some of the complexities attributed to using AI in daily activities, including healthcare systems.

More recently, Nigeria has embarked on an AI project, which is an extension of the ongoing project under NITDA. The Cocreating a National Artificial Intelligence Strategy for Nigeria project aims to engage and include top AI researchers of Nigerian descent globally. The top AI researchers are expected to support the ongoing project of crafting a national AI strategy (Tijani, 2023).

14.5.4 Russian Federation

The Law on Personal Data regulates the handling and protection of personal data, including FL systems in Russia. Another law worth mentioning is the Law on Digital Financial Assets developed by the Russian Federation. The law establishes rules for digital assets and provides a legal framework for adopting AI-based technologies in financial transactions (Steppe, 2017).

14.5.5 Republic of Singapore

The PDPA (Chik, 2013; Wong YongQuan, 2017) in Singapore regulates the harvesting, use, and disclosure of personal information, including information processed by FL systems. Additionally, rules are provided by the Monetary Authority of Singapore for the application of AI and blockchain technology in financial services. The rules capture the management of technological risks and client safety.

14.5.6 United Kingdom

The Digital Economy Act 2017 (Woods, 2017) provided by the UK government regulates several aspects of the digital economy, including electronic communications, measures for cybersecurity, and personal data protection. The act presents a regulatory framework for the application of artificial intelligence and FL systems and models in vertical industries.

14.5.7 United States of America

Currently, there is no established federal law that fully regulates FL systems in the United States. However, there are several other laws and rules that cover AI-related topics, described as follows.

14.5.7.1 Health Insurance Portability and Accountability Act

HIPAA is a federal law established to enhance the security and privacy of patients and their health records (Cohen & Mello, 2018; Dwyer et al., 2004). The act stipulates that for obtaining and storing important healthcare data, AI systems should adhere strictly to the established HIPAA regulations. Specifically, the privacy and security of PHI are regulated by HIPAA. The act provides guidelines to be followed by healthcare providers and other stakeholders in the healthcare eco-system (Atchinson & Fox, 1997; Gostin, 2001; Nosowsky & Giordano, 2006).

14.5.7.2 Food and Drug Administration

The FDA is saddled with the regulation of drugs and medical devices, which includes FL-supported healthcare devices. All FL and AI-assisted systems used for diagnosis and the treatment of patients are approved by the FDA prior to their deployment (Bosch & Lee, 1994).

14.5.7.3 Federal Trade Commission

The Federal Trade Commission (FTC) enforces consumer protection laws (Kovacic & Winerman, 2014). The FTC covers the application of artificial intelligence and associated technologies to protect personal information. The law prohibits unfair or misleading acts that violate consumer protection. Recently, state laws such as the Uniform Electronic Transactions Act and the Uniform Regulation of Virtual-Currency Businesses Act from the Uniform Law Commission have been reported (Boss, 2000; Meehan, 1999; Suwadi et al., 2023).

14.5.7.4 California Consumer Privacy Act

The California Consumer Privacy Act (CCPA) is a state law that regulates the protection of consumer rights (Bukaty, 2019; Harding et al., 2019). The law empowers and creates awareness for consumers to express their rights appropriately. Here, consumers can have control over the use and processing of their personal data. Since the CCPA applies to businesses that gather or use the personal data of California residents, it is applicable to businesses that use FL models integrated into healthcare systems to obtain or process patient data residing in the state of California, United States (Stallings, 2020).

14.6 Necessity of legal frameworks for federated learning integration

The integration of FL in digital healthcare systems holds significant potential in revolutionizing the healthcare sector. FL models can enhance data processing, improve security, and enhance trust. However, the integration of FL in healthcare systems raises important legal considerations that must be addressed to maximize the enormous benefits of this integration. Appropriate legal considerations need to be examined to ensure the responsible and ethical use of FL applications in modern healthcare systems.

FL-based systems can be trained to make informed decisions autonomously. This capability of FL systems raises questions about accountability and fairness. When developing FL for healthcare systems, appropriate legal frameworks, ethical guidelines, and societal norms should be considered. Critical questions concerning handling biases in data, preventing discrimination, and safeguarding fundamental rights and values need to be answered comprehensively. The end users of FL-empowered healthcare systems seek assurance regarding the system's accountability and transparency in critical decision-making processes.

As novel FL models and systems are being developed and deployed, understanding the reasoning behind their decisions becomes very complex. To resolve this complexity, the legal framework regulating these systems should require that FL systems are explainable, transparent, and auditable at all times. When these criteria are satisfied, the system is most likely to be accountable and create room for its users to challenge and possibly seek redress when the system presents erroneous or biased decisions.

The issue of liability and responsibility of FL systems is very critical to their acceptance and application in various domains, including the healthcare sector. To ensure that FL systems meet liability and responsibility requirements, there is a need for a comprehensive legal framework. However, one cannot rule out the possibility of FL-based systems producing erroneous results that could potentially harm or pose significant danger to the participants. The question arising from this scenario is who is to blame for such errors? The FL model, or the individual users, or the service providers? Essentially, the allocation of appropriate liability in such instances requires adequate clarification and documentation.

The working document should consider several factors, such as the nature of the decision-making process, the level of human control, and the foreseeability of harm or danger. Not until appropriate legal frameworks that spell out these requirements are put in place and enforced to ease the assignment of responsibility and required compensation for damages, users will not see FL models as fair and accountable. In addition, new legal frameworks can be developed or the existing ones can be gainfully adapted to accommodate the integration of FL in healthcare systems to maintain legal clarity and certainty. Another critical point worth mentioning is compliance. To build trust and ensure the ethical handling of sensitive user data, compliance with data protection regulations and best practices is a key requirement.

The integration of FL in digital healthcare systems also poses ethical and regulatory challenges. FL algorithms can introduce biases and produce discriminatory outcomes, especially in models with limited training data. The design of FL models should clarify the level of model training and follow the requirements of the regulatory frameworks. The development of working regulatory frameworks governing the ethical use of FL in healthcare applications would require effective collaboration among industry stakeholders, policymakers, patients, individuals, and AI technology experts.

Currently, the enormous benefits of integrating FL into digital healthcare systems have not been harnessed effectively. To explore the potential of FL in digital healthcare systems, the ensuing legal challenges need to be addressed proactively. Creating a supportive legal environment involves enforcing robust data protection measures and ensuring absolute compliance. Also, there is a necessity to foster transparency, accountability, and regular engagement in ethical and regulatory discussions. Finally, unlocking the full potential of FL in healthcare systems requires building a responsible and accountable system that follows the governing regulatory frameworks.

14.6.1 Requirements for effective legal framework for federated learning integration

The application of FL models in digital healthcare systems will continue to evolve, and as such, there is a need to put in place suitable legal frameworks that facilitate innovation while promoting ethical and responsible use at all levels. The following recommendations are suggested to guide the development of effective legal frameworks for FL integration in digital healthcare systems.

14.6.1.1 Comprehensive legal framework

A comprehensive and balanced legal framework should have a number of attributes. It should be transparent, accountable, fair, maintain consent, and guarantee the privacy and security of all stakeholders in the healthcare ecosystem. A transparent framework ensures that the FL systems considered are explainable and understandable, enabling individuals to make informed decisions. In the same vein, an accountable legal framework should specify the responsible entity for costly decisions made by the AI systems. Fairness involves examining bias, discrimination, and the social impact of the system on its authorized users. The issue of consent deserves serious consideration. Patients should be able to understand the process and procedures for handling their confidential information and understand the terms clearly. Privacy preservation is another aspect of a good legal framework. Patients' information should be rightly protected and secured against any adversaries.

14.6.1.2 Adaptive and technology-neutral approaches

Another important consideration is the establishment of adaptive and technology-neutral approaches for designing FL-supported legal frameworks. Since AI technologies are evolving and advancing rapidly, it is imperative to use flexible and adaptable rules in creating the required legal frameworks. As much as possible, technology-neutral frameworks should be considered and prioritized to save the frameworks from extinction when the technology becomes obsolete or outdated. The main point here is that the framework should be flexible and evolve with the technology without incurring substantial costs or resources for legislative updates.

14.6.1.3 Global cooperation and harmonization

Global cooperation and harmonization of resources and intellect are key to the development of effective FL frameworks. Since FL technologies transcend national boundaries, collaboration among all participants is critical to establishing consistent and interoperable legal frameworks. Harmonization efforts should focus on enhancing regulatory compliance, promoting regulatory consistency, and fostering international standardization. An ecosystem that spurs innovation and ensures the responsible use of FL models can be created through sharing best practices, experiences, and expertise across diverse jurisdictions.

14.6.1.4 Ethical guidelines and responsible federated learning practices

New ethical guidelines and responsible FL practices must be formulated and integrated into existing legal frameworks. Ethical considerations are critical to the development and deployment of AI and FL technologies in healthcare systems. Several issues, such as bias mitigation, algorithmic transparency, accountability, and human oversight, should be adequately captured in the guidelines. Policymakers can foster public trust and confidence when the right ethical principles are integrated into legal frameworks. This development will ensure that FL technologies are developed and applied efficiently to satisfy societal values and expectations.

14.6.1.5 Public–private collaboration

Public–private collaboration is critical to the development and implementation of effective policy and regulatory frameworks. The complex and interdisciplinary nature of FL requires input from various stakeholders, including government agencies, civil society organizations, academic institutions, industry players, and the public. In a collaborative setting, resources and ideas can be pooled to ensure that the right policies are formulated, leveraging diverse perspectives to address the needs and concerns of all participating stakeholders. Such diversified frameworks would promote fairness and drive healthy innovation. Finally, the role of dialogue, consultation, and public–private partnerships in driving the development of inclusive and effective legal frameworks cannot be overstated.

14.6.1.6 Cross-border collaboration

Since AI systems continuously learn and adapt, it is often difficult to define static regulations regarding the application of AI and, in particular, FL models. In FL, several participants collaborate and contribute to the training of the models, and the global nature of model development and deployment calls for appropriate harmonization and cross-border collaboration to avoid fragmented regulatory frameworks and enhance effective ethical and legal standards.

14.6.2 Legal implications of noncompliance with regulatory requirements

Tackling data privacy problems in ML-assisted applications is quite challenging. To address these problems, it is imperative to deploy FL as a viable approach to decouple data storage and processing at the local nodes. In this case, only the aggregation of a global ML model is conducted at the coordination server. There is no denying the enormous advantages of FL, enabling privacy preservation in comparison to the traditional centralized ML approaches. FL facilitates the training of an ML model and simultaneously retains personal training data on the devices of end users. Here, only the locally trained model parameters, required to update the global model, are fed as inputs to the coordination server. However, such model parameters contain sensitive features of the data that can be unlawfully exploited to redesign desirable personal information. Thus, the FL model enclosed in GDPR or any other framework must be designed in compliance with regulatory requirements (Truong et al., 2021). The question is: What happens when a breach is alerted?

Whenever a breach is detected, a user files a claim or suspicion of noncompliance, and the data protection agency (DPA) is notified immediately. The DPA critically analyzes the case and concludes whether there was a violation or not. In a scenario where no violation is concluded, no further action is required. However, if the DPA determines that a violation exists, a decision could be adopted with or without the imposition of a fine. Depending on the weight of the violation, several fines could be imposed accordingly. In the worst case scenario, the DPA can combine several possible fines. For example, a flowchart for assessing GDPR compliance and the penalty for noncompliance is given in Fig. 14.5.

14.6.3 Recommendations for effective legal frameworks for federated learning integration

In order to develop effective and adaptive AI-based regulations, the following recommendations are worthy of consideration.

First, the requisite regulations should be well crafted to maintain synergies by fostering innovation and safeguarding ethical and societal concerns. The frameworks should maintain great flexibility to cater to innovation and experimentation without compromising security and privacy.

Second, accountability and transparency in decision-making should be the hallmarks of any applicable regulatory framework. The key requirements for explainability, auditability, and documentation of AI systems should be clearly highlighted in the regulatory framework.

Third, the regulatory measures in place should provide an enabling environment for responsible data governance and address any potential privacy concerns. Also, upholding the rights of individuals should be given adequate coverage in the regulatory framework. Thus, the frameworks need to outline efficient data protection techniques and consent granting procedures to protect and preserve the rights of the participating individuals.

Fourth, the frameworks regulating FL in healthcare systems need to be harmonized and standardized to ensure consistency and facilitate interoperability in the

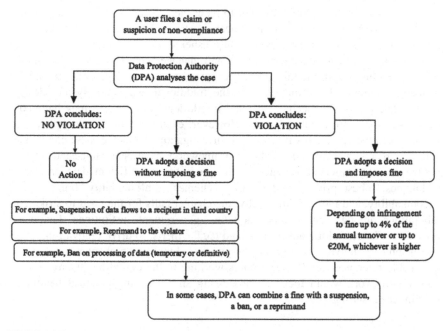

FIGURE 14.5

GDPR compliance and penalty for noncompliance. *GDPR*, General Data Protection Regulation.

global healthcare ecosystem. Certain issues, such as compliance and legal complexity, could be easily addressed when these regulations are adequately harmonized in line with global best practices. This development will usher in a fresh atmosphere for cross-border collaboration and contributions to FL and healthcare systems.

Fifth, there is a need for the regulatory authorities to actively collaborate with industry experts, academia, and civil society to develop in-depth and effective regulations. The role of stakeholder engagement in facilitating a more holistic understanding of the integration of FL in digital healthcare systems cannot be overemphasized. The legal implications of this integration call for a comprehensive and diverse perspective in the regulatory decision-making process.

Finally, continuous monitoring and evaluation of AI-based regulations are critical to examining their effectiveness, adapting to new innovations, and addressing potential risks and challenges.

14.7 Conclusions and future scope

In this chapter, legal frameworks for integrating FL in digital healthcare systems have been presented. These frameworks are critical to ensuring that FL models and systems are properly implemented to meet the privacy, security, and accountability requirements of emerging healthcare systems. The application of FL techniques in digital healthcare systems has ushered in accelerated progress and successes in healthcare system operation and management. Though FL has significantly revolutionized digital healthcare systems, several legal and regulatory challenges remain. The limitations of the traditional healthcare system design, limited data protection mechanisms, and fragile legal frameworks may have contributed to these proliferating issues. However, examining the legal implications of FL in healthcare systems is worthy of investigation. To tackle these challenges, some far reaching legal and regulatory frameworks have been suggested. In particular, the recommendations include the promotion of efficient data protection techniques, robust privacy and security schemes, confidentiality, transparency, accountability, and more. Additionally, the chapter calls for the need to strike collaborations among stakeholders, including government institutions, healthcare providers, FL technology developers, and legal experts, to drive innovation, effective healthcare delivery, and economic progress. Finally, the chapter can serve as a valuable reference for all stakeholders working in the digital healthcare ecosystem. Future work would focus on processing and analyzing practical healthcare data in an FL setting.

Acknowledgment

This work is supported in part by the Nigerian Petroleum Technology Development Fund and in part by the German Academic Exchange Service through the Nigerian-German Postgraduate Program under Grant number 57473408.

References

Ali, M., Naeem, F., Tariq, M., & Kaddoum, G. (2022). Federated learning for privacy preservation in smart healthcare systems: A comprehensive survey. *IEEE Journal of Biomedical and Health Informatics*, *27*(2), 778–789.

Allen, B., Agarwal, S., Kalpathy-Cramer, J., Dreyer, K., & Democratizing, A. I. (2019). Democratizing AI. *Journal of the American College of Radiology: JACR*, *16*(7), 961–963. Available from https://doi.org/10.1016/j.jacr.2019.04.023.

Antunes, R. S., André da Costa, C., Küderle, A., Yari, I. A., & Eskofier, B. (2022). Federated learning for healthcare: Systematic review and architecture proposal. *ACM Transactions on Intelligent Systems and Technology (TIST)*, *13*(4), 1–23.

Arikumar, K. S., Prathiba, S. B., Alazab, M., Gadekallu, T. R., Pandya, S., Khan, J. M., & Moorthy, R. S. (2022). FL-PMI: Federated learning-based person movement identification through wearable devices in smart healthcare systems. *Sensors*, *22*(4), 1377.

Atchinson, B. K., & Fox, D. M. (1997). From the field: The politics of the health insurance portability and accountability act. *Health Affairs*, *16*(3), 146–150.

Awotunde, J. B., Adeniyi, E. A., Ajagbe, S. A., Imoize, A. L., Oki, O. A., & Misra, S. (2022). Explainable artificial intelligence (XAI) in medical decision support systems (MDSS): Applicability, prospects, legal implications, and challenges. In A. L. Imoize, J. Hemanth, D.-T. Do, & S. N. Sur (Eds.), *Explainable artificial intelligence in medical decision support systems* (1st ed, pp. 45–90). The Institution of Engineering and Technology.

Ayoade, O. B., et al. (2022). Explainable artificial intelligence (XAI) in medical decision systems (MDSSs): Healthcare systems perspective. In A. L. Imoize, J. Hemanth, D.-T. Do, & S. N. Sur (Eds.), *Explainable artificial intelligence in medical decision support systems* (1st ed, pp. 1–43). The Institution of Engineering and Technology.

Azoulay, A. (2019). *Towards an ethics of artificial intelligence*. United Nations. Available at: https://www.un.org/en/chronicle/article/towards-ethics-artificial-intelligence.

Bautista, A. B., Burgos, A., Nickel, B. J., Yoon, J. J., Tilara, A. A., & Amorosa, J. K. (2009). Do clinicians use the American College of Radiology Appropriateness criteria in the management of their patients? *American Journal of Roentgenology, 192*(6), 1581–1585.

Beeler, P. E., Bates, D. W., & Hug, B. L. (2014). Clinical decision support systems. *Swiss Medical Weekly, 144*(5152), w14073.

Boscarino, N., Cartwright, R. A., Fox, K., & Tsosie, K. S. (2022). Federated learning and indigenous genomic data sovereignty. *Nature Machine Intelligence, 4*(11), 909–911.

Bosch, J. C., & Lee, I. (1994). Wealth effects of Food and Drug Administration (FDA) decisions. *Managerial and Decision Economics, 15*(6), 589–599.

Boss, A. H. (2000). The Uniform Electronic Transactions Act in a global environment. *Idaho Law Review, 37*, 275.

Boulton, G., Rawlins, M., Vallance, P., & Walport, M. (2011). Science as a public enterprise: The case for open data. *The Lancet, 377*(9778), 1633–1635.

Bujalkova, M. (2001). International guidelines on bioethics. *Bratislavske Lekarske Listy, 27*(2), 117. Available from https://doi.org/10.1136/jme.27.2.117.

Bukaty, P. (2019). The California Consumer Privacy Act (CCPA): An implementation guide. IT Governance Ltd.

Carayon, P., Hoonakker, P., Hundt, A. S., Salwei, M., Wiegmann, D., Brown, R. L., & Patterson, B. (2020). Application of human factors to improve usability of clinical decision support for diagnostic decision-making: a scenario-based simulation study. *BMJ Quality & Safety, 29*(4), 329–340.

Chik, W. B. (2013). The Singapore Personal Data Protection Act and an assessment of future trends in data privacy reform. *Computer Law & Security Review, 29*(5), 554–575.

Chik, W. B., & Pang, J. K. Y. (2014). The meaning and scope of personal data under the Singapore personal data protection act. *Singapore Academy of Law Journal, 26*(2), 354–397.

Chotchoungchatchai, S., Marshall, A. I., Witthayapipopsakul, W., Panichkriangkrai, W., Patcharanarumol, W., & Tangcharoensathien, V. (2020). Primary health care and sustainable development goals. *Bulletin of the World Health Organization, 98*(11), 792.

Cohen, I. G., & Mello, M. M. (2018). HIPAA and protecting health information in the 21st century. *JAMA: The Journal of the American Medical Association, 320*(3), 231–232.

Dong, S., Abbas, K., & Jain, R. (2019). A survey on distributed denial of service (DDoS) attacks in SDN and cloud computing environments. *IEEE Access, 7*, 80813–80828.

Dwyer, S. J., III, Weaver, A. C., & Hughes, K. K. (2004). Health insurance portability and accountability act. *Security Issues in the Digital Medical Enterprise*, 72(2), 9−18.

Farahani, B., & Monsefi, A. K. (2023). Smart and collaborative industrial IoT: A federated learning and data space approach. *Digital Communications and Networks*, 9(2), 436−447.

Geis, J. R., Brady, A. P., Wu, C. C., Spencer, J., Ranschaert, E., Jaremko, J. L., et al. (2019). Ethics of artificial intelligence in radiology: Summary of the Joint European and North American Multisociety Statement. *Journal of the American College of Radiology: JACR*, 16(11), 1516−1521. Available from https://doi.org/10.1016/j.jacr.2019.07.028.

Goddard, M. (2017). The EU General Data Protection Regulation (GDPR): European regulation that has a global impact. *International Journal of Market Research*, 59(6), 703−705.

Gostin, L. O. (2001). National health information privacy: regulations under the Health Insurance Portability and Accountability Act. *JAMA: the Journal of the American Medical Association*, 285(23), 3015−3021.

Gu, X., Sabrina, F., Fan, Z., & Sohail, S. (2023). A review of privacy enhancement methods for federated learning in healthcare systems. *International Journal of Environmental Research and Public Health*, 20(15), 6539.

Hallevy, P.G. (2015). AI V. IP − *Criminal liability for intellectual property IP offenses of artificial intelligence AI entities*. Available from: https://doi.org/10.2139/ssrn.2691923.

Harding, E. L., Vanto, J. J., Clark, R., Hannah Ji, L., & Ainsworth, S. C. (2019). Understanding the scope and impact of the California consumer privacy act of 2018. *Journal of Data Protection & Privacy*, 2(3), 234−253.

Iguoba, V., & Imoize, A. L. (2022). The psychology of explanation in medical decision support systems. In A. L. Imoize, J. Hemanth, D.-T. Do, & S. N. Sur (Eds.), *Explainable artificial intelligence in medical decision support systems* (1st ed, pp. 489−506). The Institution of Engineering and Technology.

Ilegieuno, S., Chukwuani, O., & Adaralegbe, I. (2021). *Artificial intelligence and the future of law practice in Nigeria. Internet of Things, artificial intelligence and blockchain technology* (pp. 307−326). Cham: Springer International Publishing.

Ilegieuno, S., Chukwuani, O., & Eigbobo, M. (2021). *Examining the legal issues involved in the application of blockchain technology. Internet of Things, artificial intelligence and blockchain technology* (pp. 89−109). Cham: Springer International Publishing.

Imoize, A. L., Adedeji, O., Tandiya, N., & Shetty, S. (2021). 6G enabled smart infrastructure for sustainable society: Opportunities, challenges, and research roadmap. *Sensors*, 21(5), 1709.

Imoize, A. L., Mekiliuwa, S. C., Omiogbemi, I. M. B., & Omofonma, D. O. (2020). Ethical issues and policies in software engineering. *International Journal of Information Security and Software Engineering*, 6(1), 6−17.

Imoize, A. L., et al. (2022). Blockchain technology for secure COVID-19 pandemic data handling. In C. Chakraborty, & J. J. P. C. Rodrigues (Eds.), *Smart health technologies for the COVID-19 pandemic: Internet of medical things perspectives* (1st edn, pp. 141−179). The Institution of Engineering and Technology. Available from https://doi.org/10.1049/pbhe042e_ch6.

Imoize, A. L., Balas, V. E., Solanki, V. K., Lee, C.-C., & Obaidat, M. S. (Eds.), (2023). *Handbook of security and privacy of AI-enabled healthcare systems and Internet of*

Medical Things (1st ed.). CRC Press. Available from https://doi.org/10.1201/9781003370321.

Kaur, G., Malhorta, R., & Shukla, V. K. (2022). *GDPR oriented vendors contracts in relation to data transfer: Analysis of standard clauses 2010 and 2021, . Emerging technologies in data mining and information security: Proceedings of IEMIS 2022* (Volume 2, pp. 629−635). Singapore: Springer Nature Singapore.

Köchling, A., & Wehner, M. C. (2020). Discriminated by an algorithm: A systematic review of discrimination and fairness by algorithmic decision-making in the context of HR recruitment and HR development. *Business Research, 13*(3), 795−848. Available from https://doi.org/10.1007/s40685-020-00134-w.

Kovacic, W. E., & Winerman, M. (2014). The Federal Trade Commission as an independent agency: Autonomy, legitimacy, and effectiveness. *Iowa Law Review, 100*, 2085.

Kumar, R. L., et al. (Eds.), (2021). *Internet of Things, artificial intelligence and blockchain technology* (1st edn). Switzerland AG: Springer Nature. Available from https://doi.org/10.1007/978-3-030-74150-1.

Lateef, M. A., Taiwo, L. O., & Adeyoju, A. (2022). Examining the powers of the NITDA to enforce data protection laws in Nigeria. *Global Privacy Law Review, 3*(2), 89−97.

Li, J., Meng, Y., Ma, L., Du, S., Zhu, H., Pei, Q., & Shen, X. (2021). A federated learning based privacy-preserving smart healthcare system. *IEEE Transactions on Industrial Informatics, 18*(3), 2021−2031. Available from https://doi.org/10.1109/TII.2021.3098010, March 2022.

Li, T., Sahu, A. K., Talwalkar, A., & Smith, V. (2020). Federated learning: Challenges, methods, and future directions. *IEEE Signal Processing Magazine, 37*(3), 50−60.

Lu, Y., Nakicenovic, N., Visbeck, M., & Stevance, A. S. (2015). Policy: Five priorities for the UN sustainable development goals. *Nature, 520*(7548), 432−433.

Markose, A., Krishnan, R., & Ramesh, M. (2016). Medical ethics. *Journal of Pharmacy and Bioallied Sciences, 8*(Suppl. 1), S1−S4. Available from https://doi.org/10.4103/0975-7406.191934.

McCartney, A. M., Anderson, J., Liggins, L., Hudson, M. L., Anderson, M. Z., TeAika, B., & Phillippy, A. M. (2022). Balancing openness with Indigenous data sovereignty: An opportunity to leave no one behind in the journey to sequence all of life. *Proceedings of the National Academy of Sciences, 119*(4), e2115860119.

McGuire, S. (2011). US department of agriculture and US department of health and human services, dietary guidelines for Americans, 2010, Washington, DC: US government printing office, January 2011 *Advances in Nutrition, 2*(3), 293−294.

Meehan, S. C. (1999). Consumer Protection Law and the Uniform Electronic Transactions Act (UETA): Why states should adopt UETA as drafted. *Idaho Law Review, 36*, 563.

Meshram, C., Lee, C. C., Bahkali, I., & Imoize, A. L. (2023). An efficient fractional chebyshev chaotic map-based three-factor session initiation protocol for the human-centered IoT architecture. *Mathematics, 11*(9), 2085.

Meshram, C., Obaidat, M.S., Imoize, A.L., Bahkali, I., Tambare, P., & Hsiao, K.F. (2022, October). An Efficient Authentication Technique using Convolution Chebyshev Chaotic Maps for TMIS. In *2022 International Conference on Communications, Computing, Cybersecurity, and Informatics (CCCI)* (pp. 1−6). IEEE.

Moore, W., & Frye, S. (2019). Review of HIPAA, part 1: History, protected health information, and privacy and security rules. *Journal of Nuclear Medicine Technology, 47*(4), 269−272.

Mulgund, P., Mulgund, B. P., Sharman, R., & Singh, R. (2021). The implications of the California Consumer Privacy Act (CCPA) on healthcare organizations: Lessons learned from early compliance experiences. *Health Policy and Technology, 10*(3), 100543.

Mulshine, M. (2015). *A major flaw in Google's algorithm allegedly tagged two black people's faces with the word 'gorillas'*. New York City: Business Insider.

Ni, A. Y. (2017). *Protection of personally identifiable information in government: A survey of US Regulatory Framework and Emerging Technological Challenges. Routledge handbook on information technology in government* (pp. 266–283). Routledge.

Noor, P. (2020). Can we trust AI not to further embed racial bias and prejudice? *BMJ (Clinical Research ed.), 368*. Available from https://doi.org/10.1136/bmj.m363, m363.

Nosowsky, R., & Giordano, T. J. (2006). The Health Insurance Portability and Accountability Act of 1996 (HIPAA) privacy rule: Implications for clinical research. *Annual Review of Medicine, 57*, 575–590.

Nwaneri, S. C., Yinka-Banjo, C., Uregbulam, U. C., Odukoya, O. O., & Imoize, A. L. (2022). Explainable neural networks in diabetes mellitus prediction. In A. L. Imoize, J. Hemanth, D.-T. Do, & S. N. Sur (Eds.), *Explainable artificial intelligence in medical decision support systems* (1st ed, pp. 313–334). The Institution of Engineering and Technology.

Osanaiye, T., Choo, O., Raymond Choo, K. K., & Dlodlo, M. (2016). Distributed denial of service (DDoS) resilience in cloud: Review and conceptual cloud DDoS mitigation framework. *Journal of Network and Computer Applications, 67*, 147–165.

Pawloski, P. A., Brooks, G. A., Nielsen, M. E., & Olson-Bullis, B. A. (2019). A systematic review of clinical decision support systems for clinical oncology practice. *Journal of the National Comprehensive Cancer Network, 17*(4), 331–338.

Popowicz-Pazdej, A. (2021). The new EU Standard Contractual Clauses as a type of appropriate safeguard in the international transfer of personal data. *Journal of Data Protection & Privacy, 5*(1), 61–71.

Prayitno., Shyu, C. R., Putra, K. T., Chen, H. C., Tsai, Y. Y., Hossain, K. T., & Shae, Z. Y. (2021). A systematic review of federated learning in the healthcare area: From the perspective of data properties and applications. *Applied Sciences, 11*(23), 11191.

Pyo, G. (2021). An alternate vision: China's cybersecurity law and its implementation in the Chinese courts. *Columbia Journal of Transnational Law, 60*, 228.

Rahman, A., Hossain, M. S., Muhammad, G., Kundu, D., Debnath, T., Rahman, M., & Band, S. S. (2023). Federated learning-based AI approaches in smart healthcare: Concepts, taxonomies, challenges and open issues. *Cluster Computing, 26*(4), 2271–2311.

Rani, S., Kataria, A., Kumar, S., & Tiwari, P. (2023). Federated learning for secure IoMT-applications in smart healthcare systems: A comprehensive review. *Knowledge-Based Systems, 274*, 110658.

Rieke, N., Hancox, J., Li, W., Milletari, F., Roth, H. R., Albarqouni, S., & Cardoso, M. J. (2020). The future of digital health with federated learning. *NPJ Digital Medicine, 3*(1), 119.

Rufai, A. T., Dukor, K. F., Ageh, O. M., & Imoize, A. L. (2022). XAI robot-assisted surgeries in future medical decision support systems. In A. L. Imoize, J. Hemanth, D.-T. Do, & S. N. Sur (Eds.), *Explainable artificial intelligence in medical decision support systems* (1st ed., pp. 167–195). The Institution of Engineering and Technology.

Schwartz, P. M., & Solove, D. J. (2011). The PII problem: Privacy and a new concept of personally identifiable information. *NYU Law Review, 86*, 1814.

Schwartz, P. M., & Solove, D. J. (2014). Reconciling personal information in the United States and European Union. *California Law Review, 102*, 877.

Stallings, W. (2020). Handling of personal information and deidentified, aggregated, and pseudonymized information under the California consumer privacy act. *IEEE Security & Privacy, 18*(1), 61−64.

Steppe, R. (2017). Online price discrimination and personal data: A General Data Protection Regulation perspective. *Computer Law & Security Review, 33*(6), 768−785.

Suwadi, P., Manthovani, R., & Assyifa, A. K. (2023). Legal comparison of electronic contract in electronic commerce between Indonesia and the United States Based on the United Nations Commission on International Trade Law. *Journal of Law and Sustainable Development, 11*(3), e714.

Tijani, B. (2023). *Co-creating a national artificial intelligence strategy for Nigeria.* https://www.linkedin.com/pulse/co-creating-national-artificial-intelligence-strategy-tijani?trk = public_profile_article_view. Accessed 28.08.23

Treleaven, P., Smietanka, M., & Pithadia, H. (2022). Federated learning: The pioneering distributed machine learning and privacy-preserving data technology. *Computer, 55*(4), 20−29.

Truong, N., Sun, K., Wang, S., Guitton, F., & Guo, Y. (2021). Privacy preservation in federated learning: An insightful survey from the GDPR perspective. *Computers & Security, 110*, 102402.

Voigt, P., & Von dem Bussche, A. (2017). (1st Ed.). *The EU general data protection regulation (GDPR).* A Practical Guide, (10). Cham: Springer International Publishing. (3152676), 10−5555.

Wang, W., Li, X., Qiu, X., Zhang, X., Zhao, J., & Brusic, V. (2023). A privacy preserving framework for federated learning in smart healthcare systems. *Information Processing & Management, 60*(1), 103167.

Wong YongQuan, B. (2017). Data privacy law in Singapore: The personal data protection act 2012. *International Data Privacy Law, 7*(4), 287−302.

Woods, L. (2017). Digital Economy Act 2017: Data sharing provisions. *European Data Protection Law Review, 3*, 244.

Wu, F., Lu, C., Zhu, M., Chen, H., Zhu, J., Yu, K., & Pan, Y. (2020). Towards a new generation of artificial intelligence in China. *Nature Machine Intelligence, 2*(6), 312−316.

Zhang, C., Xie, Y., Bai, H., Yu, B., Li, W., & Gao, Y. (2021). A survey on federated learning. *Knowledge-Based Systems, 216*, 106775.

Secure federated learning in the Internet of Health Things for improved patient privacy and data security

15

Kassim Kalinaki[1,2], Adam A. Alli[1], Baguma Asuman[1] and Rufai Yusuf Zakari[3]

[1]*Department of Computer Science, Islamic University in Uganda (IUIU), Mbale, Uganda*
[2]*Borderline Research Laboratory, Kampala, Uganda*
[3]*Department of Computer Science, School of Science and Information Technology (SSIT), Skyline University Nigeria, Kano, Nigeria*

15.1 Introduction

The emergence of federated learning (FL) has transformed how we approach machine learning (ML) in the era of the Internet of Things (IoT), specifically in the healthcare domain in what is currently referred to as IoHT. IoHT is a promising application domain of FL that consists of networked and internet-facing medical devices, sensors, and software applications (Kaur, Atif, & Chauhan, 2020). These interconnected devices can establish seamless data exchanges, yielding copious volumes of electronic health records (EHR) from diverse IoT devices (Fahim, Kalinaki, & Shafik, 2023). This influx of data assumes a pivotal role, empowering an array of analytical techniques to unveil valuable insights and bolster medical applications. Moreover, these IoHT devices, including wearable gadgets, have emerged as indispensable health assistants, enabling real-time health tracking, safety surveillance, rehabilitation monitoring, therapy outcome assessment, preemptive illness screening, and an array of other health metrics (Srivastava et al., 2022; Kalinaki, Fahadi, et al., 2023). However, with the exponential increase in data generation, data privacy and security concerns have become a pressing issue in the IoHT (Chemisto et al., 2023). For instance, malevolent parties and unauthorized individuals can breach the security of IoHT devices and manipulate patients' stored information, posing a grave danger to the well-being of patients (Alli et al., 2021; Alabdulatif, Thilakarathne, & Kalinaki, 2023; Ali et al., 2023; Kalinaki, Thilakarathne, et al., 2023). Moreover, these concerns are further exacerbated by the fact that we are currently navigating a data-driven world where the cost of admission to various healthcare services is the voluntary surrender of personal information (Pfitzner, Steckhan, & Arnrich, 2021).

Federated Learning for Digital Healthcare Systems. DOI: https://doi.org/10.1016/B978-0-443-13897-3.00003-5

Previous endeavors using traditional ML to guarantee healthcare data security and privacy have been fruitless since valuable user data is housed within a central server, where it is called upon to power the rigorous training and testing procedures necessary for crafting robust ML models (Yin, Zhu, & Hu, 2021). As studies have shown, relying on centralized models leads to a variety of challenges, ranging from limited processing capacity and time constraints, especially in resource-restricted environments, to particularly critical issues concerning the security and privacy of individuals' personal information, which are still prevalent and have been overlooked for an extended period (Mothukuri et al., 2021). Additionally, a centralized AI infrastructure for upcoming adaptable healthcare systems may prove unfeasible as healthcare traffic is expected to be dispersed across varied and extensive healthcare systems. Consequently, a collaborative and AI-centered scheme is imperative for promoting privacy preservation and the scalability of modern digital healthcare endeavors to ensure the effective adoption and implementation of intelligent healthcare systems.

FL has recently emerged as a prominent player in the ML domain owing to its decentralized data model. Moreover, it offers a unique approach to address the above challenges while improving the performance of ML models. Initially introduced in 2017 (McMahan et al., 2017), FL is a privacy-preserving and distributed ML approach, allowing multiple participating stakeholders to train an ML model collaboratively without sharing their raw data. Rather than openly exchanging data between parties with distrust, FL operates by exchanging model parameters that can be combined into a collective model. This exciting approach utilizes a client-server structure, where the server coordinates the training process, constructs the model, and provides it to all cooperating clients. Subsequently, the clients apply the trained model to their datasets in a distributed fashion. This technique has revolutionized how we approach ML in IoT systems, specifically in the healthcare sector, where data is generated from multiple devices with limited processing power and storage capacity (Gaba et al., 2023). IoHT applications are highly suitable for FL due to the large amount of sensitive data collected from several IoHT devices. FL's approach of facilitating the local training of models on individual devices significantly reduces the risks of data breaches, minimizes data transmission over the network, and provides better control over data ownership (Nguyen et al., 2022). Additionally, this technique enhances the training efficiency of the healthcare system by amassing copious datasets and computational resources from local IoHT devices that might otherwise be unattainable if a centralized, traditional AI approach is employed (Nguyen et al., 2021).

This study emphasizes secure FL (SFL), an extension of FL that focuses on enhancing data privacy and security in IoHT systems. This is achieved through a combination of cryptographic and privacy-preserving techniques to protect sensitive data while facilitating multiple parties to collaborate in training ML models without sharing their raw data. Accordingly, several studies have proposed various privacy-preserving methods based on SFL, such as differential privacy (DP), secure multiparty computation (SMC), homomorphic encryption (HE), secret sharing, and Blockchain, to safeguard patient data and guarantee privacy (Adnan et al., 2022; Antunes et al., 2022; Hosseini et al., 2022; Islam, Ghasemi, & Mohammed, 2022; L. Zhang et al., 2022). So far, the existing literature in this emerging and promising domain is insufficient to solve the current challenges using traditional centralized models. Additionally, due to the infancy of FL, many of the proposed privacy-preserving techniques have not been fully deployed. As the volume of sensitive health data generated by a plethora of IoHT devices continues to grow, it is crucial to develop efficient and effective SFL techniques to protect patient data. Therefore a comprehensive review of SFL techniques is essential to fill this gap by identifying the strengths and weaknesses of different recent approaches and developing more robust SFL techniques. On that basis, therefore, we present the main contributions of this chapter.

15.1.1 Chapter contributions

Driven by the growing apprehension over security and privacy issues and the inadequacy of current remedies to address these concerns, we present the critical advancements of this chapter, which include:

- A thorough examination of FL in healthcare, emphasizing the importance of privacy and data security in the IoHT.
- A detailed discussion on FL's existing security and privacy challenges in IoHT.
- A meticulous exploration of diverse and recent privacy-preserving techniques of FL in IoHT, encompassing DP, HE, SMC, secret sharing, and Blockchain.
- A glimpse into the promising future of SFL in IoHT, highlighting the potential avenues for further development in this area.

15.1.2 Chapter organization

This chapter is organized as follows: Section 15.1 consists of the introduction, emphasizing the emergence of FL in solving concerns related to the security and privacy of IoHT. Section 15.2 depicts the existing security and privacy challenges of FL in IoHT. Section 15.3 contains the different privacy-preserving techniques, including DP, HE, SMC, secret sharing, and Blockchain. Lessons learned are summarized in Section 15.4. A glimpse into the promising future of SFL in IoHT is highlighted in Section 15.5, and the conclusion is provided in Section 15.6.

15.2 Security and privacy challenges of federated learning in Internet of Health Things

FL in the IoHT involves data from multiple sources to train and model health-related problems, enabling comprehensive representations of the entire health environment where IoT devices are deployed. However, one of the primary challenges lies in ensuring the safety and ownership of data during the training and modeling process (Pfitzner, Steckhan, & Arnrich, 2021; Rahman et al., 2021). Critical questions arise regarding data control, sharing regulations, and the entities responsible for governing rules for various activities. Determining who holds power over the data and establishing clear guidelines on data sharing and access becomes essential. Additionally, the formulation of robust data usage and control policies is crucial. Underestimating the importance of these issues can lead to complications. Furthermore, challenges arise in preventing data misuse and model misuse. Consent, accountability, transparency, and fairness issues may surface throughout the FL process. These factors require careful consideration to maintain ethical standards and build stakeholder trust.

FL aims to address privacy concerns by aggregating models from multiple sources. However, this approach does not provide an absolute safeguard against data leakages, which could potentially expose information patterns and increase the risk of re-identification (Albahri et al., 2023). Data leakage can also occur during the integration of local models. To mitigate the challenge of integrating data and models from multiple sources while preserving privacy, mechanisms such as adding noise or employing techniques such as SMC or HE are utilized (Yang et al., 2023). These mechanisms introduce additional layers of privacy protection to prevent unauthorized access to sensitive information. However, implementing these privacy-preserving mechanisms comes with certain drawbacks. Firstly, such mechanisms incur computational costs and require substantial computing resources. The additional computational overhead can lead to increased processing time and resource requirements, potentially impacting the efficiency and scalability of the FL process. Secondly, adopting privacy-preserving mechanisms such as SMC or HE adds complexity to the FL system. These techniques may necessitate specialized expertise and infrastructure, making the implementation and maintenance of the FL system more challenging.

Sharing models at the edge in FL introduces vulnerabilities that adversaries can exploit to extract sensitive information through inference. Furthermore, in the FL environment, devices may participate with malicious intentions, compromising the integrity of the FL process. These devices may engage in activities such as data poisoning, launching attacks that compromise model integrity, or manipulating model updates to weaken the overall security of the FL system. The attacks pose significant safety and privacy risks, which remain ongoing challenges that must be addressed. Extracting sensitive information through inference can lead to privacy breaches and unauthorized access to confidential data (Acar et al., 2018). Adversarial participation in the FL process can disrupt the accuracy and reliability of the trained models, potentially undermining the trustworthiness of the entire FL system.

Healthcare regulations and laws have been established by governments worldwide to safeguard sensitive information in the healthcare sector. These regulations set national standards for protecting patients' sensitive information, thereby emphasizing the importance of obtaining consent before accessing such big data (Singh & Singh, 2023). However, it must be noted that different governments have different laws and regulations about healthcare data privacy and security. Adherence to these laws becomes crucial when implementing an FL system in the IoHT. Incorporating the requirements of multiple laws into the FL process presents a significant challenge. It necessitates a multifaceted approach to ensure compliance and consideration across various legal frameworks. Complying with healthcare laws requires a comprehensive understanding of the specific regulations applicable to the jurisdictions involved. Organizations deploying FL in healthcare must assess and align their processes, protocols, and technical measures with the legal requirements of each relevant jurisdiction (Chalamala et al., 2022). This may include data protection regulations, privacy laws, and industry-specific frameworks, which may not be easy to achieve.

In summary, implementing FL in the IoHT poses computational challenges that span various aspects, including establishing frameworks for data ownership, sharing regulations, and governance. This requires collaborative efforts involving healthcare organizations, data scientists, and regulatory bodies. As the IoT advances in processing power and other computing resources, users' expectations and reliance on IoT systems that require AI capabilities also increase. This growing reliance on AI-driven IoHT systems further amplifies the importance of addressing concerns related to security, privacy, confidentiality, and other related aspects. To successfully implement FL in IoHT, it is crucial to address the computational challenges by establishing robust frameworks that govern data ownership, sharing, and overall governance. This involves the collaboration of stakeholders, including healthcare organizations, data scientists, and regulatory bodies, to ensure a comprehensive data protection and privacy approach. Furthermore, as the computational capabilities of IoT devices improve, there is a need to prioritize security, privacy, and confidentiality in IoHT systems. The increasing interest in these areas reflects the growing awareness of the potential risks associated with sensitive healthcare data and the need to safeguard patient information in AI-driven healthcare applications.

15.3 Privacy-preserving techniques in federated learning for Internet of Health Things

Amidst the vast possibilities to bolster user data privacy, FL encounters its entanglement of security and privacy concerns, as discussed above. As a result, the call to action resounds louder than ever, urging the creation of ingenious solutions that enhance privacy within the emerging domain of FL techniques tailored for intelligent healthcare applications. This section details the different privacy-preserving techniques of FL and their related case studies, specifically for the IoHT.

15.3.1 Differential privacy

First introduced by Abadi et al. (2016) and later deployed in the FL setting by Geyer et al. (2017), DP is a framework for analyzing and designing privacy-preserving algorithms and systems. It provides a mathematical definition and a collection of approaches that aim to safeguard individual privacy and security while enabling the beneficial extraction of insights from data. DP has shown immense promise in mitigating gradient data leakage of sensitive information. The core idea behind DP is to introduce controlled randomness or noise into the computation or output of a data analysis algorithm. This noise helps to obscure the contribution of any individual data point and makes it difficult for a computationally overpowered adversary to link specific data points to their corresponding outputs. In other words, the goal is to provide plausible deniability for any given individual's data (Abadi et al., 2016). Fueled by this remarkable edge, numerous scholarly endeavors have explored differentially private federated learning (DPFL) frameworks, encompassing various domains, from cutting-edge technologies to intelligent healthcare solutions. For instance, Wu et al. (2021) introduced a novel DPFL framework that simultaneously addresses task expenditure and privacy considerations by introducing artificial noise into the local dataset. Through meticulous analysis, the researchers investigated the contribution, privacy costs, computation, and communication linked with each data owner, resulting in an inherent information asymmetry. To optimize the model owner's utility within this information asymmetry, they proposed a sophisticated 3D contract approach for incentive mechanism design. Rigorous simulations conducted in the study confirmed the efficacy of the proposed mechanism within the DPFL framework, surpassing the performance of various baseline mechanisms. Their scheme, however, overlooks the impact of communication overhead on the outcomes.

In another study, Choudhury et al. (2019) presented a novel FL framework based on DP designed to learn a global model from decentralized health data among various sites. The framework adopts two key measures to protect patient privacy: one that refrains from sharing raw data and the other that utilizes a DP scheme to fortify the model against privacy breaches. With the backdrop of two real-world healthcare applications encompassing the electronic health data of a staggering one million patients, the proposed framework unveiled its true potential. A meticulously conducted evaluation showcased its effectiveness and revealed a groundbreaking revelation: the FL framework upheld the sacred tenets of privacy while preserving the utility of the global model.

More recently, an innovative privacy protection framework for IoHT, aptly named block-free gradient (BFG), explicitly designed for decentralized FL employing a combination of blockchain technology, DP, and the transformative power of a generative adversarial network (GAN), was presented (Liu et al., 2023). Their innovative framework not only effectively tackles the crucial obstacle of mitigating a solitary vulnerability and thwarting inference attacks but also astoundingly surpasses prevailing projections, constraining the prevalence of poisoning attacks to an astonishingly low threshold of less than 26%. Furthermore, BFG successfully tackles the storage problem associated with Blockchain, attains an optimal equilibrium between the privacy budget and the proficiency of the global model, and remarkably withstands the potential repercussions of node withdrawal. Through a meticulous array of simulation experiments encompassing various image datasets, the authors indisputably establish the unparalleled effectiveness of the BFG framework, unveiling a groundbreaking standard for decentralized FL systems by virtue of its extraordinary amalgamation of precision, resilience, and safeguarding of privacy. However, all the benefits of this scheme come with a cost of computational efficiency when compared with traditional ML-based methods, underscoring its weakness.

Moreover, a cutting-edge FL framework that prioritizes privacy in IoHT systems was introduced (Nair, Sahoo, & Raj, 2023). This advanced paradigm revolves around fortifying privacy by mitigating susceptibilities in data transfer and incorporating user anonymity and load reduction functionalities, all achieved through the seamless fusion of an edge computing-driven multitier system architecture. Furthermore, to fortify the security of the transmitted data against potential attacks, a differential privacy technique utilizing Laplacian noise has been implemented on the shared attributes. This approach adds an extra layer of confidentiality to the data and ensures its protection even in the face of adversarial circumstances. However, the proposed scheme's computational complexity was not investigated, making it challenging to ascertain which resources would be constrained during its applications, especially on real-life datasets.

Finally, to overcome limitations in healthcare due to the scarcity of large-scale medical datasets caused by privacy concerns, a novel approach called DPFL and a case study using histopathology images as a testbed were proposed (Adnan et al., 2022). The authors investigated the impact of independent and nonindependent data distributions, as well as the number of participating healthcare providers and their dataset sizes, using the publicly available TCGA (The Cancer Genome Atlas) dataset. The study demonstrates that private, distributed training using the DPFL framework performs similarly to conventional training methods while ensuring strong privacy guarantees.

While the above studies depict the effectiveness of DP in preserving patient information security and privacy in the healthcare sector, one limitation of DP is that its effectiveness dwindles when dealing with a small amount of data. This is because injecting noise into a diminutive dataset while training the model can negatively impact the outcome (Li et al., 2023). Hence, with colossal datasets, DP is guaranteed to provide the necessary privacy in IoHT. Table 15.1 depicts the summary of the studies deploying DP in modern digital healthcare systems, along with their corresponding datasets.

15.3.2 Homomorphic encryption in Internet of Health Things

Encryption is a mechanism that is used to protect the confidentiality of data. This means that if data is to be accessed, it should be accessed by authorized users with a key. The drawback of traditional encryption systems is that there must be

Table 15.1 Summary of studies deploying differential privacy (DP) in modern digital healthcare systems.

Summary of proposed DP technique	Datasets	References
A DPFL framework that simultaneously addresses task expenditure and privacy considerations by introducing artificial noise into the local dataset was proposed	MNIST-IID, MNIST NON-IID	Wu et al. (2021)
DP framework, which protects patient privacy through refraining from sharing raw data and fortifying the model against privacy breaches	MIMIC III data and Limited MarketScanExplorys Claims-EMR Data (LCED)	Choudhury et al. (2019)
Proposed the BFG framework combining DP, decentralized Blockchain, and GAN	MNIST dataset and CIFAR-10 dataset	Liu et al. (2023)
Proposed an edge computing-driven multi-tier system architecture for privacy preservation by incorporating user anonymity and load reduction functionalities	eICU Collaborative Research Database and the standard MNIST dataset.	Nair, Sahoo, and Raj (2023)
A DPFL framework was proposed using histopathology images	TCGA dataset	Adnan et al. (2022)

BFG, *Block-free gradient;* DPFL, *differentially private federated learning;* GAN, *generative adversarial network;* MIMIC, *Medical Information Mart for Intensive Care;* TCGA, *The Cancer Genome Atlas.*

a process of sharing keys, and no computational approach can happen on encrypted data (Singh & Singh, 2023). In addition, traditional encryption does not take care of the sensitivity of data, which arises from many applications, such as in health (Acar et al., 2018; Alli & Alam, 2019). To solve the problems that arise from traditional encryption, homomorphic mechanisms that enable encrypted data to be processed without being decrypted have been established (Ogburn, Turner, & Dahal, 2013). This paradigm attracts applications involving intermediate computational steps before arriving at the destination where it can be decrypted. In HE, addition, multiplication, or a combination of addition and multiplication can be done on encrypted data (Acar et al., 2018; W. Yang et al., 2023). Given these unique characteristics, HE is bound to attract applications in secure computational off-loading, ML, and data manipulation of sensitive data in IoHT. One of the categorizations of HE is partially homomorphic encryption (PHE), in which the cloak of secrecy can be draped over sensitive data as only a chosen set of mathematical operations can be conducted on encrypted values. Another categorization is somewhat homomorphic encryption (SWHE), which allows for a restricted set of computations, such as addition or multiplication, to be performed up to a specific degree of robustness but with reduced repetitions. Finally, fully homomorphic encryption (FHE) possesses additive and multiplicative isomorphism properties that empower an inexhaustible array of operations and facilitate the implementation of any feasible computable function. Moreover, FHE significantly improves the efficiency of secure MPC. Notably, FHE surpasses other HE methods as it can competently handle the encryption and computation of any ciphertext (Chen, Iliashenko, & Laine, 2021; Meftah et al., 2021; Wu, Zhao, & Zhang, 2021).

HE provides an advantage to the FL environment since computation is done on multiple platforms and requires that trusted systems be set. Currently, most FL solutions operate by deploying the training models and computational processes on the client devices. After training, the different models are redeployed to the server for integration. The success of such models depends solely on the client device's structural efficiency in terms of battery life, processing power, and storage capacity. Several studies have deployed HE to improve patient information security and privacy in IoHT. For instance, an ingenious approach to FL that employs homomorphic re-encryption and enlists IoHT devices to collect data, with model training conducted via fog nodes and data aggregation done on servers, was introduced (Ku et al., 2022). This novel scheme effectively protects data from IoT devices and resolves steep computational and storage costs. Furthermore, by utilizing homomorphic re-encryption for model aggregation, their solution achieves a remarkable level of resilience that remains unaffected even if some fog nodes become inoperative. However, the time cost of re-encryption becomes longer as the length of the secret key increases, leading to a higher computational overhead.

To safeguard local medical models and prevent adversaries from accessing private medical information through model reconstruction attacks in IoHT, Zhang et al. (2022) introduced sophisticated cryptographic techniques such as masks and HE. Instead of exclusively emphasizing the magnitude of datasets, the investigators discerned the intrinsic attributes encompassed within datasets possessed by diverse participants, serving as the paramount determinant for quantifying the contribution rate of local models to the global model during each training epoch. Moreover, they proffered a resilient schema, impervious to dropout, to safeguard the unchecked progression of FL, notwithstanding instances where the count of active clients descends below a preestablished threshold. The scheme's ability to maintain data privacy is thoroughly analyzed, with theoretical analyses of the associated computation and communication costs. To demonstrate its efficacy in healthcare applications, the researchers conducted skin lesion classification using the HAM10000 medical dataset's training images. The results from their experiments indicate the promising outcomes of the proposed scheme while maintaining privacy and outperforming existing approaches. Nevertheless, their strategy failed to account for the challenges posed by heterogeneous clients equipped with resource-constrained hardware and the complexities inherent in asynchronous FL.

Table 15.2 Summary of studies deploying homomorphic encryption (HE) in modern digital healthcare systems.

Summary of proposed HE technique	Datasets	References
A homomorphic re-encryption framework was proposed for IoHT	Real-life datasets from the University of California, Irvine Machine Learning Repository	Ku et al. (2022)
Proposed a framework based on DL, masks, and HE to prevent model reconstruction attacks in IoHT	HAM10000 medical datasets	Zhang et al. (2022)
Proposed a HE architecture for IoHT and implemented an encrypted query technique within the framework	Encrypted heart disease dataset from Kaggle	Hsu and Huang (2022)

IoHT, *Internet of Health Things.*

Furthermore, authors (Hsu & Huang, 2022) devised a privacy-preserving FL architecture grounded on HE for IoHT to guarantee individual participant data and local models' privacy. Moreover, to ensure data privacy conformance, they implemented an encrypted query technology within this architecture, which enables data providers to explore encrypted data in ciphertext, identify encrypted data fulfilling task criteria, and execute the model training methodology without giving concessions to any requirements of the task originator. Table 15.2 summarizes the studies deploying HE in modern digital healthcare systems, and Fig. 15.1 shows the scheme of HE.

15.3.3 Secure multiparty computation in Internet of Health Things

SMC, a subdomain of cryptography, safeguards confidential data by enabling collaborative computing without any party having access to other participants' information. This technique facilitates secure calculation of functions for multiple participants without relying on third-party trust or revealing inputs (Khalid et al., 2023). Nevertheless, SMC's computational overhead and high communication costs stem from the added encryption and decryption operations (Ma et al., 2022). Several studies have been undertaken to address security and privacy concerns using SMC. For instance, a novel scheme was proposed based on SMC to mitigate the significant privacy challenges that arise when handling sensitive medical data of patients during the diagnostic process (Li et al., 2020). To achieve their objectives, Li et al. (2020) introduced a methodology that encrypts registered patients' healthcare data before transmitting it to the healthcare facility server.

Through the application of HE, a novel approach to FL in the IoHT was introduced in this study, emphasizing privacy preservation (Wibawa et al., 2022). To combat potential threats from malicious entities, a secure MPC protocol was

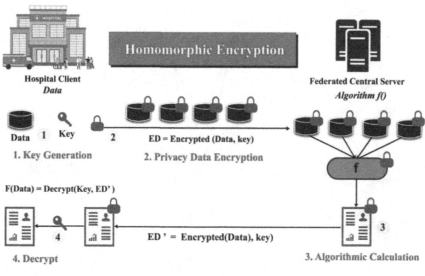

FIGURE 15.1

Homomorphic encryption scheme.

utilized to protect the integrity of the DL model. In another study, Kalapaaking et al. (2022), proposed a novel FL framework for 6G and IoMT, based on a convolutional neural network and incorporating both SMC and encrypted inference methods. Their framework addresses privacy and security challenges by enabling several healthcare facilities consisting of clusters of varied IoMT and edge devices to train locally and encrypt their models before transmitting them for encryption and aggregation in the cloud based on SMC. Their innovative approach yields an encrypted global model distributed back to individual healthcare institutions for further refinement through localized training to optimize the model's efficacy. Additionally, their method enables hospitals to execute encrypted inference seamlessly on their edge servers or within the cloud, ensuring the utmost confidentiality of both data and model throughout the entire procedure. Fig. 15.2 depicts the scheme for SMC.

Although these approaches provide an additional layer of data privacy and security, they do entail a trade-off between communication efficiency and model performance. For instance, DP incorporates random noise into client training data, enhancing privacy at the potential cost of diminished model proficiency. On the other hand, HE encrypts model parameters exclusively, which safeguards data but also has performance implications. Lastly, SMC maintains client input privacy, which is computationally demanding and requires substantial communication among the involved parties (Dang et al., 2022). Fig. 15.2 depicts the SMC scheme.

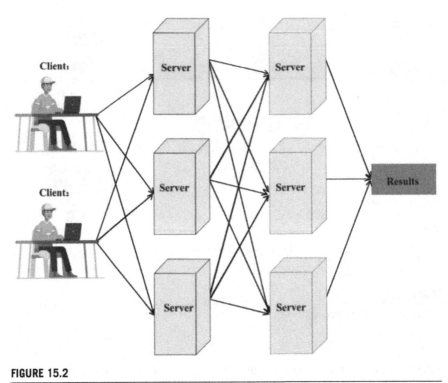

FIGURE 15.2

SMC scheme. *SMC*, Secure multiparty computation.

15.3.4 Secret sharing

The application of secret sharing, an exquisite cryptographic methodology, facilitates the seamless dispersion of a clandestine entity amidst many stakeholders, ensuring the reconstitution of the secret solely through the harmonious collaboration of an adequate cohort (Luo et al., 2018). To fortify the electronic health system against cunning impostors, an innovative fusion of a perceptron-based session key and a modified logistic map-based intermediate key was introduced (Dey, Bhowmik, & Karforma, 2022). The objective was to ensure the protection of patients' clinical reports and data through an uncompromising lossless secret sharing method that required the active involvement of all recipients in the report generation process. To validate the effectiveness of their technique amidst the New Normal COVID-19 E-Health era, the researchers conducted a comprehensive set of mathematical tests, including graphical examination, brute force analysis, statistical randomization, and performance time assessment. The conclusive results unambiguously demonstrate the efficacy of their approach in safeguarding patient data. Another study assessed a decentralized security module explicitly designed for protecting the extensive collection of clinical images, constituting approximately 80% of the overall health data (Sarosh et al., 2021). Their approach involved utilizing

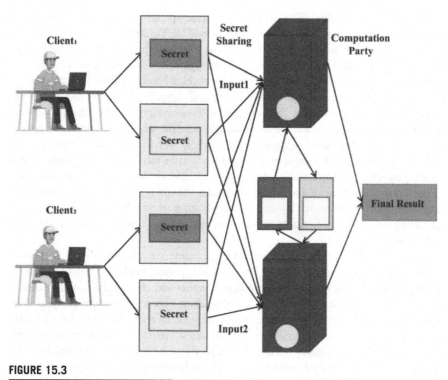

FIGURE 15.3

Secret sharing scheme.

the highly regarded Rivest Cipher 6 (RC6) encryption scheme, augmented by a computational secret sharing scheme, to effectively store medical images across distributed systems. Fig. 15.3 shows the secret sharing scheme.

15.3.5 Blockchain technology

The advent of decentralized networks, leveraging Blockchain technology, has significantly enhanced the privacy of IoT networks (Ali, Karimipour, & Tariq, 2021). The underlying principles of Blockchain networks align seamlessly with the requisites of data privacy, trustworthy security, and decentralization, as envisioned by FL. FL merged with Blockchain leads to a promising avenue for swift enhancements in the IoHT (Stephanie et al., 2023). FL aggregates local model parameters on a central server and distributes a global model to every participating device. However, this centralized server becomes vulnerable to attackers, who may disrupt or manipulate FL training data. To mitigate such risks, the application of Blockchain as a protective mechanism offers the potential to manage data access and decentralize the orchestration mechanisms within FL, creating an unchanging and accountable/traceable ledger. Consequently, recent investigations have embraced the merging of FL and Blockchain, facilitating the realization of a secure and privacy-preserving intelligent healthcare framework (Chang et al., 2021; Kalapaaking, Khalil, & Yi, 2023).

A novel health management framework was presented grounded in the IoHT paradigm, employing blockchain technology within edge nodes (Rahman et al., 2020). The authors advocate replacing the conventional blockchain-based gradient aggregator solution, specifically a consensus-driven distributed aggregator. Such an innovative approach effectively mitigates the concerns of bias and privacy breaches associated with a centralized aggregation entity. A notable example that aligns with these principles is the fog-IoT-based approach for medical services (Baucas, Spachos, & Plataniotis, 2023), which integrates a private blockchain mechanism to regulate access rights, guaranteeing interactions between trusted devices and the local training and global knowledge base of the predictive model. Similarly, an FL-based system for fraud detection, empowered by Blockchain (Lakhan et al., 2023), applies privacy-preserving measures and detection of fraud algorithms across various Fog-Cloud endpoints.

Moreover, the pioneering work introduces an innovative agent architecture for intelligent healthcare systems, leveraging the combined power of FL and blockchain technology, primarily focusing on the proposed detection and classification of skin cancer patients (Połap, Srivastava, & Yu, 2021). The scheme securely encrypts personal data within the dataset utilizing blockchain methodologies. In a related study, Kumar et al. (2021) proposed a novel approach wherein a modest volume of data is gathered from diverse sources, encompassing various hospitals. This data is subsequently employed to effectively train a global DL model using a blockchain-based FL paradigm to identify patients afflicted with COVID-19. Through blockchain technology, the authenticity of the data is diligently upheld, while FL ensures the model's global training, effectively safeguarding the privacy of the organizations involved. Furthermore, an innovative blockchain-FL framework designed specifically for healthcare applications (Liu et al., 2022) aims to facilitate accurate and reliable FL aggregation results while concurrently incentivizing healthcare facilities, such as hospitals, to willingly collaborate and share their respective local data purposely for training in FL undertakings.

In summary, merging FL with Blockchain within IoHT applications is a paramount strategy to improve the security and privacy of the global prediction model (Rehman et al., 2022). By leveraging Blockchain, decentralized coordination of global models becomes feasible through consensus among participants, thus ensuring a distributed approach. The Blockchain-based framework not only fosters seamless cooperation and partnerships among healthcare service providers but also establishes an entirely trustworthy, distributed, and adaptable environment (Abou El Houda et al., 2022), propelling it as an emerging area of research.

15.4 Lessons

This chapter offers a comprehensive introduction to the burgeoning field of FL within contemporary digital healthcare systems, emphasizing its potential in rejuvenating the healthcare sector. One of the salient insights gained pertains to FL's transformative capability in personalized medicine, enabling model adaptation to individual patient characteristics through a collaborative framework. However, it also critically scrutinizes this context's security and privacy challenges. In response, exploring privacy-preserving techniques within FL emerges as a central theme of discussion. The chapter underscores the paramount importance of real-world datasets as essential benchmarks for evaluating the feasibility and robustness of privacy-preserving methodologies in the context of FL for healthcare applications. These datasets encompass a wide range, including MNIST-IID, MNIST NON-IID (Wu et al., 2021), Medical Information Mart for Intensive Care (MIMIC III) data and Limited MarketScan Explorys Claims-EMR Data (LCED) (Choudhury et al., 2019), MNIST dataset and the CIFAR-10 dataset (Liu et al., 2023), eICU Collaborative Research Database and the standard MNIST dataset (Nair, Sahoo, & Raj, 2023), the Cancer Genome Atlas (TCGA) dataset (Adnan et al., 2022), University of California, Irvine Machine Learning Repository (Ku et al., 2022), and encrypted heart disease dataset from Kaggle (Hsu & Huang, 2022). These tangible testbeds faithfully mirror the intricacies and heterogeneity of healthcare data, compellingly demonstrating FL's adaptability and efficacy within the healthcare domain. As elucidated, numerous researchers have leveraged these datasets to develop and assess privacy-preserving techniques within FL, including HE, DP, SMC, Blockchain, and secret sharing. The seamless integration of these techniques into FL workflows showcases their applicability and robustly responds to the critical security and privacy challenges that pervade healthcare settings. These meticulous evaluations, grounded in real-life datasets, underscore a steadfast commitment to scientific rigor and empirical validation. In summary, this research paradigm accentuates the pivotal role of FL in harmonizing personalized healthcare with stringent data protection protocols, thus charting a path toward efficient and secure collaborative healthcare practices.

15.5 Future research directions on secure federated learning in modern digital healthcare systems

The emerging and promising field of FL is in its infancy, presenting exciting opportunities for personalized healthcare that can adapt to individual patient characteristics and transform how healthcare institutions can share data while safeguarding patient data security and privacy. However, the application of FL in healthcare faces significant limitations due to privacy concerns associated with directly sharing personalized models. To overcome these persistent challenges, future research efforts should focus on exploring privacy-preserving techniques within FL, such as secure model aggregation and secure model distillation (Chen et al., 2023; Jahani-Nezhad et al., 2023). Implementing these approaches makes it possible to strike a balance between personalized healthcare and data privacy.

Furthermore, since FL relies on data sharing and collaboration among healthcare institutions, it is imperative to investigate techniques that enable secure data sharing. This includes exploring methods for secure data aggregation (Yu & Cui, 2023), secure model update transmission (Zhang et al., 2022; Li et al., 2023), and secure parameter sharing (Ahmed et al., 2023). These techniques must account for healthcare systems' privacy and security requirements while reaping FL's collaborative benefits. Moreover, the proposed strategies for privacy preservation in FL are computationally expensive, posing challenges in terms of scalability within healthcare settings. Therefore it is crucial for future research to prioritize the development of efficient privacy-preserving algorithms equally. These algorithms should balance privacy guarantees and computational efficiency, allowing FL to scale up effectively for large-scale healthcare datasets.

Also, when adopting FL models for clinical decision-making, it is essential to ensure explainability and trustworthiness. Consequently, future research endeavors should focus on techniques that enhance the interpretability and transparency of FL models while preserving privacy (Chaddad et al., 2022). This necessitates the development of privacy-preserving explainability methods and the establishment of trust frameworks specifically tailored for FL in the context of healthcare. By addressing these critical aspects, FL can reach its full potential in revolutionizing personalized medicine while maintaining the highest standards for privacy and data security, contributing significantly to advancing healthcare practices.

15.6 Conclusion

In conclusion, intelligent healthcare has undergone a momentous revolution driven by rapid advancements in communication technologies, AI, and the IoHT. This transformative journey has been marked by tremendous strides in improving healthcare outcomes and revolutionizing patient care. However, as the demand for data-driven solutions in healthcare grows, traditional AI approaches face critical scalability and privacy limitations. The vast amounts of sensitive medical data and stringent privacy regulations create formidable challenges for AI implementation within the healthcare domain. Fortunately, FL has emerged as a beacon of hope, offering a promising solution to overcome these hurdles. By allowing ML models to be trained on decentralized data, FL facilitates collaboration and knowledge sharing across multiple healthcare institutions while safeguarding the privacy and security of sensitive information. This paradigm shift in AI methodology has the potential to unlock groundbreaking possibilities for the future of smart healthcare. This chapter has shed light on these hurdles, comprehensively examining the various techniques to preserve privacy within this context. Techniques such as HE, DP, SMC, secret sharing, and innovative blockchain-based approaches have been explored and discussed in detail, offering insights into their effectiveness and limitations. Employing these privacy-preserving techniques will ensure the data integrity of healthcare organizations while harnessing the power of FL for collaborative research and analysis. The lessons learned from the deployment of FL in the IoHT landscape are invaluable. They serve as a compass, guiding researchers, students, and stakeholders in their quest to navigate the intricate terrain of secure FL within the healthcare realm.

Furthermore, these insights also illuminate the pressing research opportunities that lie ahead, beckoning researchers to explore new horizons and develop novel methodologies that push the boundaries of secure FL. The knowledge shared in this chapter holds immense significance in this ever-evolving field of smart healthcare. Its impact extends far beyond the confines of academic discourse, reaching the hearts and minds of those dedicated to advancing healthcare technologies. By harnessing the potential of FL and prioritizing security and privacy, we can forge a future where intelligent healthcare systems not only deliver optimal patient care but also uphold and safeguard the privacy rights of individuals. Together, we can embark on a transformative journey toward a brighter, healthier future for all.

References

Abadi, M., Chu, A., Goodfellow, I., McMahan, H. B., Mironov, I., Talwar, K., & Zhang, L. (2016). Deep learning with differential privacy, *Proceedings of the ACM Conference on Computer and Communications Security*, 24−28-October-2016, pp. 308−318. Available from https://doi.org/10.1145/2976749.2978318.

Abou El Houda, Z., Hafid, A. S., Khoukhi, L., & Brik, B. (2022). When collaborative federated learning meets blockchain to preserve privacy in healthcare. *IEEE Transactions on Network Science and Engineering [Preprint]*. Available from https://doi.org/10.1109/TNSE.2022.3211192.

Acar, A., Aksu, H., Uluagac, A. S., & Conti, M. (2018). A survey on homomorphic encryption schemes: Theory and implementation. *ACM Computing Surveys*, *51*(4). Available from https://doi.org/10.1145/3214303.

Adnan, M., Kalra, S., Cresswell, J. C., Taylor, G. W., & Tizhoosh, H. R. (2022). Federated learning and differential privacy for medical image analysis. *Scientific Reports*, *12*(1), 1−10. Available from https://doi.org/10.1038/s41598-022-05539-7, 2022 12:1.

Ahmed, J., Nguyen, T. N., Ali, B., Javed, M. A., & Mirza, J. (2023). On the physical layer security of federated learning based IoMT networks. *IEEE Journal of Biomedical and Health Informatics*, *27*(2), 691−697. Available from https://doi.org/10.1109/JBHI.2022.3173947.

Alabdulatif, A., Thilakarathne, N. N., & Kalinaki, K. (2023). A novel cloud enabled access control model for preserving the security and privacy of medical big data. *Electronics*, *12*(12), 2646. Available from https://doi.org/10.3390/electronics12122646.

Albahri, A. S., Duhaim, A. M., Fadhel, M. A., Alnoor, A., Baqer, N. S., Alzubaidi, L., Albahri, O. S., Alamoodi, A. H., Bai, J., Salhi, A., & Santamaría, J. (2023). A systematic review of trustworthy and explainable artificial intelligence in healthcare: Assessment of quality, bias risk, and data fusion. *Information Fusion*, *96*, 156−191. Available from https://doi.org/10.1016/J.INFFUS.2023.03.008.

Ali, M., Naeem, F., Tariq, M., & Kaddoum, G. (2023). Federated learning for privacy preservation in smart healthcare systems: A comprehensive survey. *IEEE Journal of Biomedical and Health Informatics, 27*(2), 778−789. Available from https://doi.org/10.1109/JBHI.2022.3181823.

Ali, M., Karimipour, H., & Tariq, M. (2021). Integration of blockchain and federated learning for Internet of Things: Recent advances and future challenges. *Computers & Security, 108*, 102355. Available from https://doi.org/10.1016/J.COSE.2021.102355.

Alli, A. A., Kassim, K., Mutwalibi, N., Hamid, H., & Ibrahim, L. (2021). Secure fog-cloud of things: Architectures, opportunities and challenges. In M. Ahmed, & P. Haskell-Dowland (Eds.), *Secure edge computing* (1st edn, pp. 3−20). CRC Press. Available from https://doi.org/10.1201/9781003028635-2.

Alli, A. A., & Alam, M. M. (2019). SecOFF-FCIoT: Machine learning based secure offloading in Fog-Cloud of things for smart city applications. *Internet of Things, 7*, 100070. Available from https://doi.org/10.1016/j.iot.2019.100070.

Antunes, R. S., André da Costa, C., Küderle, A., Yari, I. A., & Eskofier, B. (2022). Federated learning for healthcare: Systematic review and architecture proposal. *ACM Transactions on Intelligent Systems and Technology (TIST), 13*(4). Available from https://doi.org/10.1145/3501813.

Baucas, M. J., Spachos, P., & Plataniotis, K. N. (2023). Federated learning and blockchain-enabled fog-IoT platform for wearables in predictive healthcare. *IEEE Transactions on Computational Social Systems*, 1−10. Available from https://doi.org/10.1109/TCSS.2023.3235950.

Chaddad, A., Lu, Q., Li, J., Katib, Y., Kateb, R., Tanougast, C., Bouridane, A., & Abdulkadir, A. (2022). *Explainable, domain-adaptive, and federated artificial intelligence in medicine.* Available from https://doi.org/10.1109/JAS.2023.123123.

Chalamala, S. R., Kummari, N. K., Singh, A. K., Saibewar, A., & Chalavadi, K. M. (2022). Federated learning to comply with data protection regulations. *CSI Transactions on ICT, 10*(1), 47−60. Available from https://doi.org/10.1007/S40012-022-00351-0, *2022 10:1*.

Chang, Y., Fang, C., & Sun, W. (2021). *A blockchain-based federated learning method for smart healthcare*, downloads.hindawi.com [Preprint]. Available from https://doi.org/10.1155/2021/4376418.

Chen, H., Iliashenko, I., & Laine, K. (2021) When HEAAN Meets F.V. A New Somewhat Homomorphic Encryption with Reduced Memory Overhead, *Lecture Notes in Computer Science (including subseries Lecture Notes in Artificial Intelligence and Lecture Notes in Bioinformatics)*, 13129 LNCS, pp. 265−285. Available from https://doi.org/10.1007/978-3-030-92641-0_13.

Chen, K., Zhang, X., Zhou, X., Mi, B., Xiao, Y., Zhou, L., Wu, Z., Wu, L., & Wang, X. (2023). Privacy preserving federated learning for full heterogeneity, *ISA Transactions* [Preprint]. Available from https://doi.org/10.1016/J.ISATRA.2023.04.020.

Choudhury, O., Gkoulalas-Divanis, A., Salonidis, T., Sylla, I., Park, Y., Hsu, G., & Das, A. (2019). *Differential privacy-enabled federated learning for sensitive health data.* Available from https://arxiv.org/abs/1910.02578v3 (Accessed 17.05.23).

Chemisto, Musa, Gutu, Tar J. L., Kalinaki, Kassim, Mwebesa, Darlius Bosco, Egau, Percival, Kirya, Fred, Oloya, Ivan Tim, & Kisitu, Rashid (2023). Artificial intelligence for improved maternal healthcare: A systematic literature review. *IEEE Xplore*, 1−6. Available from https://doi.org/10.1109/AFRICON55910.2023.10293674.

Dang, T. K., Lan, X., Weng, J., & Feng, M. (2022). Federated learning for electronic health records. *ACM Transactions on Intelligent Systems and Technology, 13*(5), 72. Available from https://doi.org/10.1145/3514500.

Dey, J., Bhowmik, A., & Karforma, S. (2022). 'Neural perceptron & strict lossless secret sharing oriented cryptographic science: Fostering patients' security in the "new normal" COVID-19 E-Health. *Multimedia Tools and Applications, 81*(13), 17747−17778. Available from https://doi.org/10.1007/s11042-022-12440-y.

Fahim, K. E., Kalinaki, K., & Shafik, W. (2023). *'Electronic devices in the artificial intelligence of the Internet of Medical Things (AIoMT). Handbook of security and privacy of AI-Enabled Healthcare Systems and Internet of Medical Things* (1st Edition). CRC Press. Available from https://doi.org/10.1201/9781003370321-3.

Gaba, S., Buddhiraja, I., Kumar, V., & Makkar, A. (2023). Federated learning based secured computational offloading in cyber-physical IoST systems. *Communications in computer and information science*, 344−355. Available from https://doi.org/10.1007/978-3-031-23599-3_26.

Geyer, R. C., Klein, T., & Nabi, M. (2017). *Differentially private federated learning: A client level perspective.* Available from https://arxiv.org/abs/1712.07557v2 (Accessed 17.05.23).

Hosseini, S. M., Sikaroudi, M., Babaei, M., & Tizhoosh, H. R. (2022). Cluster based secure multi-party computation in federated learning for histopathology images, *Lecture Notes in Computer Science (including subseries Lecture Notes in Artificial Intelligence and Lecture Notes in Bioinformatics)*, 13573 LNCS, pp. 110−118. Available from https://doi.org/10.1007/978-3-031-18523-6_11.

Hsu, R. H., & Huang, T. Y. (2022). Private data preprocessing for privacy-preserving federated learning, *Proceedings of the 2022 5th IEEE International Conference on Knowledge Innovation and Invention, ICKII 2022*, pp. 173−178. Available from https://doi.org/10.1109/ICKII55100.2022.9983518.

Islam, T. U., Ghasemi, R., & Mohammed, N. (2022). Privacy-preserving federated learning model for healthcare data, *2022 IEEE 12th Annual Computing and Communication Workshop and Conference, CCWC 2022*, pp. 281−287. Available from https://doi.org/10.1109/CCWC54503.2022.9720752.

Jahani-Nezhad, T., Maddah-Ali, M. A., Li, S., & Caire, G. (2023). SwiftAgg + : Achieving asymptotically optimal communication loads in secure aggregation for federated learning. *IEEE Journal on Selected Areas in Communications [Preprint].* Available from https://doi.org/10.1109/JSAC.2023.3242702.

Kalapaaking, A. P., Stephanie, V., Khalil, I., Atiquzzaman, M., Yi, X., & Almashor, M. (2022). SMPC-based federated learning for 6G-enabled Internet of Medical Things. *IEEE Network, 36*(4), 182−189. Available from https://doi.org/10.1109/MNET.007.2100717.

Kalapaaking, A. P., Khalil, I., & Yi, X. (2023). Blockchain-based federated learning with SMPC model verification against poisoning attack for healthcare systems. *IEEE Transactions on Emerging Topics in Computing [Preprint].* Available from https://doi.org/10.1109/TETC0.2023.3268186.

Kalinaki, K., Fahadi, M., Alli, A. A., Shafik, W., Yasin, M., & Mutwalibi, N. (2023). *'Artificial intelligence of Internet of Medical Things (AIoMT) in smart cities: A review of cybersecurity for smart healthcare. Handbook of security and privacy of AI-enabled healthcare systems and Internet of Medical Things* (1st Edition). CRC Press. Available from https://doi.org/10.1201/9781003370321-11.

Kalinaki, K., Thilakarathne, N. N., Mubarak, H. R., Malik, O. A., & Abdullatif, M. (2023). *Cybersafe capabilities and utilities for smart cities. Cybersecurity for smart cities* (pp. 71−86). Cham: Springer. Available from https://doi.org/10.1007/978-3-031-24946-4_6.

Kaur, H., Atif, M., & Chauhan, R. (2020). *An Internet of Healthcare Things (IoHT)-based healthcare monitoring system*, in, pp. 475–482. Available from https://doi.org/10.1007/978-981-15-2774-6_56.

Khalid, N., Qayyum, A., Bilal, M., Al-Fuqaha, A., & Qadir, J. (2023). Privacy-preserving artificial intelligence in healthcare: Techniques and applications. *Computers in Biology and Medicine*, *158*, 106848. Available from https://doi.org/10.1016/J.COMPBIOMED.2023.106848.

Ku, H., Susilo, W., Zhang, Y., Liu, W., & Zhang, M. (2022). Privacy-preserving federated learning in medical diagnosis with homomorphic re-encryption. *Computer Standards & Interfaces*, *80*, 103583. Available from https://doi.org/10.1016/J.CSI.2021.103583.

Kumar, R., Khan, A. A., Kumar, J., Golilarz, N. A., Zhang, S., Ting, Y., Zheng, C., & Wang, W. (2021). Blockchain-federated-learning and deep learning models for COVID-19 detection using CT imaging. *IEEE Sensors Journal*, *21*(14), 16301–16314. Available from https://doi.org/10.1109/JSEN.2021.3076767.

Lakhan, A., Mohammed, M. A., Nedoma, J., Martinek, R., Tiwari, P., Vidyarthi, A., Alkhayyat, A., & Wang, W. (2023). Federated-learning based privacy preservation and fraud-enabled blockchain IoMT system for healthcare. *IEEE Journal of Biomedical and Health Informatics*, *27*(2), 664–672. Available from https://doi.org/10.1109/JBHI.2022.3165945.

Li, D., Liao, X., Xiang, T., Wu, J., & Le, J. (2020). Privacy-preserving self-serviced medical diagnosis scheme based on secure multi-party computation. *Computers & Security*, *90*, 101701. Available from https://doi.org/10.1016/J.COSE.2019.101701.

Li, H., Li, C., Wang, J., Yang, A., Ma, Z., Zhang, Z., & Hua, D. (2023). Review on security of federated learning and its application in healthcare. *Future Generation Computer Systems*, *144*, 271–290. Available from https://doi.org/10.1016/J.FUTURE.2023.02.021.

Liu, W., He, Y., Wang, X., Duan, Z., Liang, W., & Liu, Y. (2023). BFG: privacy protection framework for internet of medical things based on blockchain and federated learning. *Connection Science*, *35*(1), 2199951. Available from https://doi.org/10.1080/09540091.2023.2199951.

Liu, Y., Yu, W., Ai, Z., Xu, G., Zhao, L., & Tian, Z. (2022). A blockchain-empowered federated learning in healthcare-based cyber physical systems. *IEEE Transactions on Network Science and Engineering [Preprint]*. Available from https://doi.org/10.1109/TNSE.2022.3168025.

Luo, E., Bhuiyan, M. Z. A., Wang, G., Rahman, M. A., Wu, J., & Atiquzzaman, M. (2018). PrivacyProtector: Privacy-protected patient data collection in IoT-based healthcare systems. *IEEE Communications Magazine*, *56*(2), 163–168. Available from https://doi.org/10.1109/MCOM.2018.1700364.

Ma, X., Liao, L., Li, Z., Lai, R. X., & Zhang, M. (2022). Applying federated learning in software-defined networks: A survey. *Symmetry*, *14*(2), 195. Available from https://doi.org/10.3390/SYM14020195, *2022, Vol. 14, Page 195*.

McMahan, B., Moore, E., Ramage, D., Hampson, S., & y Arcas, B. A. (2017). Communication-efficient learning of deep networks from decentralized data. *PMLR*, 1273–1282. Available from https://proceedings.mlr.press/v54/mcmahan17a.html.

Meftah, S., Tan, B. H. M., Mun, C. F., Aung, K. M. M., Veeravalli, B., & Chandrasekhar, V. (2021). DOReN: Toward efficient deep convolutional neural networks with fully homomorphic encryption. *IEEE Transactions on Information Forensics and Security*, *16*, 3740–3752. Available from https://doi.org/10.1109/TIFS.2021.3090959.

Mothukuri, V., Parizi, R. M., Pouriyeh, S., Huang, Y., Dehghantanha, A., & Srivastava, G. (2021). A survey on security and privacy of federated learning. *Future Generation Computer Systems*, *115*, 619–640. Available from https://doi.org/10.1016/J.FUTURE.2020.10.007.

Nair, A. K., Sahoo, J., & Raj, E. D. (2023). Privacy preserving federated learning framework for IoMT based big data analysis using edge computing. *Computer Standards & Interfaces*, *86*, 103720. Available from https://doi.org/10.1016/J.CSI.2023.103720.

Nguyen, D. C., Ding, M., Pathirana, P. N., Seneviratne, A., Li, J., Niyato, D., & Poor, H. V. (2021). Federated learning for industrial internet of things in future industries. *IEEE Wireless Communications*, *28*(6), 192−199. Available from https://doi.org/10.1109/MWC.001.2100102.

Nguyen, D. C., Pham, Q. V., Pathirana, P. N., Ding, M., Seneviratne, A., Lin, Z., Dobre, O., & Hwang, W. J. (2022). Federated learning for smart healthcare: A survey. *ACM Computing Surveys (CSUR)*, *55*(3). Available from https://doi.org/10.1145/3501296.

Ogburn, M., Turner, C., & Dahal, P. (2013). Homomorphic encryption. *Procedia Computer Science*, *20*, 502−509. Available from https://doi.org/10.1016/J.PROCS.2013.09.310.

Pfitzner, B., Steckhan, N., & Arnrich, B. (2021). Federated learning in a medical context: A systematic literature review. *ACM Transactions on Internet Technology (TOIT)*, *21* (2). Available from https://doi.org/10.1145/3412357.

Połap, D., Srivastava, G., & Yu, K. (2021). Agent architecture of an intelligent medical system based on federated learning and blockchain technology. *Journal of Information Security and Applications*, *58*, 102748. Available from https://doi.org/10.1016/J.JISA.2021.102748.

Rahman, M. A., Hossain, M. S., Islam, M. S., Alrajeh, N. A., & Muhammad, G. (2020). Secure and provenance enhanced internet of health things framework: A blockchain managed federated learning approach. *IEEE Access*, *8*, 205071−205087. Available from https://doi.org/10.1109/ACCESS.2020.3037474.

Rahman, M. A., Hossain, M. S., Showail, A. J., Alrajeh, N. A., & Alhamid, M. F. (2021). A secure, private, and explainable IoHT framework to support sustainable health monitoring in a smart city. *Sustainable Cities and Society*, *72*, 103083. Available from https://doi.org/10.1016/J.SCS.2021.103083.

Rehman, A., Abbas, S., Khan, M. A., Ghazal, T. M., Adnan, K. M., & Mosavi, A. (2022). A secure healthcare 5.0 system based on blockchain technology entangled with federated learning technique. *Computers in Biology and Medicine*, *150*, 106019. Available from https://doi.org/10.1016/J.COMPBIOMED.2022.106019.

Sarosh, P., Parah, S. A., Bhat, G. M., Heidari, A. A., & Muhammad, K. (2021). Secret sharing-based personal health records management for the Internet of Health Things. *Sustainable Cities and Society*, *74*, 103129. Available from https://doi.org/10.1016/J.SCS.2021.103129.

Singh, M., & Singh, A. K. (2023). A comprehensive survey on encryption techniques for digital images. *Multimedia Tools and Applications*, *82*(8), 11155−11187. Available from https://doi.org/10.1007/S11042-022-12791-6.

Srivastava, J., Routray, S., Ahmad, S., & Waris, M. M. (2022). Internet of Medical Things (IoMT)-based smart healthcare system: Trends and progress. *Computational Intelligence and Neuroscience*, *2022*, 1−17. Available from https://doi.org/10.1155/2022/7218113.

Stephanie, V., Khalil, I., Atiquzzaman, M., & Yi, X. (2023). Trustworthy privacy-preserving hierarchical ensemble and federated learning in Healthcare 4.0 with blockchain. *IEEE Transactions on Industrial Informatics*, *19*(7), 7936−7945. Available from https://doi.org/10.1109/TII.2022.3214998.

Wibawa, F., Catak, F. O., Sarp, S., & Kuzlu, M. (2022). BFV-based homomorphic encryption for privacy-preserving CNN models. *Cryptography*, *6*(3), 34. Available from https://doi.org/10.3390/CRYPTOGRAPHY6030034, *2022, Vol. 6, Page 34.*

Wu, M., Ye, D., Ding, J., Guo, Y., Yu, R., & Pan, M. (2021). Incentivizing differentially private federated learning: A multidimensional contract approach. *IEEE Internet of Things Journal*, *8*(13), 10639−10651. Available from https://doi.org/10.1109/JIOT.2021.3050163.

Wu, T., Zhao, C., & Zhang, Y. J. A. (2021). Privacy-preserving distributed optimal power flow with partially homomorphic encryption. *IEEE Transactions on Smart Grid*, *12*(5), 4506−4521. Available from https://doi.org/10.1109/TSG.2021.3084934.

Yang, H., Ge, M., Xue, D., Xiang, K., Li, H., & Lu, R. (2023). Gradient leakage attacks in federated learning: Research frontiers, taxonomy and future directions. *IEEE Network [Preprint]*. Available from https://doi.org/10.1109/MNET.001.2300140.

Yang, W., Wang, S., Cui, H., Tang, Z., & Li, Y. (2023). A review of homomorphic encryption for privacy-preserving biometrics. *Sensors*, *23*(7), 3566. Available from https://doi.org/10.3390/S23073566, *2023, Vol. 23, Page 3566*.

Yin, X., Zhu, Y., & Hu, J. (2021). A comprehensive survey of privacy-preserving federated learning. *ACM Computing Surveys (CSUR)*, *54*(6). Available from https://doi.org/10.1145/3460427.

Yu, S., & Cui, L. (2023). *Secure data aggregation in federated learning*, pp. 99−107. Available from https://doi.org/10.1007/978-981-19-8692-5_7.

Zhang, L., Xu, J., Vijayakumar, P., Sharma, P. K., & Ghosh, U. (2022). Homomorphic encryption-based privacy-preserving federated learning in IoT-enabled healthcare system. *IEEE Transactions on Network Science and Engineering [Preprint]*. Available from https://doi.org/10.1109/TNSE.2022.3185327.

Zhang, P., Hong, Y., Kumar, N., Alazab, M., Alshehri, M. D., & Jiang, C. (2022). BC-EdgeFL: A defensive transmission model based on blockchain-assisted reinforced federated learning in IIoT environment. *IEEE Transactions on Industrial Informatics*, *18*(5), 3551−3561. Available from https://doi.org/10.1109/TII.2021.3116037.

Index

Note: Page numbers followed by "*f*" and "*t*" refer to figures and tables, respectively.

Printed in the United States
by Baker & Taylor Publisher Services